Harry Van den Akker, Robert F. Mudde

Mass, Momentum and Energy Transport Phenomena

Also of interest

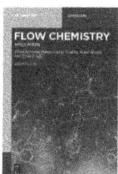

Harry Van den Akker, Robert F. Mudde

Mass, Momentum and Energy Transport Phenomena

A Consistent Balances Approach

2nd edition

DE GRUYTER

Authors
Prof. Dr Harry E.A. Van den Akker
School of Engineering
Bernal Institute
University of Limerick
Limerick V94 T9PX
Ireland
E-mail: harry.vandenakker@ul.ie

Prof. Dr Robert F. Mudde
Chemical Engineering Department
Faculty of Applied Sciences
Delft University of Technology
Delft 2600 AA
The Netherlands

ISBN 978-3-11-124623-9
e-ISBN (PDF) 978-3-11-124657-4
e-ISBN (EPUB) 978-3-11-124715-1

Library of Congress Control Number: 2023937381

Bibliographic information published by the Deutsche Nationalbibliothek
The Deutsche Nationalbibliothek lists this publication in the Deutsche Nationalbibliografie;
detailed bibliographic data are available on the Internet at http://dnb.dnb.de.

Preface to the second edition

The authors are indebted to De Gruyter for their willingness to continue the availability of our textbook that enjoys a continued international interest. At several leading American universities, it has been selected as the core textbook for one-term courses on Transport Phenomena, not in the least because it is much more compact than the classical *Transport Phenomena* by Bird, Stewart, and Lightfoot. Another difference is that for didactic reasons we have opted for a different order of mass, energy (heat), and momentum balances in the first chapter. In the remainder of the book, we once more use a different order. We refer to a 2022 paper in *Physics of Fluids* (Vol. 34, art. #037106; doi.org.10.1063/5.0084383) for a more extensive discussion on these differences.

Furthermore, our textbook distinguished itself by the consistent usage of the art of drawing up balances, in the derivation of the differential equations which are key to Transport Phenomena as well as when solving problems in flow, transport, and transfer problems. Teaching students how to solve transport problems by applying the technique of drawing up balances is a very important objective of our textbook. This strategy is illustrated in some 80 worked examples. In addition, our derivation of the Navier–Stokes equations at the end of the textbook pursues the same balance approach, and in this way is a perfect preparation towards the technique of computational fluid dynamics (CFD).

In this second edition, we further fine-tuned the explanation of the theory at several places, and we corrected printing errors, particularly in equations, partly due to useful suggestions by Dr Klaas Besseling. We are also indebted to Mr Krook for his assistance in the smooth transition to De Gruyter. Of course, we are open for suggestions for further improvements to our textbook.

Delft, 25 June 2023

Harry E.A. Van den Akker
Robert F. Mudde

https://doi.org/10.1515/9783111246574-202

Preface

In 1956, Hans Kramers in Delft published his first lecture notes (in Dutch) on 'Fysische Transportverschijnselen' – to the best of our knowledge the first systematic treatment of the emerging discipline of Transport Phenomena. In 1958, Robert B. Bird spent a period in Delft as a guest of Hans Kramers. This visit gave the two professors the excellent opportunity to explore and improve the way of teaching Transport Phenomena. Bird published his *Notes on Transport Phenomena* in the fall of 1958, followed in 1960 by the first Wiley edition of the famous *Transport Phenomena* textbook by Bird, Stewart, and Lightfoot.

In Delft, the Dutch students kept using Kramers' shorter lecture notes in Dutch, which in the course of the years were continuously improved, also by Kramers' successors Wiero J. Beek and John M. Smith. All those years, the analogy of momentum, heat, and mass transport remained the leading theme, just like in Bird's textbook. An essential element in the way Transport Phenomena has been taught in Delft has always been the emphasis on developing the students' ability of solving realistic engineering problems. Over the years, hundreds of challenging exam problems were devised.

In 1996, the current authors published a new version of the Delft lecture notes on Transport Phenomena for various reasons. Students, their interests, their prior education, and the way they prepare for exams were changing. In many curricula, the course got a different role and place. New applications in biotechnology, biomedical, smart materials, and solar evolved – remote from the traditional chemical industry – and computational fluid dynamics (CFD) was developed into a real analytical tool. All these required different didactic methods for teaching as well as different examples and exam problems.

Our new version of the Delft textbook *Transport Phenomena* (still in Dutch) was built on the earlier Delft lecture notes but was still based on the classical analogy of momentum, heat, and mass transport, although the order of treatment was changed: fluid mechanics largely moved to the end, provoked by ideas developed by Kees Rietema at Eindhoven University of Technology. Most importantly, however, we put a much stronger emphasis on the basic method of drawing up balances, either about a particular device (a macro-balance) or about a differential element anywhere in a material or fluid (a micro-balance). In most cases, such a balance turns into a differential equation. We believe that teaching students as to how to draw up balances and solve differential equations is an excellent preparation for exploiting modern CFD techniques. The exam requirement that students should be capable of solving original problems was maintained.

Our textbook was quite successful: a second edition was released in 2003–2005 and a third edition in 2008. In recent years, however, increasing numbers of foreign students arrived at Delft University of Technology for various MSc programs. This development has prompted the idea of publishing an English version of our Dutch textbook, simultaneously updating and improving a fourth edition in Dutch.

https://doi.org/10.1515/9783111246574-203

This textbook – *Transport Phenomena: The Art of Balancing* – is the result. We hope it will find its way to foreign universities as well.

We like to express our sincere appreciation for all suggestions for improvements we received over the years. In particular we like to acknowledge the contributions from A.G.N. Boers, C. Ouwerkerk, G.C.J. Bart, C.R. Kleijn, J.J.Ph. Elich, L.M. Portela, J.A. Battjes, R.B. Bird, and J.E. Schievink. We remain open for suggestions for further improvements.

Delft, August 2014

Harry E.A. Van den Akker
Robert F. Mudde

Contents

1 Balances

1.1 The balance: recipe and form

The field of transport phenomena covers the transport of the three most important quantities – mass, energy, and momentum – in any (physical or chemical) process. The addition of the words 'in any process' in particular is an indication of one of the most important features of the field: transport phenomena is, above all, an engineering field with a wide range of applications.

Nonetheless, the field is also fundamental, given that it forms the basis for many other chemical engineering disciplines such as reactor engineering, separation technology, and fluid mechanics. This makes transport phenomena a must for any chemical engineer. A good knowledge of the subject is also very useful to those in other professions such as mechanical, mining, civil, and building engineers, physicists, chemists, and materials scientists.

The area covered by the field of transport phenomena and the discipline of chemical engineering is considerable. There are, for example, all kinds of processes in the chemical and petro-chemical industry, the flow of one or more phases through a pipeline, the behaviour of bubbles in a bioreactor, or the filling of a casting mould with liquid metal. At the other end of the scale, the field is also very important to everyday matters, such as the heat emission of a radiator and the associated air flows in the room, and transport of oxygen by blood flow. Fortunately, these very different processes can be clearly understood and described with a limited number of rules.

Flow phenomena and heat and mass transfer are described in this field in terms of continuum properties, with only occasional references to molecular processes. This is how the basis is laid for chemical engineering: the expertise of designing and improving processes in which substances are transported, transformed, processed, or shaped. It is important here to fully understand the essence of a process – that is, to identify the essential stages in the transport of mass, heat, and/or momentum. The transport of these three quantities can, as it happens, be described in exactly the same way. Transport phenomena lays the basis for physical technology and provides the necessary tools. This textbook is about these tools.

First and foremost, transport phenomenon is a subject of **balances** and **concepts** by which physical processes and phenomena can be described. In many cases, the subject is about deviations from a state of equilibrium and the subsequently occurring **resistances** to heat and mass transport. It frequently concerns a quantitative description of cause and effect. With the help of these still somewhat vague terms, it is possible to gain an outlined, but also very detailed, understanding and description of the aforementioned and countless other processes. This chapter will discuss the term **balance** in extensive detail.

https://doi.org/10.1515/9783111246574-001

For the description of the transport of any quantity, such as the transport of oxygen from bubbles to the liquid phase in a fermenter or the transport of heat through the wall of a furnace, the balance is an essential tool. The basic principle of the balance is the bookkeeping of a selected physical quantity. This concept is of particular importance when working with what are known as conserved quantities; these are quantities (like mass and energy) that are not lost during a process, but conserved.

The field of transport phenomena deals with steady-state or transient (time-dependent) processes in which mass, energy, and momentum are exchanged between domains as a result of driving forces (differences in concentrations of mass, energy and momentum, and/or in pressure). Transport phenomena is therefore primarily about the 'bookkeeping' of the three physical quantities: mass, energy, and momentum.

This bookkeeping is done by drawing up balances over control volumes. A control volume is a domain (or system) with boundaries and can be closed or open, usually allowing for transport or exchanges across the boundaries. Such bookkeeping can refer to large control volumes, which involve **macro-balances;** however, balances can also be drawn up in relation to very small control volumes – these are known as **micro-balances**, which provide information at a local scale. In almost all cases, solving problems, such as about transport or transfer rates or about changes in concentrations or temperatures, starts with drawing up one or more of such balances.

The next step is to derive from such balances proper equations, in many cases differential equations, the latter requiring initial and/or boundary conditions of course. The final step is about solving these (differential) equations to find the answer to the problem under consideration. In this approach, it is essential to denote all quantities with symbols!

The **general recipe** for drawing up a balance and solving the problem can be summarised as follows:

1) Make a sketch of the situation. Use symbols rather than numerical values to indicate quantities.
2) Select the quantity G that is being transported or transferred in the process under consideration.
3) Select the **control volume** V about which information is to be obtained.
4) Find out whether and if so, how, the quantity of G in the control volume V changes during a brief period of time Δt. Draw up the balance (using symbols).
5) Solve the (differential) equation resulting from the balance.

The quantity of G in V can change in all kinds of ways. These should be examined systematically and, if applicable, included in the balance. For example, during Δt, G can flow into V from outside. As a result, the quantity of G inside V increases. It is also possible for G to flow outwards, from inside V. In this case, the quantity of G in V falls. We refer to **inflow** and **outflow**, respectively. Of course, it is also possible for the **production** of G to occur inside V during period Δt: as a result, the total quantity of G in V increases. Negative production (= destruction, consumption, annihilation) is also possible, for example, if G stands for the mass of a reagent that is being trans-formed into another substance in a chemical process.

Bear in mind that G may not necessarily be the quantity in which you are inter-ested. In order to calculate temperature T, for example, a thermal energy balance has to be drawn up, and the thermal energy U must be selected for G rather than T be-cause thermal energy (not temperature) can be transported and transferred.

The general structure for a balance is now as follows (see also Figure 1.1):

The change of G in V during Δt = G at time $(t + \Delta t)$ in V – G at time t in V
$\quad\quad\quad$ = quantity of G that flows from outside into V during Δt +
$\quad\quad\quad$ – quantity of G that flows outside from inside V during Δt +
$\quad\quad\quad$ + net quantity of G that is produced in V during Δt

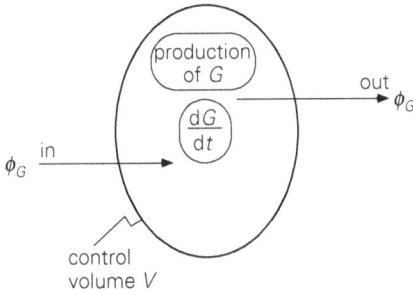

Figure 1.1

From now on, the symbol ϕ will be used to denote a transport (rate), with the dimen-sion 'quantity of G per unit of time'. Instead of transport rate, the term **flow rate** is used. The letter P stands for net production per unit of time or net production rate. With the help of this notation, the quantity of G that flows 'inwards' (= from the out-side to the inside) **during** the period of time Δt can, if Δt is very short, be written as the product of the flow rate 'in' at time t and the period of time Δt:

$$\phi_{G,in}(t) \cdot \Delta t$$

The same applies to the flow of G from the inside to outside and to the net production during Δt:

$$\phi_{G,out}(t) \cdot \Delta t \quad \text{and} \quad P_G(t) \cdot \Delta t$$

This means the balance is

$$G^{inV}(t + \Delta t) - G^{inV}(t) = \phi_{G,in} \cdot \Delta t - \phi_{G,out} \cdot \Delta t + P_G \cdot \Delta t \tag{1.1}$$

Dividing both sides of the equation (1.1) by Δt and taking the limit $\Delta t \to 0$ produces

$$\frac{d}{dt} G^{\text{in}V} = \phi_{G,\text{in}} - \phi_{G,\text{out}} + P_G \tag{1.2}$$

Equation (1.2) is the basic form of the balance and is called the **balance equation**. The left-hand side therefore stands for the incremental change of the total quantity of G in V, while the three ways in which the total quantity of G in V can change are given on the right-hand side. The left-hand side is also known as 'unsteady-state term'. All terms in equation (1.2) have the same dimension – *should* have the same dimension: quantity of G per unit of time.

If, for a given quantity G, the net production *always* equals zero, in the case of mass, for example, then equation (1.2) is simplified to

$$\frac{d}{dt} G^{\text{in}V} = \phi_{G,\text{in}} - \phi_{G,\text{out}} \tag{1.3}$$

Equation (1.3) is called a **conservation law**. Some quantities are indeed conserved – that is, they are only transported and/or (re)distributed.

It is very important to understand that the flows 'in' and 'out' actually have to cross the outer boundary of the control volume V.

Example 1.1. Population.
The population of the Netherlands is known on 31 December of a particular year. Based on this number, describe a method for determining the number of people living in the Netherlands on 31 December of the following year, without counting them all over again.

Solution
The way of keeping track of the population is to use the balance method. To that end, use the general formula: first, the control volume has to be selected, with the number of people being the quantity for which the balance has to be drawn up. In this case, it is the Netherlands (see Figure 1.2).

Figure 1.2

For the inflows and outflows during the year of interest, all the borders of the control volume have to be taken into consideration. This means that not only do the external borders with Belgium, Germany, and the North Sea have to be involved in the drawing up of the balance, but also individual transit points like airports and harbours. The borders should be defined here in the traditional sense of 'passing through Customs' (the fact that checks are no longer made on the internal borders of the European Union is not relevant here). Finally, the question of 'production' must not be overlooked: all births and deaths have to be counted up in order to arrive at the net production term in the balance equation, equation (1.2).

Notice that a balance method of this kind also works for the number of people in the Netherlands aged between 20 and 30 years. However, extra care should be taken with regard to all the terms on the right-hand side of equation (1.2): in the case of the inflows and outflows, the proportion of 20–30-year-olds in the flows should be used, while the net production term should include all those who have turned 20 (counting positive) and 30 (counting negative) in period Δ*t* as well as those who died between these ages.

Summary

The balance is an essential instrument in describing and calculating transport phenomena. Five basic rules are important when drawing up a balance and solving a problem. Together they may be considered as the recipe for solving transport issues:

1) make a sketch of the situation including all transports in and out, and the production inside, and use symbols;
2) select the quantity for which a balance has to be drawn up;
3) select the control volume about which the balance is to be drawn up;
4) draw up the balance the form of which always is

$$\frac{d}{dt} = \text{in} - \text{out} + \text{production}$$

5) solve the resulting (differential) equation.

1.2 The mass balance

1.2.1 The total mass balance and the species balance

We will first deal with the balance for the total mass M. The total mass is typically a conserved quantity: the production or destruction of mass is impossible (except in the case of nuclear reactions), but it can be transported. This is why a conservation law applies to M. By substituting M for G, equation (1.3) then becomes

$$\frac{d}{dt}M = \phi_{m,\text{in}} - \phi_{m,\text{out}} \tag{1.4}$$

The units of the different variables in equation (1.4) offer a useful tool for checking the balance that has been drawn up. In equation (1.4), all four terms have the unit kg/s.

Next is the mass balance for one of the species of a multiple-species system (see Figure 1.3). The bookkeeping is now concerned with mass M_A of substance A, which is present in control volume V. The general form of such a species-mass balance is

$$\frac{d}{dt}M_A = \phi_{m,A,\text{in}} - \phi_{m,A,\text{out}} + P_A \tag{1.5}$$

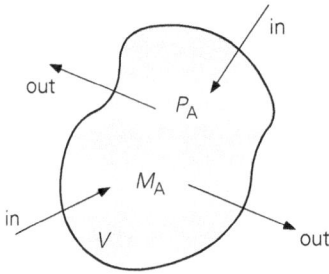

Figure 1.3

In the case of equation (1.5), unlike equation (1.4), we cannot talk of a conservation law anymore, on account of the production term P_A which is due to one or more chemical reactions. As in the case of chemical reactions working in terms of moles is to be preferred, all four terms in such a species-mass balance, equation (1.5), have the unit mol/s or kmol/s.

It is often much more useful to draw up the balance with the help of concentrations. Later on, it will be shown frequently that it is concentrations and differences in concentration that determine transport and not so much the quantity (the mass, in this case) itself. In general, a concentration is defined as the quantity G per unit of volume. Put another way, the concentration (g) of G is the quantity of G in volume V divided by the size of volume V; as a formula, this is

$$g = \frac{G}{V} \tag{1.6}$$

The concentration of the total mass is equivalent to density ρ, while the concentration of a species is represented by a c. Therefore, for a multi-component system consisting of species A, B, . . . etc.:

$$c_A = \text{mass of A per unit of volume} = M_A/V,$$
$$c_B = \text{mass of B per unit of volume} = M_B/V, \text{ etc.} \tag{1.7}$$

Naturally, for the total (mass) concentration ρ, the following applies:

$$\rho = \frac{M}{V} = \frac{M_A + M_B + \cdots}{V} = c_A + c_B + \cdots \tag{1.8}$$

In order to draw up a mass balance for species A, which is present in the reactor in Figure 1.4, it is strongly advisable to use the general recipe consistently, as summarised at the end of Section 1.1:
1) make a sketch and indicate all quantities and transports in symbols (see Figure 1.4)
2) quantity → mass M_A of A in reactor
3) control volume → whole reactor volume V
4) systematically check whether, by what means and how M_A changes in V, and draw up the **balance** for M_A in V.

Step 5 of the recipe is not yet appropriate at this time.

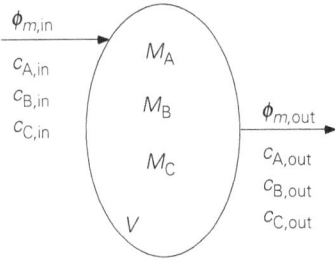

Figure 1.4

It is therefore a matter of systematically checking which effects can contribute to the increase or decrease of M_A in V. These effects will then become part of the right-hand part of the balance equation, while dM_A/dt is on the left. The different terms on the right-hand side will be, in this case:

flow in: a mass flow $\phi_{m,in}$ goes into the reactor, which contains species A with a concentration $c_{A,in}$

$$\phi_{m,A,in} = \text{mass of A that flows in per unit of time } V$$
$$= (\text{volume that flows in per unit of time } V) \times (\text{mass of A}$$
$$\text{contained in a unit of volume of the ingoing flow})$$
$$= \phi_{V,in} \cdot c_{A,in} \tag{1.9}$$

The symbol $\phi_{V,in}$ stands for the volumetric flow rate that goes into the reactor – that is, the number of cubic metres that flows into the reactor per second.

flow out: for this, by analogy, the following applies

$$\phi_{m,A,out} = \phi_{V,out} \cdot c_{A,out} \tag{1.10}$$

production: as a result of a chemical reaction

$$P_A = r_A \cdot V \tag{1.11}$$

where r_A is the production of A per unit of volume (in mol/m^{3}s). Because of this chemical reaction term, it is more useful to express all the concentrations in this balance equation in mol/m^3.

The species-mass balance for A in the reaction is therefore

$$\frac{d}{dt}M_A = \phi_{m,A,in} - \phi_{m,A,out} + P_A \tag{1.12}$$

and, as long as V is constant in time and with the help of equations (1.7), (1.9), (1.10) and (1.11), can be rewritten as

$$V\frac{d}{dt}c_A = \phi_{V,in} \cdot c_{A,in} - \phi_{V,out} \cdot c_{A,out} + r_A V \tag{1.13}$$

Equation (1.13) is equivalent to

$$\frac{d}{dt}c_A = \frac{\phi_{V,in}}{V} \cdot c_{A,in} - \frac{\phi_{V,out}}{V} \cdot c_{A,out} + r_A \tag{1.14}$$

Notice that all the terms in equation (1.14) now have the unit mol/m³s.

If the corresponding equations (1.13) for all the species present are written down and then counted up, the mass balance for the overall mass in V (the overall mass balance for short) is obtained:

$$V\frac{d}{dt}c_A = \phi_{V,in} \cdot c_{A,in} - \phi_{V,out} \cdot c_{A,out} + r_A V$$

$$V\frac{d}{dt}c_B = \phi_{V,in} \cdot c_{B,in} - \phi_{V,out} \cdot c_{B,out} + r_B V$$

$$\vdots \qquad \vdots \qquad \vdots \qquad \vdots \tag{1.15}$$

$$\text{————————————————} +$$

$$V\frac{d}{dt}(c_A + c_B + \cdots) = \phi_{V,in}(c_A + c_B + \cdots)_{in} +$$

$$- \phi_{V,out}(c_A + c_B + \cdots)_{out} + (r_A + r_B + \cdots)V$$

The left-hand side can be written as $V\,d\rho/dt = dM/dt$. The final term of the right-hand side is zero, as production of one species as a result of reactions entails the annihilation of other species. If $\phi_V \cdot \rho = \phi_m$ is also used, then equation (1.15) does indeed cover the overall mass balance, see equation (1.4).

Example 1.2. A chemical reactor under steady conditions.
A chemical reaction takes place in a reactor. Substance A reacts, becoming a new product B. An aqueous solution of A flows into the reactor. The volumetric flow rate is 1 l/s, of which the concentration of A is $c_{A,in} = 100$ kg/m³. The reaction in the reactor is not complete so that A is still present in the reactor exit. The volume flow at the exit is also 1 l/s, and the concentration here is $c_{A,out} = 20$ kg/m³. The situation is **steady** (i.e. does not change over time).

How great is the (negative) production of A in the reactor?

Solution
Clearly, in order to answer the question, a balance for the mass of A in the reactor will have to be drawn up. To do this, apply the recipe:
1) sketch → see Figure 1.5.
2) quantity → M_A
3) control volume → whole reactor
4) steady → '$\frac{d}{dt}$ = 0'; the volumetric flow rates in and out are exactly the same → $\phi_{V,in} = \phi_{V,out} = \phi_V$

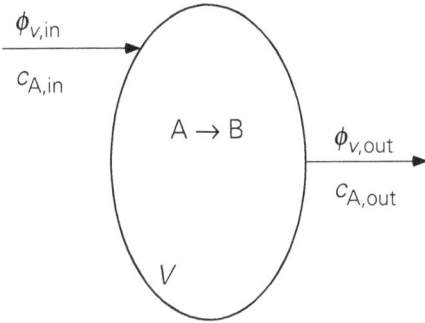

$\phi_{v,in}$
$c_{A,in}$

A → B

$\phi_{v,out}$
$c_{A,out}$

V

Figure 1.5

This means the balance is as follows:

$$0 = \phi_V \cdot c_{A,in} - \phi_V \cdot c_{A,out} + P_A$$

For the production, this gives

$$P_A = \phi_V \cdot c_{A,out} - \phi_V \cdot c_{A,in} = 10^{-3} \cdot (20 - 100) = -8.0 \cdot 10^{-2} \text{ kg/s}$$

Note we use the numerical values for concentrations and flow rate only at the very end of the solution: this is a general rule! By working in symbols, we can see and check (the units inclusively) how the variable of interest depends on the independent variables, in this case how the production rate relates to concentrations and flow rate, irrespective of the numerical values. The numerical values should be substituted as the last step of the solution only.

Incidentally, it is also possible to work with **mass fractions** instead of concentrations. These are written as an x and defined as the fraction of the species under consideration in relation to the overall mass:

$$x_A = \frac{M_A}{M} \tag{1.16}$$

In terms of the mass fraction, the mass balance for species A is

$$\frac{d}{dt}(Mx_A) = \phi_{m,in} \cdot x_{A,in} - \phi_{m,out} \cdot x_{A,out} + P_A \tag{1.17}$$

Notice that x has no dimension – that is, the unit is kg/kg.

To simplify the balance equations and to arrive at (differential) equations that can be solved easily, some simplifying concepts have become quite commonplace in the chemical engineering world. First, the concept of an **ideally stirred tank**, or **ideally mixed tank**, has been introduced: this denotes a (flow) device in which the composition of its contents is always the same everywhere in the tank. This means that at all times the composition of the liquid that flows out of the tank is the same as that inside the tank. In the case of an ideally stirred tank being a chemical reactor operated with a continuous inflow and outflow, the term **continuous stirred tank reactor** (CSTR) is widely used.

For the first time in this textbook, first-order linear differential equations have to be solved in the next example. The technique of solving this type of differential equation is presented in Appendix 1A.

Example 1.3. Coconut oil in an ideally stirred tank.

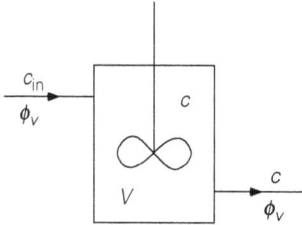

Figure 1.6

Look now at the situation in Figure 1.6 for an ideally stirred tank. There is a continuous and steady volumetric flow rate ϕ_V going into and out of the tank. This flow contains a certain concentration of palm seed oil c_{P0}. The tank is and remains completely full of palm seed oil. This is therefore a steady situation: flow rates and composition of the liquid do not change over time. At time $t = 0$, however, the supply of palm seed oil is stopped abruptly while at the same moment in time a liquid with a coconut oil concentration $c_{C,in}$ is supplied to the tank at the same volumetric flow rate.

The question now is how do the concentrations of palm seed oil and coconut oil change at the exit as a function of time?

Solution
To find the answer, a mass balance has to be drawn up, in accordance with the formula – first, for the palm seed oil:
1) sketch → see Figure 1.6
2) quantity → mass of palm seed oil → $M_P = V \cdot c_P$
3) control volume → volume of the tank V
4) mass balance (for period $t \geq 0$):

$$V\frac{d}{dt}c_P = \phi_V \cdot 0 - \phi_V \cdot c_P + 0 \tag{1.18}$$

From $t = 0$ the entrance concentration $c_{P,in}$ of the palm seed oil equals zero. Equation (1.18) can now be rewritten as

$$\frac{d}{dt}c_P = -\frac{\phi_V}{V}c_P \tag{1.19}$$

Equation (1.19) now has to be solved as follows (as a boundary condition)

$$t = 0 \rightarrow c_P = c_{P0}$$

This produces

$$c_P(t) = c_{P0} \cdot \exp\left(-\frac{\phi_V}{V}t\right) \tag{1.20}$$

According to the same recipe, a mass balance can also be drawn up for the coconut oil for the period $t \geq 0$:

$$V \frac{d}{dt} c_C = \phi_V c_{C,\text{in}} - \phi_V \, c_C + 0 \tag{1.21}$$

Equation (1.21) can be rewritten as

$$\frac{d}{dt} c_C = \frac{\phi_V}{V} - (c_{C,\text{in}} - c_C) \tag{1.22}$$

Solving this differential equation along with the initial condition $c_C = 0$ at $t = 0$ results in

$$c_C(t) = c_{C,\text{in}} \cdot \left[1 - \exp\left(-\frac{\phi_V}{V} \, t \right) \right] \tag{1.23}$$

The latter equation expresses at which rate the concentration of coconut oil in the tank becomes equal to the concentration $c_{C,\text{in}}$ at the inlet after the switch.

Summary

The general form of the species-mass balance for species A is

$$\frac{d}{dt} M_A = \phi_{m,A,\text{in}} - \phi_{m,A,\text{out}} + P_A$$

This balance can also be written in terms of concentrations and chemical reaction rate r_A:

$$\frac{d}{dt} V c_A = \phi_{V,\text{in}} \cdot c_{A,\text{in}} - \phi_{V,\text{out}} \cdot c_{A,\text{out}} \pm r_A V$$

The concentration of species A is denoted as c_A and, because of the chemical reaction(s), is in moles of A per unit of volume, with the rate r_A in moles of A per unit of volume and per unit of time.

Some examples (without chemical reactions) have shown how balance equations lead to first-order linear differential equations which can easily be solved by the separation of variables method.

1.2.2 Chemical reactors

In Example 1.3, there was no production term. However, in chemical reactors, production (and/or consumption) of one or more species is an essential feature. In such cases, the production term will have to be specified in order to be able to solve the balance.

This method will be illustrated for a **first-order chemical reaction** in which the rate of transformation of species A per unit of volume, denoted as r_A (in mol/m^{3}s), is directly proportional to the concentration c_A of A (in mol/m^3):

$$r_A = k_r \, c_A \tag{1.24}$$

Here, k_r is the **first-order reaction rate constant** (unit: s^{-1}). For a reaction in which species A is consumed, the production rate of A is given by $P_A = -r_A \cdot V$ and is negative. For the simple reaction A → B, the positive production rate of species B depends on just the concentration c_A of species A and is given by $P_B = +r_A \cdot V$.

We will now examine different types of reactor.

The batch reactor

Consider first an ideally stirred tank that is operated **batch-wise**: that is, the tank is filled in one go, after which nothing flows in or out. A is transformed according to the first-order reaction. This kind of reactor is known as a **batch reactor**. Given that there are no flows in or out of the reactor, the mass balance for substance A in the case of a constant reactor (control) volume is as follows:

$$V \frac{d}{dt} c_A = 0 - 0 + r_A V = r_A V = -k_r c_A V \tag{1.25}$$

The solution to equation (1.25) is obtained by using the pertinent boundary condition $c_A(t = 0) = c_{A0}$ representing the addition in one go of all reactive species to the tank:

$$\frac{c_A}{c_{A0}} = \exp(-k_r t) \tag{1.26}$$

As expected, quantity A in the tank reduces over time. After all, no A is supplied into the tank or withdrawn from the tank during the period in which A was being transformed.

The plug flow reactor

It is now the turn of the **ideal tubular reactor** for a chemical reaction to occur. In an ideal tubular reactor of this kind, each liquid package moves as quickly as all the others. This is known as **plug flow** and the reactor is denoted as a **plug flow reactor**. In our case, the reaction is again first order (with reaction rate constant k_r), where A is transformed into B.

Assuming that the concentration of substance A at the start of the reactor is c_{A0}, what will the concentration be at the tube exit?

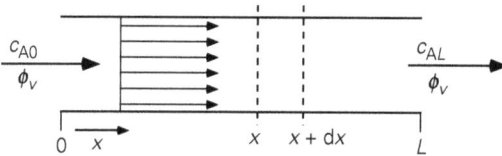

Figure 1.7

For the sake of simplicity, diffusion (will be covered later) is ignored and it is assumed that the liquid (in spite of the reaction) has a constant density. As a result of the chemical reaction, the concentration of A will steadily decrease towards the direction of the exit. To establish this pattern (the **profile**), a small slice of the tube somewhere in the reactor between x and $x + dx$ should be examined where x is the coordinate along the reactor in the direction of the flow (see Figure 1.7). Again, it is assumed that the condition is steady – that is, there are no changes over time. A mass flow of A, quantity $\phi_v \cdot c_A$, comes into the control volume through the left-hand boundary plane. Of course, the concentration taken is the one as it is at position x. This is shown as $c_A|_x$. Meanwhile, $\phi_V \cdot c_A$ flows out through the right-hand boundary plane, but now with c_A as it is at position $x + dx$ is $c_A|_{x+dx}$. Finally, the consumption of A inside the slice as a result of the chemical reaction is again represented by $r_A V$ where now $V = A dx$. The mass balance over the slice is now

$$0 = \phi_V \cdot c_A|_x - \phi_V \cdot c_A|_{x+dx} - k_r\, c_A \cdot A dx \tag{1.27}$$

in which A is the surface of the cross section of the tube (perpendicular to the direction of the flow). Notice that the left-hand side of the equation is zero, given that the state is assumed to be steady; in the control volume, it is therefore not possible for the mass of A to change. By dividing all terms of the right-hand side of the equation by dx and taking the limit $dx \to 0$, equation (1.27) can be written as follows:

$$\phi_V \frac{dc_A}{dx} = -k_r \cdot A \cdot c_A \tag{1.28}$$

By using the method of separation of variables and thanks to the boundary condition $c_A(x = 0) = c_{A0}$, the solution to this is

$$\frac{c_A(x)}{c_{A0}} = \exp\left(-\frac{k_r \cdot A \cdot x}{\phi_V}\right) \tag{1.29}$$

This means the concentration at the exit is

$$\frac{c_A(L)}{c_{A0}} = \exp\left(-\frac{k_r \cdot AL}{\phi_V}\right) = \exp\left(-\frac{k_r \cdot V}{\phi_V}\right) \tag{1.30}$$

in which $V = A \cdot L$ is the volume of the tubular reactor.

The above technique for drawing up a balance over a thin slice, of which the dimension goes to zero (limit $dx \to 0$), leads to a **micro-balance**, see equation (1.27), and provides detailed information (i.e. a **profile**). As the local rate of the chemical reaction depends on the local concentration that decreases along the tubular reactor, this approach of drawing up a micro-balance for a thin slice really is a must, although the original question was related only to the concentration at the reactor exit.

The continuous stirred tank reactor (CSTR)
Example 1.3 will now be expanded to include a chemical reaction. The reactor, a CSTR, contains a catalyst that accelerates the conversion of species A. The catalyst cannot leave the reactor. At first, a flow rate ϕ_V that does not contain any A passes through the reactor. From time $t = 0$, the incoming flow ϕ_V contains a concentration c_{A0} of A.

How does the concentration of A at the reactor exit progress as a function of time?

Again, a mass balance has to be drawn up and the resulting differential equation solved. Mass A is the quantity to be considered for the balance; and the control volume is the volume of the reactor. The balance is as follows:

$$\frac{d}{dt}Vc_A = \underbrace{\phi_V c_{A0}}_{\text{in}} - \underbrace{\phi_V c_A}_{\text{out}} - \underbrace{k_r c_A V}_{\text{production}} \qquad (1.31)$$

For constant reactor volume and constant ϕ_V this balance equation can be rewritten as

$$\frac{d}{dt}c_A = -\left(\frac{\phi_V}{V} + k_r\right)c_A + \frac{\phi_V}{V}c_{A0} \qquad (1.32)$$

This differential equation is a first-order linear differential equation of the type

$$\frac{dy}{dx} = \alpha y + \beta \qquad (1.33)$$

the solution of which is discussed in Appendix 1A.

With the initial condition that $c_A = 0$ at $t = 0$ (the reactor did not contain any A at the start), the solution to equation (1.31) is

$$c_A(t) = \frac{\phi_V/V}{\phi_V/V + k_r}c_{A0} \cdot \left\{1 - \exp\left[-\left(\frac{\phi_V}{V} + k_r\right)t\right]\right\} \qquad (1.34)$$

Notice that both ϕ_V/V and k_r have the unit s^{-1}. The eventual steady-state situation may be considered as a limit case and is obtained by substituting $t \to \infty$ into equation (1.34):

$$c_A = \frac{1}{1 + k_r\frac{V}{\phi_V}}c_{A0} \qquad (1.35)$$

This result can be directly derived from equation (1.31) too: in the eventual steady state the d/dt-term is zero.

When comparing the performances of the ideal tubular reactor and the ideally stirred tank, both with the same ϕ_V/V – ratio, it is immediately clear that the transformation in the tubular reactor is much greater. This is caused by a difference in the average concentration $\langle c_A \rangle$ between both reactors. For a reactant that depletes the following applies: in the tubular reactor, $\langle c_A \rangle$ is always greater than c_{AL}, while in the tank reactor $\langle c_A \rangle$ is actually always equal to the concentration at the exit. Given that the transformation is proportional to the concentration in the case of a first-order reaction, the transformation in the tubular reactor will therefore be greater.

Example 1.4. A plug flow reactor with recirculation.
For the purpose of exercise, we will now look at a practical situation in which part of the outgoing flow is fed back into the reactor. This is done in order to gain a better use of a reactant, if only a proportion of that reactant is transformed in a single passage; this means the length of the reactor does not have to be so long. It also means that in the case of an exothermic reaction (very common), the temperature at the

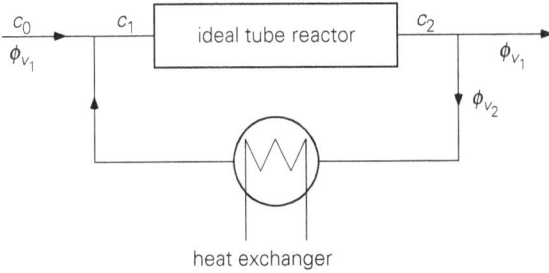

Figure 1.8

entrance and consequently also the conversion of A in the reactor can be jacked up 'cheaply' using part of the heat in the exit flow (after all, most reactions take place more quickly with higher temperatures). In other cases, a heat exchanger is often placed in the recirculation duct that allows some of the reaction heat to be extracted from the flow and used elsewhere. This situation is illustrated in Figure 1.8.

In the ideal tubular reactor, species A is transformed according to a first-order chemical reaction. The volumetric flow rate of the inflow and outflow is ϕ_{V_1}. In order to remove the reaction heat, a volumetric flow rate ϕ_{V_2} is recirculated by a heat exchanger. The concentration of A in inflow ϕ_{V_1} is c_0. The concentration that flows into the reactor is shown as c_1 ($\neq c_0$), while the concentration that leaves the reactor is referred to as c_2. For the sake of simplicity, it is assumed that the reaction only occurs in the reactor. The volume of the reactor is V and the situation is steady. How great is c_2 compared with c_0?

Solution
The flow rate $\phi_{V_1} + \phi_{V_2}$ going into the reactor can be found by drawing up a total mass balance for the T-junction in the duct before the plug flow reactor. Similarly, the concentration c_1 follows from drawing up the mass balance for A for the same T-junction:

$$0 = \phi_{V_1} c_0 + \phi_{V_2} c_2 - \left(\phi_{V_1} + \phi_{V_2}\right) c_1 \tag{1.36}$$

rearranging this equation and introducing the **recirculation ratio** $a \equiv \dfrac{\phi_{V_2}}{\phi_{V_1} + \phi_{V_2}}$; all this gives

$$c_1 = \frac{\phi_{V_1} c_0 + \phi_{V_2} c_2}{\phi_{V_1} + \phi_{V_2}} = (1 - a)\, c_0 + a\, c_2 \tag{1.37}$$

From the earlier discussion on the plug flow reactor, it follows that for the current configuration the exit concentration c_2 obeys to

$$\frac{c_2}{c_1} = \exp\left(-k_r \frac{V}{\phi_{V_1} + \phi_{V_2}}\right) \tag{1.38}$$

Eliminating c_1 from equation (1.38) with the help of equation (1.37) results into

$$\frac{c_2}{c_0} = \frac{1 - a}{\exp\left[(1 - a)k_r \frac{V}{\phi_{V_1}}\right] - a} \tag{1.39}$$

Two limit cases merit closer examination:

1) $a = 0$, that is, no recirculation: for this situation, it follows from equation (1.39) that

$$\frac{c_2}{c_0} = \exp\left(-k_r \frac{V}{\phi_{V_1}}\right) \tag{1.40}$$

Of course, this is the result for the ideal tubular reactor of equation (1.30): after all, there is now no recirculation.

2) $a \to 1$, that is, the recirculation flow is much greater than ϕ_{V_1}. Now, with the help of the series expansion $\exp x \to 1 + x$ for $x \to 0$, equation (1.39) can be approximated as follows:

$$\frac{c_2}{c_0} = \frac{1}{1 + k_r \frac{V}{\phi_{V_1}}} \tag{1.41}$$

This is precisely the result that was obtained before – see equation (1.35) – for an ideally stirred tank in a steady-state situation!

! **Summary**

In this section, several chemical reactors have been examined: the batch reactor, the plug flow reactor (being the ideal tubular reactor), the continuous stirred tank reactor (CSTR), and the plug flow reactor with recirculation. A first-order reaction was assumed in every case.

Every time, one or more mass balances have to be drawn up for a suitable control volume. In many cases, this control volume was the full reactor vessel and the balance was a macro-balance. In the case of a tubular reactor, however, you have to draw up a micro-balance for a thin slice somewhere in the reactor, the slice thickness being dx with $dx \to 0$; this procedure leads to a differential equation from which the concentration profile $c(x)$ inside the reactor follows.

1.2.3 Residence time distribution

Consider a small liquid package which, as part of a volume flow ϕ_V, flows into and through a tube of volume V, cross-sectional area A, and length L. At time $t = 0$ the package (or element) has just entered the tube. At time t it has covered a distance of $\ell = v \cdot t$ (see Figure 1.9); therefore, at time $\tau = L/v$, the element is at the end of the tube. Time τ is referred to as the **residence time**.

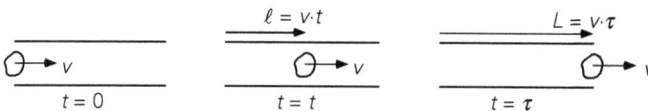

Figure 1.9

If every package in a flow has the same velocity $v = \phi_V/A$, then the same residence time τ applies to every package:

$$\tau = \frac{L}{v} = \frac{AL}{\phi_V} = \frac{V}{\phi_V} \tag{1.42}$$

A flow in which every liquid package (or element) has the same residence time is described as **plug flow** (see also Figure 1.7).

In reality, the packages do not generally have one and the same residence time. After all, not every package has the same velocity and nor do they all cover the same route. In reality, the highest velocity in a tube is at the axis, while along the wall it is actually zero; this makes plug flow an ideal type of flow that in fact never occurs. In a stirred tank, the packages follow very different paths precisely because of the mixing. The result is that in reality, there is always a residence time distribution in every device, as well as a **mean residence time** τ, shown by

$$\tau = \frac{V}{\phi_V} \tag{1.43}$$

Knowledge of this residence time distribution can be important for a successful operation of the process under consideration. An example that comes to mind is that of the sedimentation ('settling') of fixed particles in a sewage treatment plant. The residence time of *all* the packages of liquid has to be sufficiently long for all the particles in them to settle.

There are several functions by which the residence time distribution can be characterised; they will be discussed one-by-one below. The non-dimensional time $\theta = t/\tau$ is used in every case, with τ defined through equation (1.43).

The *E*-function

First of all, the residence time distribution can be characterised with the help of the 'ages' of the packages when they pass the exit: this way the age distribution at the exit can be obtained, which is known as the *E*-function. The definition of the *E*-function ('exit age function') is

$$E(\theta)\, d\theta = \text{volume fraction of the outgoing flow with a} \atop \text{non-dimensional residence time of between } \theta \text{ and } \theta + d\theta \tag{1.44}$$

The recipe for determining $E(\theta)$ should be: take a sample from the outgoing flow and determine the fraction with a non-dimensional residence time of between θ and $\theta + d\theta$. Do this for every θ between 0 and ∞. An example of an *E*-function is given in Figure 1.10.

Notice that the integral of $E(\theta)$ for interval $[0,\infty]$ is equal to 1. This can be understood by realising that

$$\int_0^{\theta_1} E(\theta)\, d\theta = \text{volume fraction of the sample of the outgoing} \atop \text{flow with residence time } \theta \leq \theta_1 \tag{1.45}$$

Every package that leaves the device via the outflow has only had a finite residence time in the device. For this reason, equation (1.45) means

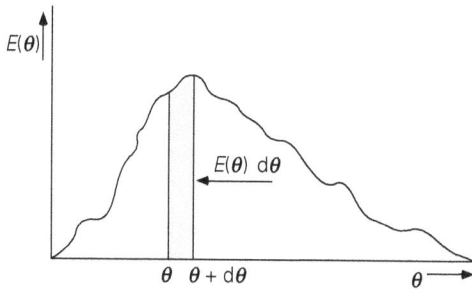

Figure 1.10

$$\int_{0}^{\infty} E(\theta)\, d\theta = 1 \tag{1.46}$$

The mean residence time τ can be calculated from the E-function according to

$$\tau = \int_{0}^{\infty} t\, E(\theta)\, d\theta \tag{1.47}$$

while the standard deviation σ of the residence time distribution can be described by

$$\sigma^2 = \int_{0}^{\infty} (t'-1)^2\, E(t')\, dt' \tag{1.48}$$

This method of describing the residence time distribution is generally not very practical – after all, how do you determine how long a package has stayed in a device?

In practical situations, usually experiments are therefore conducted in which a known change is applied to the inflow of a device and the response to this at the exit is measured as a function of time in order to find the residence time distribution that goes with that particular device. Incidentally, this procedure is not only used for a single device but also for a combination of devices such as in a plant.

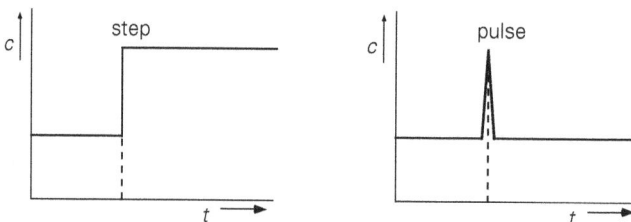

Figure 1.11

The most commonly occurring entrance changes are the **step** and the **pulse** (see Figure 1.11). In the case of the step, a sudden change is applied to the concentration of some marked substance (marker, or tracer) in the inflow. This new level is then maintained. The pulse scenario involves a brief 'injection' of the tracer in the inflow (the injection is so small in size and short in time that the flow rate ϕ_V remains more or less constant).

The C-function

The *C*-function shows the time-dependent response at the exit of a device to **pulse**-induced tracer material at the entrance. This process is illustrated in Figure 1.12.

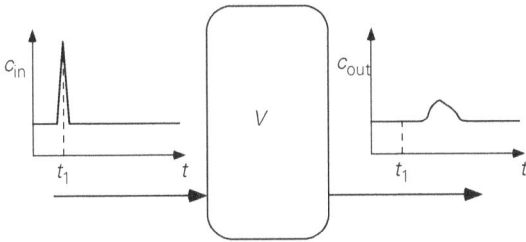

Figure 1.12

The total quantity of material A injected into the device amounts to M_A. For the average concentration c_0 of A in the device, if all of A is inside it, it then follows that $c_0 = M_A/V$ (where V is the volume of the device). The *C*-function is now defined as

$$C(\theta) = \frac{c_{tr,out}}{c_0} \tag{1.49}$$

$c_{tr,out}$ is the concentration of the tracer (denoted by index tr) at the exit. The mass balance (for $t > 0$) for the tracer material in the device is

$$\frac{d}{dt} M_{tr} = \phi_V\, c_{tr,in} - \phi_V\, c_{tr,out} = -\phi_V\, c_{tr,out} \tag{1.50}$$

In equation (1.50), $c_{tr,in} = 0$ given that the tracer material is applied in a pulse at $t = 0$ and the balance is drawn up for $t > 0$.

The integral of the *C*-function from 0 to ∞ is also equal to 1. Proof of this follows from the integration of the left-hand side of equation (1.50) from $t = 0$ to $t \to \infty$:

$$\int_0^\infty \frac{d}{dt} M_{tr}\, dt = M_{tr}\big|_0^\infty = -M_{tr}(t = 0) = -V\, c_0 \tag{1.51}$$

where $M_{tr}(t \to \infty) = 0$ is used since all the tracer material will exit the device sooner or later. Combining equations (1.50) and (1.51) produces

$$V c_0 = \int_0^\infty \phi_V \, c_{tr,out} \, dt \tag{1.52}$$

Dividing both the right-hand side and the left-hand side of this equation by $V c_0$ gives

$$1 = \int_0^\infty \frac{\phi_V}{V} \frac{c_{tr,out}}{c_0} \, dt = \int_0^\infty C(\theta) \, d\theta \tag{1.53}$$

The recipe for determining the C-function is: follow the tracer concentration change at the exit of the device after a tracer pulse has been applied at the entrance. The resulting curve should satisfy the condition of equation (1.53).

It is possible to demonstrate that in a steady situation the E- and C-functions are identical: $C(\theta) = E(\theta)$. The methods by which they are determined are essentially different, however: the E-function is determined from a single sample, while in the case of the C-function, the response at the exit has to be monitored for a sufficiently long period of time.

The F-function

The F-function is another function that can be found by following the concentration in the exit as a function of time. The F-function represents the response of the exit to a step function at the entrance – for example, $c = 0$ to $c = c_{in}$ (see Figure 1.13):

$$F(\theta) = \frac{c_{out}}{c_{in}} \tag{1.54}$$

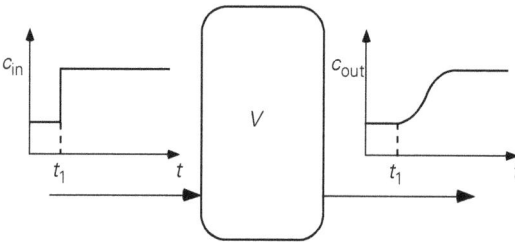

Figure 1.13

The F-function has some characteristic properties. For example, $F \to 1$ for $t \to \infty$, that is, for $\theta \to \infty$: ultimately, the concentration in the exit is equal to that at the entrance. A possible shape of the F-function is shown in Figure 1.14.

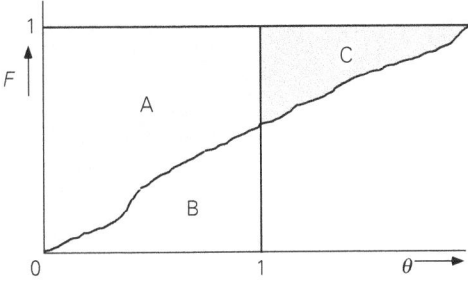

Figure 1.14

There is another characteristic that every F-function has: the shaded areas B and C in Figure 1.14 are of equal size. This can be derived as follows: the mass balance for the tracer material in the device is $(t \geq 0)$

$$\frac{d}{dt} M_{tr} = \phi_V \, c_{in} - \phi_V \, c_{out} \tag{1.55}$$

Integration from $t = 0$ to $t \to \infty$ produces

$$\int_0^\infty \frac{d}{dt} M_{tr} \cdot dt = \int_0^\infty \phi_V (c_{in} - c_{out}) dt \tag{1.56}$$

The left-hand side of this equation can again be rewritten as

$$\int_0^\infty \frac{d}{dt} M_{tr} \cdot dt = [M_{tr}]_0^\infty = V \, c_{in} \tag{1.57}$$

After all, at $t = 0$ the device does not contain any tracer yet, so $M_{tr}\,(t = 0) = 0$. Combining equations (1.56) and (1.57) produces

$$1 = \int_0^\infty \frac{\phi_V}{V} \left(1 - \frac{c_{out}}{c_{in}} \right) dt = \int_0^\infty [1 - F(\theta)] \, d\theta \tag{1.58}$$

In other words, the sum of the areas A and C in Figure 1.14 is exactly equal to 1: $A + C = 1$. At the same time, the sum of areas A and B is equal to 1 (the length of both sides of the square AB is 1). The conclusion is therefore that areas B and C are equal, regardless of the form of the F-function.

The physical significance of the F-function is that F is equal to the volume fraction in the outgoing flow with a non-dimensional residence time of less than θ. From the definitions of C and E, it follows that $C(\theta) \, d\theta$ is the volume fraction in the outgoing flow with a residence time of between θ and $\theta + d\theta$; the following therefore applies:

$$F(\theta) = \int\limits_0^\theta C(\theta')\, d\theta' \; = \; \int\limits_0^\theta E(\theta')\, d\theta' \tag{1.59}$$

The second equality applies only in a steady-state situation. An extremely practical relationship follows from equation (1.59):

$$C(\theta) = \frac{dF(\theta)}{d\theta} \tag{1.60}$$

It is very useful since the discrete character of the pulse function prevents its use as a boundary condition when trying to calculate a C-function for an ideal type of flow device from the differential equation that results from a species mass balance. The step function, however, does make the use of a boundary condition in the calculation of an F-function feasible. The C-function can then be calculated from the F-function thanks to the property of equation (1.60). See Examples 1.5 and 1.6 below.

Real-world and ideal flow devices
E-, C-, and F-functions characterise, each in their own way, the state of mixing in a vessel, or system, and in this way how a system responds to changes from outside. In the real world, it is possible and rather easy to just measure the F- and C-functions. In Figure 1.15 some C-functions are presented. Figure 1.15(a) shows a response in which the distribution is rather nicely centred around the mean residence time τ as a result

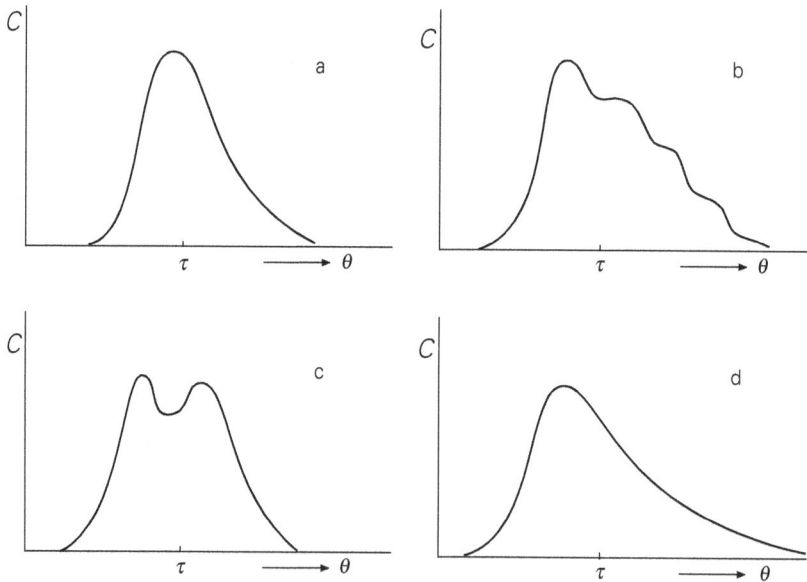

Figure 1.15

of a rather sound flow or mixing state of the tank under review. In Figure 1.15(b), however, the stepwise drop in the response for times longer than τ may be indicative of an internal circulation. A dual peak such as in Figure 1.15(c) may suggest that the liquid flow may get split upon entering the vessel with the two liquid streams reaching the exit via separate paths of unequal length. In Figure 1.15(d), the maximum of the distribution function arriving earlier than τ as calculated with equation (1.43) may be due to short-circuiting between entrance and exit, some parts of the vessel acting as 'dead zones' which do not really take part in the through flow.

By comparing a measured response with a set of calculated residence distribution functions, it might be possible to roughly characterise the flow or mixing state of a tank or system as ideally stirred, a plug flow, a number of ideally stirred tanks in series, or a combination of plug flow and ideally stirred tanks.

We will therefore now turn to various types of ideal flow devices – such as one of more ideally stirred tanks or a plug flow – for which F- and C-functions can be calculated. Remember that, since these functions characterise the residence time distribution, chemical reactions can and indeed must be disregarded. After all, the residence time distribution in a particular device is determined solely by the flow or mixing state.

Example 1.5. F- and C-functions for an **ideally stirred tank**.
Calculate the F- and C-functions for the ideally stirred tank. See Figure 1.6 and also Example 1.3.

Solution
In Example 1.3, coconut oil replaces palm seed oil at the entrance at, say, $t = 0$. In other words, this is a 'step' from 0 to c_{in} in the concentration of coconut oil at the entrance. In order for the F- and C-functions to be determined, it is necessary to determine the concentration of the tracer (the coconut oil) at the exit as a function of the time. The mass balance for coconut oil is given by equation (1.21) and can be shown to result in the expression for the exit concentration given by equation (1.23). With $\tau = V/\phi_v$ and $\theta = t/\tau$ it follows for the F-function that

$$F(\theta) = \frac{c_{out}}{c_{in}} = 1 - \exp(-\theta) \tag{1.61}$$

and therefore, thanks to equation (1.60), for the C-function that

$$C(\theta) = \frac{dF}{d\theta} = \exp(-\theta) \tag{1.62}$$

Example 1.6. F- and C-functions for two ideally stirred tanks in series.
Consider two ideally stirred tanks in series, with inflow and outflow, connected as shown in Figure 1.16. The tanks have identical volumes V. When $t = 0$, and with a constant volumetric flow rate ϕ_v, the concentration of some tracer at the entrance is suddenly changed stepwise from $c = 0$ to $c = c_0$. Determine the F- and C-functions.

Solution
In order to be able to determine the F- and C-functions of the system of these two tanks, it is necessary to know what the concentration at the exit is – that is, $c_2(t)$.

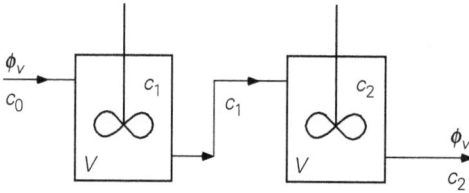

Figure 1.16

Mass balance of tank 1:

$$V\frac{dc_1}{dt} = \phi_V(c_0 - c_1) \quad \text{with } c_1(0) = 0 \tag{1.63}$$

Mass balance of tank 2:

$$V\frac{dc_2}{dt} = \phi_V(c_1 - c_2) \quad \text{with } c_2(0) = 0 \tag{1.64}$$

The solution to equation (1.63) is

$$c_1 = c_0\left[1 - \exp\left(-\frac{\phi_V}{V}t\right)\right] \tag{1.65}$$

Equation (1.64) is not so easy to solve because c_1 is a function of the time. Substituting the solution for c_1 – equation (1.65) – into equation (1.64) results in:

$$V\frac{dc_2}{dt} = \phi_V c_0\left[1 - \exp\left(-\frac{\phi_V}{V}t\right)\right] - \phi_V c_2 \quad \text{with } c_2(0) = 0 \tag{1.66}$$

This is an inhomogeneous linear differential equation. Several methods for solving such an inhomogeneous differential equation are presented in Appendix 1B. The solution to equation (1.66) is

$$c_2(t) = c_0\left[1 - \left(1 + \frac{\phi_V}{V}t\right)\cdot\exp\left(-\frac{\phi_V}{V}t\right)\right] \tag{1.67}$$

With definitions: $\tau = 2V/\phi_V$ (as the overall volume of this system is what matters, viz. $2V$) and $\theta = t/\tau$, it follows for the F-function that

$$F(\theta) = \frac{c_2(\theta)}{c_0} = 1 - (1 + 2\theta)\exp(-2\theta) \tag{1.68}$$

and for the C-function that

$$C(\theta) = \frac{dF}{d\theta} = 4\theta\exp(-2\theta) \tag{1.69}$$

Example 1.7. F-function in the case of a plug flow.
The ideal plug flow reactor was discussed in Section 1.2.2 (see Figure 1.7). We are concerned here with the F-function of the ideal plug flow, without a chemical reaction taking place. In a plug flow, every liquid package has velocity v. Now, from $t \geq 0$, a marker substance is injected into the inflow (concentration c_0). Each liquid element has exactly the same residence time τ as given by equation (1.42).

Solution
There is therefore no residence time distribution involved: no marker substance comes out before $t < \tau$, while for $t \geq \tau$ the marker substance always comes out with concentration c_0. For the *F*-function, therefore, the following applies:

$$F(\theta) = 0 \quad \text{for} \quad \theta < 1$$
$$F(\theta) = 1 \quad \text{for} \quad \theta \geq 1$$

(1.70)

Summary

In a device through which a flow is passing, the packages do not take the same length of time from the point of entry to the point of exit. In this context, the terms residence time, mean residence time, non-dimensional (residence) time, and residence time distribution have been introduced. Knowledge of this residence time distribution is important in order to be able to properly operate the device. Three functions have been discussed for the purpose of showing the residence time distribution:

the *E*-function: shows the residence times of the parts of a sample from the outgoing flow **at one point in time**;

the *C*-function: shows the reaction at the exit to a **pulse** at the entrance;

the *F*-function: shows the reaction at the exit to a **step** at the entrance.

A relation between the *F*- and *C*-functions can be derived:

$$C(\theta) = \frac{dF}{d\theta}$$

For a system consisting of one or more ideal devices, expressions can be derived for the *F*-function of the system. In the real world, *C*- and *F*-functions can be measured only.

1.2.4 Multiple tanks in series

The ideal tubular reactor of Section 1.2.2 (Figure 1.7) can be mimicked by putting a large number of ideally stirred tank reactors in series (see Figure 1.17) and operating them in a steady state. This will be proven below by using balances.

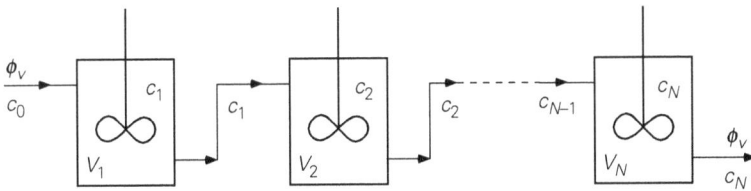

Figure 1.17

After the first tank, the time-independent concentration is (see Section 1.2.2)

$$\frac{c_1}{c_0} = \left(1 + k_r \frac{V_1}{\phi_v}\right)^{-1} = \left(1 + \frac{k_r V/\phi_v}{N}\right)^{-1}$$

(1.71)

$V_1 = V_2 = \ldots = V/N$ (N is the number of tanks in series and V the volume of all the tanks combined). Concentration c_1 is the concentration of the inflow into tank 2, so the following applies

$$\frac{c_2}{c_1} = \left(1 + \frac{k_r V/\phi_v}{N}\right)^{-1} \tag{1.72}$$

or with the help of equation (1.71)

$$\frac{c_2}{c_0} = \left(1 + \frac{k_r V/\phi_v}{N}\right)^{-2} \tag{1.73}$$

Finally, the concentration after the Nth tank is found as follows:

$$\frac{c_N}{c_0} = \left(1 + \frac{k_r V/\phi_v}{N}\right)^{-N} \tag{1.74}$$

For greater N, the tank reactors in series will increasingly resemble the ideal tubular reactor. Each slice of a plug flow reactor may be conceived as a small ideally mixed reactor. Ultimately, in the limit $N \to \infty$, both systems are the same:

$$\lim_{N \to \infty} \left(1 + \frac{k_r \tau}{N}\right)^{-N} = \exp(-k_r \tau) \tag{1.75}$$

In Figure 1.18, the curves of c_N are shown as a function of $k_r\tau$, with $\tau = V/\phi_v$, for a number of values of N. The result of the ideal plug flow reactor is also shown.

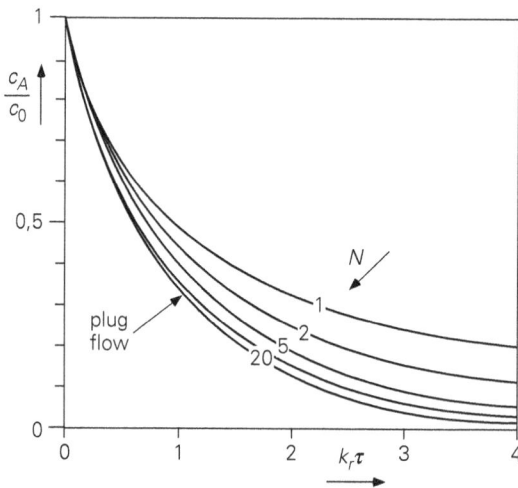

Figure 1.18

Example 1.8. F- and C-functions for N ideally stirred tanks in series.
In order to determine the F- and C-functions of N tanks in series, N transient mass balances must be considered without a source term for chemical reactions. This situation is illustrated in Figure 1.19.

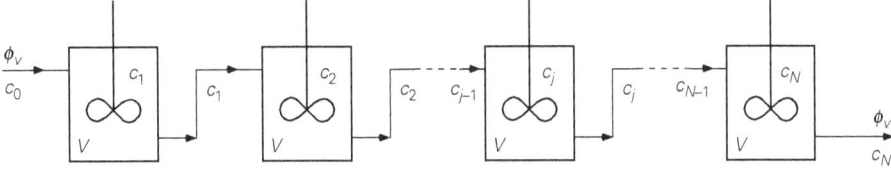

Figure 1.19

The tanks all have an equal volume, V. When $t = 0$, and with a constant volume flow, the inflow at the entrance is suddenly changed from $c = 0$ to $c = c_0$ (step method). Determine the F- and C-functions.

Solution
In order to be able to determine the F-function, it is necessary to know what the concentration at the exit is – that is, $c_N(t)$.

Mass balance of tank 1:

$$V\frac{dc_1}{dt} = \phi_V(c_0 - c_1) \quad \text{with } c_1(0) = 0 \tag{1.76}$$

Mass balance of tank j:

$$V\frac{dc_j}{dt} = \phi_V(c_{j-1} - c_j) \quad \text{with } c_j(0) = 0 \tag{1.77}$$

or:

$$\frac{dc_j}{dt} - \frac{N}{\tau}c_{j-1} + \frac{N}{\tau}c_j = 0 \quad \text{in which} \quad \tau = \frac{NV}{\phi_V} \tag{1.78}$$

The solution to equation (1.78) is given by:

$$c_j = \exp\left(-\frac{Nt}{\tau}\right)\int_0^t \frac{N}{\tau}c_{j-1}\exp\left(\frac{Nt}{\tau}\right)dt \tag{1.79}$$

In principle, applying this to a random tank in the series is not difficult, though it is laborious – after all, the concentration of all the tanks before it must first be calculated. By substituting $\zeta = \exp(Nt/\tau)$ and dividing by c_0, equation (1.79) is rewritten as

$$\frac{c_j}{c_0} = \frac{1}{\zeta}\int_1^\zeta \frac{c_{j-1}}{c_0}d\zeta \tag{1.80}$$

Making the calculation for one tank is easy:

$$\frac{c_1}{c_0} = \frac{1}{\zeta}\int_1^\zeta d\zeta = \frac{\zeta-1}{\zeta} = 1 - \frac{1}{\zeta} \tag{1.81}$$

It is not difficult for two tanks either:

$$\frac{c_2}{c_0} = \frac{1}{\zeta} \int_1^\zeta \frac{c_1}{c_0} d\zeta = \frac{1}{\zeta} \int_1^\zeta \left(1 - \frac{1}{\zeta}\right) d\zeta = \frac{1}{\zeta} (\zeta - 1 - \ln \zeta) = 1 - \frac{1}{\zeta} (1 + \ln \zeta) \qquad (1.82)$$

For three tanks, the task is a little more laborious and produces this:

$$\frac{c_3}{c_0} = \frac{1}{\zeta} \int_1^\zeta \frac{c_2}{c_0} d\zeta = \frac{1}{\zeta} \int_1^\zeta \left[1 - \frac{1}{\zeta}(1 + \ln \zeta)\right] d\zeta = \frac{\zeta - 1}{\zeta} - \frac{1}{\zeta} \ln \zeta - \frac{1}{2\zeta}(\ln \zeta)^2 = 1 - \frac{1}{\zeta}\left[1 + \ln \zeta + \frac{1}{2}(\ln \zeta)^2\right] \qquad (1.83)$$

For N tanks, the pattern continues, fortunately, and the following applies:

$$\frac{c_N}{c_0} = 1 - \frac{1}{\zeta}\left[1 + \ln \zeta + \frac{1}{2}(\ln \zeta)^2 + \ \dots \ + \frac{1}{(N-1)!}(\ln \zeta)^{N-1}\right] \qquad (1.84)$$

The F-function follows by substituting $\theta = t/\tau$ and $\zeta = \exp(N\theta)$ into equation (1.84):

$$F(\theta) = 1 - \exp(-N\theta) \cdot \left[1 + N\theta + \frac{1}{2}(N\theta)^2 + \ \dots \ + \frac{1}{(N-1)!}(N\theta)^{N-1}\right] \qquad (1.85)$$

The C-function follows from $dF(\theta)/d\theta$.

In Figure 1.20, $F(\theta)$ has been plotted as a function of θ; from which it appears that $F(\theta)$ becomes steeper and steeper as N increases. In the event that N is infinitely large, there is no longer any distribution in residence time: the F-function becomes equal to that for a tube with plug flow. The volume of every tank has to be infinitely small in that case; otherwise the residence time itself is infinite.

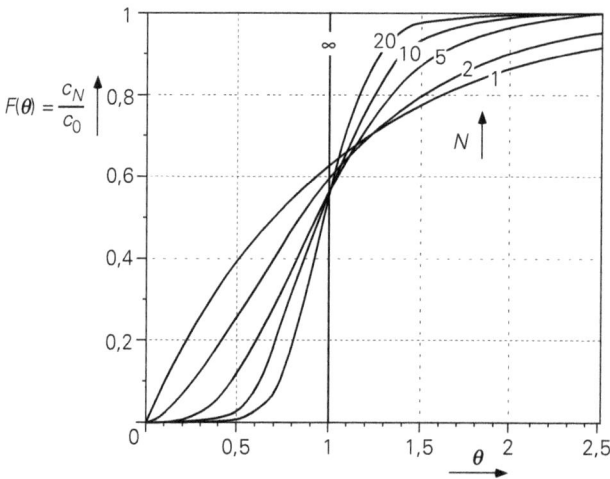

Figure 1.20

Summary

A plug flow reactor can be approximated by a large number of ideally stirred tank reactors in series. Similarly, the F-function of a series of infinitely small ideally stirred tanks approaches the F-function of a plug flow.

1.3 The energy balance

1.3.1 Introduction

The main topic in Section 1.2 was mass. In this section, the focus will be on energy. Yet, we will exploit here the same technique and the same recipe of drawing up a balance about a control volume. In addition, an energy balance exhibits the same, familiar basic form:

$$\frac{d}{dt} = \text{in} - \text{out} + \text{production}$$

Energy manifests itself in many forms, which can often be easily converted into each other, although in such transformation processes some energy is usually lost in the form of useless 'heat'. To some extent, such an energy transformation is similar to a chemical reaction that converts one particular species into one of more other ones.

We will use capital letters to denote amounts of energy present in a control volume, just like we used M for the total mass (in kg) in a control volume. Here, we will use capitals for the so-called extensive variables: E for the total amount of energy in a control volume, H for the total enthalpy, U for the total amount of thermal energy, and S for total entropy, all in J.

On the analogy of working with density ρ and with species concentrations c, all in kg/m^3, in the context of mass balances, with a view to energy balances we will use small letters for **energy concentrations** such as e, h, and u, all with the unit J/kg. We explicitly choose here, for working on a mass basis, to bypass effects of temperature and pressure on the control volume.

The table below gives an overview of the most commonly occurring forms of energy in terms of such energy concentrations:

Form	Energy concentration (expressed per unit of mass)
Kinetic energy	$\frac{1}{2}v^2$
Potential energy	gz
Internal energy	u
Pressure energy	p/ρ
Enthalpy	$h \, (= u + p/\rho)$

The internal energy (also known as **thermal** energy) stands for:
(i)   the thermal movements that individual molecules make given their degrees of freedom for translation, rotation (including internal), and vibration and
(ii) the short-range molecular interaction between individual molecules.

This means the internal energy also depends on the form and composition of the molecules. Internal energy can therefore be regarded as a physical property.

Where there is a constant volume, $du = c_v\, dT$, and $c_v T$ is often filled in for u (see the intermezzo below). For liquids (because of their low thermal expansion coefficient), a good approximation is $c_v = c_p$, and therefore also $u = c_p T$.

Intermezzo

According to the first law of thermodynamics, the following applies to incremental reversible changes in a system of constant composition and mass:

$$dU = T\, dS - p\, dV$$

The entropy, S, of a system can be regarded as a function of $\{T,V\}$. This means that dS can be written as follows:

$$dS = \left(\frac{\partial S}{\partial T}\right)_V dT + \left(\frac{\partial S}{\partial V}\right)_T dV$$

So for changing the overall internal energy, U, we can also write:

$$dU = T\left(\frac{\partial S}{\partial T}\right)_V dT + \left[T\left(\frac{\partial S}{\partial V}\right)_T - p\right] dV$$

The first term on the right-hand side represents the specific heat at constant volume, defined as:

$$c_v = T\left(\frac{\partial s}{\partial T}\right)_V$$

where we have switched from the extensive variable S (in J/K) to the specific entropy s (in J/kg K). For a system with a constant volume, it is of course only the specific heat that contributes towards to the change in specific internal energy:

$$du = c_v\, dT$$

Remember that for a system with constant volume this is an exact term, even if the specific heat is in principle a function of the temperature, for example. In integral form, the change in internal energy, again with a constant volume, is shown by:

$$u - u_{ref} = \int_{T_{ref}}^{T} c_v dT \approx c_v\,(T - T_{ref}) \quad \text{for constant } V$$

The first equality is again exact, while in the case of the second, a calculation is used whereby the specific heat is constant in the first order if T is not very different from the reference temperature T_{ref}.

For the internal energy concentration, u, we may simply use $u = c_v T$ from now on, as in an energy balance about a control volume we are always interested in a change of u from inlet to outlet and then the reference values may cancel. Further, when taking a (time or spatial) derivative of u, the constant u_{ref} drops out. If a phase transition is involved, caution should be exercised however: the latent heat (Δh, again in J/kg) of the phase transition should be taken into consideration in such cases.

The p/ρ term from the above table allows for a sort of 'potential' energy and indicates the capacity of a gas or liquid to flow under the influence of differences in pressure. In terms of energy, pressure energy can be transformed into kinetic energy for example:

a liquid (or gas) starts flowing under influence of differences in pressure in the liquid (or the gas). This flow runs naturally from areas of high pressure to those of low pressure.

It can also be seen in the first law of thermodynamics that pressure is a kind of energy concentration: $dU = TdS - pdV$. The change to the overall internal energy (dU) of the system is caused in part by the work term, pdV. In other words, p can be regarded as an energy concentration per unit of volume and p/ρ therefore as energy concentration per unit of mass. By using intensive variables ($T, p, \ldots$) and extensive variables ($U, S, V, \ldots$), it is easy to derive in thermodynamics that $U = TS - pV$. So for the internal energy per unit of mass we find $u \equiv \frac{U}{M} = TS - \frac{p}{\rho}$, which confirms the above interpretation of p/ρ.

In many cases, the total energy of a system is the sum of the internal, potential, and kinetic energy. In terms of energy concentrations, we can define the following:

$$e = u + gz + \frac{1}{2}v^2 \tag{1.86}$$

Remember that the energy concentrations are determined in relation to a suitably selected reference level. In the case of balances, however, it generally concerns **changes** to energy concentrations; reference levels are often not relevant here. Nonetheless, it is necessary to be prepared for phase transitions and their associated specific energy (enthalpy) changes Δh (in J/kg). In addition, all energy concentrations in equation (1.86) should preferably be expressed in the unit J/kg; this also applies to the kinetic and potential energy.

In most 'transport phenomena', the above definition of e is sufficient. Where chemical reactions play a role, consideration must also be given to the **chemical enthalpy** (often in J/mol) that accounts for the atomic structure of a molecule and the pertinent interactions of nucleus and electrons. When methane is burnt, for example, chemical enthalpy is released in the form of heat: the energy 'stored' in a CH_4 mole becomes available, as the products CO_2 and H_2O are ranked lower down the energy ladder. Although chemical enthalpy changes are often dominant in chemical reactors, this form of energy is not included in our definition of e in equation (1.86) and will be disregarded from now on.

Another important concept has to be introduced: **energy dissipation**. Dissipation stands for the fact that the transformation from one form of energy into another is always accompanied by friction losses, resulting in a (minor) temperature increase which is no longer useful (*note*: a perpetuum mobile does not exist). The useful forms, like kinetic, potential, and (partly, see below) pressure energy, all ultimately end up as internal energy. In more general terms, dissipation is the transformation of (for us as people the mostly useful) 'mechanical energy' into 'thermal energy' (i.e. to chaotic kinetic energy of individual molecules). The mechanism for dissipation is **friction**.

Example 1.9. The bouncing ball.
If a tennis ball is released at a particular height above the ground, it loses potential energy on its way down and gains kinetic energy. At the same time, a small amount of the ball's energy is transferred to the air, which then starts to flow and therefore also gains kinetic energy. In addition, energy is dissipated as a result of friction with the surrounding air. When the ball 'collides' with the ground, the kinetic energy is transformed into 'elastic energy' and some of it is dissipated into useless heat.

After the bounce, the elastic energy is again transformed into kinetic energy which in turns changes again into potential energy (and a small part of which again 'disappears' into the air) until all the kinetic energy is exhausted and the ball reaches its new 'highest' point. The cycle repeats itself several times of course, until the ball eventually rests motionless on the ground and all the kinetic and potential energy has been dissipated.

! **Summary**
There are many manifestations of energy which can all be converted into each another. The focus was here on internal, or thermal, energy u and on pressure energy p/ρ. The physical backgrounds of these two 'types' of energy have been discussed. The concept of 'energy dissipation' has also been highlighted – the one-way traffic through which mechanical energy is converted into useless thermal energy (heat) by friction.

As a preparation to drawing up energy balances, the concept of 'energy concentrations' has been introduced and the specific total energy concentration has been defined $e = u + \frac{1}{2}v^2 + gz$.

We use small letters for energy concentrations (in J/kg), while capital letters denote extensive variables (in J).

1.3.2 The total energy balance

Energy is a **conserved quantity**, like mass. This is an important starting point in physics. For the balance equation it means that the production term equals zero, providing that the total energy really is considered – in order words, including types of energy (chemical, electrical, nuclear, . . .) that are not part of equation (1.86). If we therefore consider the total energy of a system with a given volume, the energy content can only change through the inflow or outflow of energy through the boundaries of this volume:

$$\frac{dE}{dt} = \text{energy flow in} - \text{energy flow out}$$

There is therefore no production term in this equation because energy cannot be produced from nothing. The above equation is a **conservation law**, entirely comparable with equation (1.3) for mass.

When drawing up an energy balance for a specific **type** of energy (or a combination of types) it is important not just to take the transport terms of the selected energy type into account, as in the energy conservation law above, but also to include the transformation (the production) of the type selected into other types (and vice versa) in the balance. The general form of a balance for a specific type of energy has the same structure as the species mass balance, equation (1.5):

$$\frac{d}{dt} = \text{flow in} - \text{flow out} + \text{production} \tag{1.87}$$

From now on, we will concentrate on the total energy of a system defined as the sum of the internal, the potential, and the kinetic energy: $E = U + E_{pot} + E_{kin}$, with the related energy concentration $e = u + gz + \frac{1}{2}v^2$ Figure 1.21 gives an illustration of a control volume, with the possible energy flows associated with it.

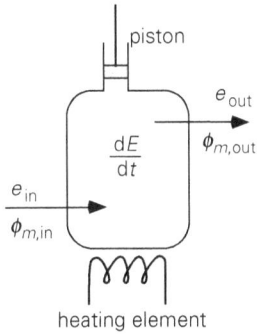

Figure 1.21

The unsteady-state (or transient) term of the left-hand side of the total energy balance which shows how quickly the total energy within the control volume V changes is

$$\frac{d}{dt} E$$

With the help of the energy concentration e (per unit of mass), this term can be written as:

$$\frac{d}{dt} E = \frac{d}{dt} (\rho V e) \tag{1.88}$$

Several forms of **energy transport** are now possible:
- An energy flow as a result of **mass** flow rate into (or out of) the control volume. This mass itself contains energy. The mass flow carries the energy 'under its arm', as it were. Analogous to the transport terms 'in' and 'out' of species A in equations (1.9) and (1.10), we also write the transport terms 'in' and 'out' for energy (in J/s) as the product of a flow rate and a concentration, that is, as a mass flow rate (in kg/s) times energy concentration (J/kg):

$$\phi_{m,in} \cdot e_{in} \text{ and } \phi_{m,out} \cdot e_{out}$$

- An energy flow as a result of **heat** (via a heating element, e.g.) flowing into the control volume (or flowing out with the help of a cooling element). This energy flow does not involve a mass flow. We denote this flow as follows:

$$\phi_q$$

- An energy flow, which also crosses the boundaries of the control volume, as a result of **work** performed per unit of time by or on the 'outside world'. Such work may be performed from outside on the control volume by means of a piston, stirrer, pump, or compressor – that is, via piston rods or revolving axles. Work can also be performed by the outside world on the control volume as a result of a mass flow rate flowing inwards across one of the boundaries of the control volume against the prevailing pressure there. Conversely, work can also be performed from the control volume on the outside world by means of a piston or a turbine (that operates a mill, for example, or generates electric power via a dynamo) or by a mass flow rate that flows across one of the boundaries of the control volume against the outside pressure. The overall (net) 'flow' as a result of **work** (on and/or from the control volume) is written as W.

Finally, **energy production** (per unit of time) P_e in the interior of V is possible. This could involve heat production, for example, in a chemical reactor as a result of an exothermic chemical reaction or in a metal wire as a result of electric current. Actually, the presence of the production term P_e is the result of our own limitation that E does not cover all types of energy – only internal, potential, and kinetic energy – ignoring e.g. chemical enthalpy (as mentioned before). Without such a limitation, there would be no production term, of course: it is impossible to produce energy from nothing. *After all, energy is a conserved quantity.* If from now on we really limit ourselves to the types of energy listed in the table at the start of this section, the production term P_e in the energy balance is equal to zero.

By substituting the above expressions for the various contributions into equation (1.87) we arrive at the general form for the total energy balance:

$$\frac{d}{dt}E = \frac{d}{dt}(\rho Ve) = \phi_{m,in} \cdot e_{in} - \phi_{m,out} \cdot e_{out} + \phi_q + W \tag{1.89}$$

It is also useful to divide the energy flow resulting from work, W, into two parts:
- one part ϕ_w, which stands for the work as a result of the pistons, stirrers, pumps, compressors, and/or turbines, and
- one part that is related to the mass flow rates that flow into and/or out of the control volume and then performs work against the prevailing local pressure:

    If a 'package' of fluid flows into the control volume at velocity v in a time interval dt through a surface A, then during this dt, a force $p \cdot A$ will be exercised locally on the package, which will move over a distance $dx = v \cdot dt$. In other words, this involves an amount of energy $dE = F \cdot dx = pA \cdot vdt$. In the balance, we need the energy *per unit of time* (i.e. the power) and that is exactly the same as $pA \cdot v$, or as $\phi_V \cdot p = \phi_m \cdot p/\rho$.

With respect to W, distinguishing between two types of work leads to three terms in the total energy balance [equation (1.89)]:

$$W = \phi_w + \left[\phi_{m,\text{in}} \frac{p}{\rho}\bigg|_{\text{in}} - \phi_{m,\text{out}} \frac{p}{\rho}\bigg|_{\text{out}} \right] \tag{1.90}$$

This means that equation (1.89) can be rewritten as

$$\frac{d\rho eV}{dt} = \phi_{m,\text{in}} \left(e + \frac{p}{\rho} \right)_{\text{in}} - \phi_{m,\text{out}} \left(e + \frac{p}{\rho} \right)_{\text{out}} + \phi_q + \phi_w \tag{1.91}$$

The work (per unit of time), ϕ_w, is therefore now just the work performed by the outside world through mechanical devices such as turbines and pumps.

In many cases, especially if the pressure in the system under consideration can be taken as a constant (using a reasonable approximation) and certainly if **phase transition**s are involved with the accompanying density changes and work performed, it is useful to work in terms of the specific **enthalpy** $h = u + p/\rho$ in the 'in' and 'out' flows (with all variables in J/kg).

Example 1.10. Transport of water through a tube.
A constant flow of water flows through a horizontal cylindrical tube (see Figure 1.22) as a result of an imposed pressure difference. This difference in pressure amounts to 2 bar (= $2 \cdot 10^5$ N/m²) and the mass flow rate is 1 kg/s. The tube has a (constant) diameter D of 2 cm and is well insulated. The water has a density ρ of 10^3 kg/m³, a specific heat at constant volume c_v of 4.2 kJ/kg K and, as usual under this type of conditions, can be regarded as incompressible.

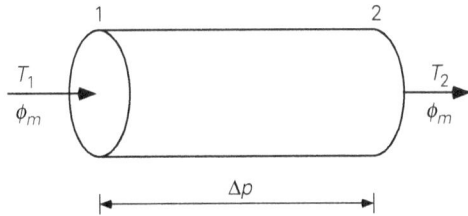

Figure 1.22

Answer the following questions:
- Give the energy balance and calculate the rise in temperature of the water.
- What power must a pump have in order to be able to cause this rise in temperature?

Solution
Consider a control volume that extends from a point right after the pump, denoted as point 1, until point 2 at the open exit of the tube (see Figure 1.22). Draw up a mass balance and a (total) energy balance for this control volume. Given that the situation is steady, the following applies

$$\phi_{m,in} = \phi_{m,out} = \phi_m \quad \text{and} \quad \frac{dE}{dt} = 0$$

Additionally, $\phi_w = 0$ (no pump) and $\phi_q = 0$ (no heat exchange between tube and outside world). This means that the energy balance for this stretch of the tube is

$$0 = \phi_m \left[u_1 + \frac{p_1}{\rho_1} + \frac{1}{2}v_1^2 + gz_1 - \left(u_2 + \frac{p_2}{\rho_2} + \frac{1}{2}v_2^2 + gz_2 \right) \right] \tag{1.92}$$

Equation (1.92) becomes much simpler when: $z_1 = z_2$ (horizontal tube), $\rho_1 = \rho_2$ (water is incompressible), and $v_1 = v_2$ (providing the diameter of the tube and the water density ρ are constant).

The energy balance then becomes:

$$u_1 + \frac{p_1}{\rho_1} = u_2 + \frac{p_2}{\rho_2} \tag{1.93}$$

Due to the imposed pressure difference $p_1 - p_2 > 0$, it follows that $u_1 < u_2$. In other words: **pressure energy is transformed into internal energy**. As we have seen, with a constant ρ, the following approximation applies according to thermodynamics:

$$u_1 - u_2 = c_V (T_1 - T_2) \tag{1.94}$$

From equations (1.93) and (1.94) it then follows:

$$T_2 - T_1 = \frac{p_1 - p_2}{\rho c_V} \tag{1.95}$$

Substituting the numerical data produces a rise in temperature of 0.048 K which is useless!

The rise in the temperature of the water is therefore an example of dissipation of mechanical energy. This energy (per unit of mass) is provided by the pump that raises the pressure of the water at the start of the tube. An energy balance for just an ideal pump[1] in the case of incompressible liquid runs therefore as

$$0 = \phi_m \left(\frac{p_0}{\rho} - \frac{p_1}{\rho} \right) + \phi_w \tag{1.96}$$

where p_0 denotes the pressure in the suction line immediately upstream of the pump and p_1 stands for the pressure immediately after the pump.

Consider now pump plus tube as the control volume. If water flows through the tube just due to the action of the pump and the suction line (upstream of the pump) is really short, then the suction pressure p_0 is equal to the pressure p_2 at the open end of the tube: $p_0 = p_2$. With this information, it follows that

$$\phi_w = \phi_m \frac{p_1 - p_2}{\rho} = 1 \cdot \frac{2.10^5}{10^3} = 200 \text{ W} \tag{1.97}$$

This result expresses that the pump just produces the pressure p_1 needed to overcome the friction in the tube. This result is of course also equal to $\phi_m (u_2 - u_1)$: the power provided by the pump ultimately ends up as internal energy.

Example 1.11. Compression of a gas.
With the help of a compressor (an 'air pump'), air is compressed from 1 to 5 bar under steady-state conditions. The power consumption of the compressor is $6 \cdot 10^3$ W. The mass flow rate through the

---

1 An ideal pump only raises the pressure of a liquid without heat development inside the pump; assume also $v_0 = v_1$.

compressor is 108 kg/h and the diameter of the entrance and exit tubes is 5 cm. Room temperature is 20 °C. How hot is the gas that leaves the compressor?

Solution
Because of the steady-state situation in the control volume (= the compressor, see Figure 1.23), dM/dt and dE/dt are both 0.

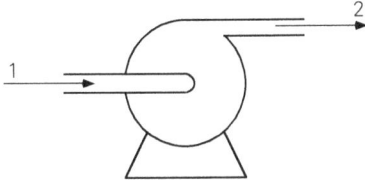

Figure 1.23

From the mass balance it follows that

$$\phi_{m,1} = \phi_{m,2} = \phi_m \tag{1.98}$$

If we assume that the compressor does not give off any heat via the wall directly to the ambient air (i.e. $\phi_q = 0$), then the energy balance is

$$0 = \phi_m\left[u_1 + \frac{p_1}{\rho_1} + \frac{1}{2}v_1^2 - \left(u_2 + \frac{p_2}{\rho_2} + \frac{1}{2}v_2^2\right)\right] + \phi_w \tag{1.99}$$

where v_1 and v_2 can be expressed in ϕ_m, the local densities ρ_1 and ρ_2, and the cross-sectional area $A (= \pi D^2/4)$ of the entrance and exit:

$$v_1 = \frac{\phi_m}{A}\frac{1}{\rho_1} \text{ and } v_2 = \frac{\phi_m}{A}\frac{1}{\rho_2} \tag{1.100}$$

In these circumstances, air can be regarded as an ideal gas (with molar mass m):

$$p = \rho\frac{RT}{m}$$
$$u = \frac{5}{2}\frac{RT}{m} = \frac{5}{2}\frac{p}{\rho} \tag{1.101}$$

This latter equality has its origins in statistical physics. With temperatures that are not extremely high, that is, where the vibrations inside the molecules do not yet play a major role, any diatomic molecule, such as the oxygen and nitrogen molecules in air, has five degrees of freedom – three for translation and two for rotation. According to the equipartition theorem, every degree of freedom has an energy of ½ RT per mole or ½ RT/m per kilogram.
    After some substitutions, the energy balance is now:

$$0 = \phi_m\left[\frac{7}{2}\left(\frac{p_2}{\rho_2} - \frac{p_1}{\rho_1}\right) + \frac{1}{2}\left(\frac{\phi_m}{A}\right)^2\left(\frac{1}{\rho_2^2} - \frac{1}{\rho_1^2}\right)\right] - \phi_w \tag{1.102}$$

Substituting the numerical data then produces: $\rho_1 = 1.19$ kg/m³, $\rho_2 = 3.54$ kg/m³, $T_2 = 492$ K, and $\Delta T = 199$ K! In practice, the outgoing gas flow is therefore often cooled – which again implies energy dissipation!

Example 1.12. The kettle.
In this example, the balance technique will be used to develop an elementary description of the heating up and boiling of water in a kettle. Look at the kettle in Figure 1.24 and assume that the heating up process occurs through a constant heat supply ϕ_{q1}. For the sake of simplicity, it is also assumed that the heat loss to the surroundings (ϕ_{q2}) is constant. We will disregard the heating up of the kettle itself for the purpose of convenience.

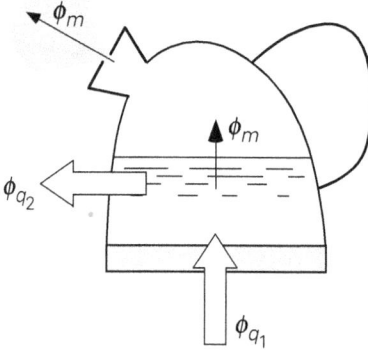

Figure 1.24

First look at the situation in which the water is being heated up (not boiling yet). In these circumstances, evaporation can be ignored; this involves only a minor error, depending on the water temperature. We will take the contents of the kettle as the control volume and look only at the liquid, ignoring the vapour. (Incidentally, the mass of all the water in the kettle including the vapour in the space above the liquid and the mass of the liquid are more or less equal.) This means that the mass balance for the water (shown by subscript l for 'liquid'):

$$\frac{d}{dt}M_l = 0 \tag{1.103}$$

As for an incompressible liquid $c_v = c_p$, and providing that c_p is independent of temperature, the energy balance for the liquid in rest in the kettle is

$$\frac{dE}{dt} = \frac{dU}{dt} = \frac{d}{dt}(M_l c_p T) = M_l c_p \frac{dT}{dt} = \phi_{q1} - \phi_{q2} \tag{1.104}$$

In other words, the temperature will rise at a constant rate:

$$\frac{dT}{dt} = \frac{\phi_{q1} - \phi_{q2}}{M_l c_p} \tag{1.105}$$

With $T = T_0$ at $t = 0$, the above differential equation results in an expression for the liquid temperature as a function of time:

$$T = T_0 + \frac{\phi_{q1} - \phi_{q2}}{M_l c_p} t \tag{1.106}$$

Then consider the situation in which the water is boiling. There is now a substantial mass flow rate (mostly vapour) coming out of the kettle that we really cannot ignore (you can hear the whistle!). This implies that we should no longer ignore the quantity (the mass) of vapour in the space above the liquid either. The mass balance for the control volume, that is, the contents of the kettle, now becomes (with subscript v for 'vapour', the vapour phase):

$$\frac{dM_l}{dt} + \frac{dM_v}{dt} = 0 - \phi_m = -\phi_m \qquad (1.107)$$

Even if M_v were to be ignored, the result of equation (1.107) is still of some value because it is not unreasonable to assume that in the short term, the quantity of water vapour in the kettle remains constant during the boiling process: at each moment, there is as much vapour leaving the kettle as there is liquid water changing into vapour.

This transition from liquid to vapour is accompanied by a change in volume. The vapour that is formed must therefore exercise work on the surroundings (the outside air), while the pressure in the kettle remains constant. These considerations play a role when we come to analyse the energy management of boiling. Again, we use what we know of thermodynamics. For a closed system (a system without mass flow rates in and/or out), thermodynamics tells us:

$$\delta Q = dU + p\, dV \qquad (1.108)$$

Instead of working with internal energy, U, we can also use enthalpy, H:

$$H \equiv U + pV \qquad (1.109)$$

This means that the conservation of energy can be written as

$$\delta Q = dH - Vdp \qquad (1.110)$$

We therefore see that, for processes at constant volume ($dV = 0$) and constant pressure ($dp = 0$), as is the case in our kettle, enthalpy is a very useful quantity because $dU/dt = dH/dt$. This is even more so given that on the right-hand side of equation (1.91), enthalpy also occurs in the in and out terms (as a result of mass flow rates in and out). For the boiling process in the kettle (an open system, due to the outgoing flow of vapour), we arrive, based on equation (1.91) and with $\phi_w = 0$, at the following enthalpy balance for the overall content of the kettle (liquid plus vapour phase):

$$\frac{dH}{dt} = \phi_{m,in}\, h_{in} - \phi_{m,out}\, h_{out} + \phi_{q_1} - \phi_{q_2} \qquad (1.111)$$

In equation (1.111), again all the effects of kinetic and potential energy are disregarded – both in the d/dt term on the left-hand side and in the ingoing and outgoing mass flow rates on the right-hand side of the equation – as they are very small in relation to the various enthalpy terms. The ingoing mass flow rate is zero and the outgoing enthalpy is h_v. For the overall enthalpy in the control volume, we can write:

$$H = h_l M_l + h_v M_v \qquad (1.112)$$

Because both pressure and temperature are constant during boiling, the enthalpy concentrations, h_l and h_v, are also constant. Equation (1.111) can therefore be written as follows:

$$h_l \frac{dM_l}{dt} + h_v \frac{dM_v}{dt} = -\phi_m h_v + \phi_{q_1} - \phi_{q_2} \qquad (1.113)$$

Combining equations (1.107) and (1.113), and using the notation $\Delta h_v = h_v - h_l$ for the enthalpy of evaporation (in kJ/kg) results in

$$-\frac{dM_l}{dt} \Delta h_v = \phi_{q_1} - \phi_{q_2} \qquad (1.114)$$

Starting from the original mass M_{l0} of liquid in the kettle, the expression for the mass of liquid in the kettle as a function of time then becomes

$$M_l = M_{l0} - \frac{\phi_{q_1} - \phi_{q_2}}{\Delta h_v} t \qquad (1.115)$$

In other words, supplying heat to a kettle of boiling water only causes – not surprisingly – a reduction of the water mass through the phase transition when the atmospheric boiling point is constant and the atmospheric pressure is constant (if we ignore the very small, but constant pressure drop over the exit opening): this results (primarily because we are ignoring kinetic and potential energy contributions) in the removal of vapour via the discharge flow rate ϕ_m.

Bear in mind that throughout the analysis we have not taken the heat capacity of the kettle itself into account.

Example 1.13. The refrigerator.
The air in the food storage area of a refrigerator is cooled with the help of a coolant (or refrigerant) that is circulated in a closed circuit. This cooling can only take place when the refrigerant in the heat exchanger is at a lower temperature than the air in the food storage area. The coolant picks up heat from the food storage area and passes it to the air outside the refrigerator and realises this with the help of phase transitions (evaporation, condensation): really a complex transport process.

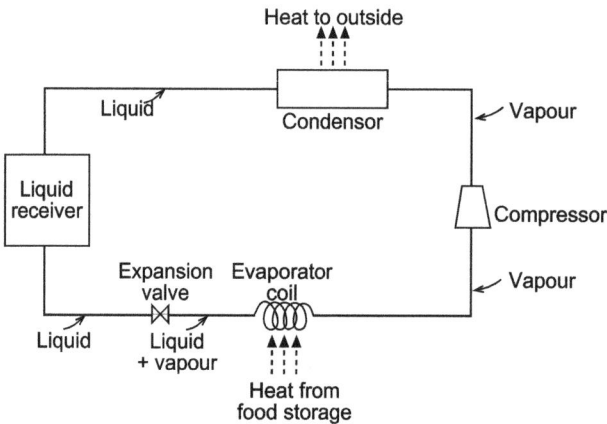

Figure 1.25

The closed cooling circuit (see Figure 1.25) comprises five essential components. The first component is a pressure vessel (also denoted as the liquid receiver) in which the refrigerant is a liquid due to a high pressure and temperature (say, 7 bar and 30 °C). The liquid then expands through an expansion valve: downstream of the valve, the pressure is lower (e.g. 1.8 bar) and, as a result, the liquid momentarily flashes (evaporates) partly by cooling to the boiling-point temperature (−15 °C) at this lower pressure. The resulting vapour-liquid mixture then passes through the heat exchanger (also denoted as evaporator coil) that is in the food storage area to be cooled. By withdrawing heat from the air in the food storage area, the remaining liquid coolant may also evaporate, still at the low pressure and pertinent boiling-point temperature. The cold refrigerant emerging from the coil is then brought back to the original high pressure by a compressor; in this compression step, temperature also rises of course, for example, to 45 °C. The hot vapour then passes through a heat exchanger, the so-called condenser (outside and at the rear side of the refrigerator), where it gives up heat to the ambient air such that the vapour condenses before it is returned to the liquid receiver.

Let's now focus on the expansion valve and draw up an energy balance for this valve during a steady-state operation. This valve is a passive component (no work ϕ_w performed) and we will ignore heat exchange with the ambient world ($\phi_q = 0$). That means equation (1.91) reduces to

$$0 = \phi_{m,in}\left(e + \frac{p}{\rho}\right)_{in} - \phi_{m,out}\left(e + \frac{p}{\rho}\right)_{out} \tag{1.116}$$

For the valve, obviously $\phi_{m,in} = \phi_{m,out}$. As a phase change is involved, with a substantial change in density and with expansion work, a description in terms of enthalpy is needed. Then equation (1.116) leads to

$$0 = \left(h + \frac{1}{2}v^2 + gz\right)_{in} - \left(h + \frac{1}{2}v^2 + gz\right)_{out} \tag{1.117}$$

As a result of the flashing, the vapour-liquid mixture leaving the valve with a much lower (mixture) density should have a much higher velocity than the liquid refrigerant entering the valve (to satisfy the above mass balance). The resulting change in kinetic energy, however, is still small in comparison with the enthalpy of the entering liquid refrigerant. Ignoring therefore – just like in the preceding example – the changes in kinetic and potential energy turns equation (1.117) into the simple relation

$$h_{l,in} = [xh_v + (1-x)h_l]_{out} \tag{1.118}$$

in which x denotes the fraction of the refrigeration that has flashed (evaporated) – subscripts v and l denote vapour and liquid, respectively. By finding enthalpy data of the refrigerant of interest at the prevailing temperatures and pressures from thermodynamic property tables, one can calculate this fraction x; in the above case, x amounts to some 27%.

Summary
In principle, a total energy balance for a control volume has the same basic form as a species mass balance and comprises transport and production terms. In drawing up a total energy balance, all three types of energy (internal or thermal, kinetic, and potential energy) should be taken into account:

$$e = u + \frac{1}{2}v^2 + gz$$

As a matter of fact, there is the convective transport of energy by mass flows into and out of the control volume. Also **enthalpy** $h = u + p/\rho$ plays a role in energy balances, with the flows entering and leaving the control volume as well as when a phase transition is involved. Further, work may be done by or upon the outside world that also may exchange heat with the control volume – all such effects have to be accommodated in an energy balance.

Through an energy balance, conversions of one type of energy in one or more other types can easily be demonstrated and accommodated. The general form of a total energy balance runs as:

$$\frac{d\rho Ve}{dt} = \phi_{m,in}\left(e + \frac{p}{\rho}\right)_{in} - \phi_{m,out}\left(e + \frac{p}{\rho}\right)_{out} + \phi_q + \phi_w$$

The concept of dissipation of mechanical energy into heat due to friction was illustrated with the help of some examples.

1.3.3 The thermal energy balance

So far, we have dealt with the total energy, but it is sometimes more helpful to look solely at a specific type of energy, such as thermal or internal energy U. The balance for U is

$$\frac{d}{dt}U = \frac{d}{dt}(\rho V u) = \phi_{m,in} \cdot u_{in} - \phi_{m,out} \cdot u_{out} + \phi_q + P_u \qquad (1.119)$$

As usual, the left-hand side of the equation represents the change over time in the internal energy U in the control volume. The first two terms of the right-hand side are the difference between inflow and outflow of internal energy by mass flow rates. The third term denotes the net heat flow ϕ_q that goes into or out of the volume. Finally, the last term P_u stands for production of internal (thermal) energy in the control volume as a result of **dissipation** (effects of friction) and for production of internal energy by **work** that is performed on the system.

Take note: thermal energy is certainly *not* a conserved quantity and therefore equation (1.119) contains a production term P_u. The balance of equation (1.119) is therefore not a **conservation law** either. The question now is to find an expression for production term P_u. Pistons, pumps, compressors, and turbines add or extract 'mechanical' energy to or from the control volume rather than heat (thermal energy): a term resulting from the actions of pistons, pumps, compressors, and turbines is therefore not part of a thermal energy balance. The production term P_u can be found via the following derivation:

Consider again the overall energy balance, equation (1.91) or, for a steady-state situation, equation (1.92), in which the terms $\phi_m \cdot [u + p/\rho]_{in}$ and $\phi_m \cdot [u + p/\rho]_{out}$ occur, among others. These terms stand for the enthalpy difference between entrance and exit. For this enthalpy decrease, the following applies:

$$dh = du + p \cdot d\left(\frac{1}{\rho}\right) + \frac{1}{\rho} \cdot dp \qquad (1.120)$$

The first two terms on the right-hand side are related to heat and should feature explicitly in a thermal energy balance: after all, according to the first law of thermodynamics, the following applies, formulated on a mass basis

$$\delta Q = du + p \cdot d\left(\frac{1}{\rho}\right) \qquad (1.121)$$

in which δQ stands for the change to the 'heat content' (per kg) as a result of a heat transfer process: adding heat raises the temperature and reduces the density. It is also clear from this law that work performed on the system by a constant pressure (the 2nd term at the right-hand side) under adiabatic conditions ($\delta Q = 0$), affects the internal energy of the system. Remember that equation (1.121) relates to a **reversible process**: the kinetic energy of the molecules can be used to bring them closer together and vice versa.

Owing to equation (1.121), the term $\phi_m \cdot pd(1/\rho)$ should feature explicitly in a thermal energy balance such as equation (1.119), in addition to the term $\phi_m \cdot du$. The term is the result of the difference in density between ingoing and outgoing mass flow rates. In a similar way to how energy production W due to work was divided into two terms in advance of equation (1.90), in equation (1.119) we now remove the **reversible work term** $\phi_m \cdot pd(1/\rho)$ from production term P_u and write it with the inflows and outflows. What remains of P_u is an **irreversible** energy transformation: the **dissipation** of mechanical energy into heat as a result of friction within the control volume. This friction, the production of thermal energy (heat), is the result of the resistance that is offered within the control volume to the flow *from entrance to exit* and is therefore in proportion with the mass flow rate. For this dissipation term, we therefore write $\phi_m\, e_{\text{diss}}$, in which

$e_{\text{diss}} \equiv$ the quantity of mechanical energy that is irreversibly transformed into thermal energy (heat) per unit of mass.

For a steady-state situation, in which the ingoing and outgoing mass flow rate are of course equal ($\phi_{m,\text{in}} = \phi_{m,\text{out}} = \phi_m$), equation (1.119) can then be rewritten in the following form:

$$0 = \phi_m \left[u_{\text{in}} - u_{\text{out}} - \int_{\text{in}}^{\text{out}} p \cdot d\!\left(\frac{1}{\rho}\right) \right] + \phi_q + \phi_m\, e_{\text{diss}} \tag{1.122}$$

which is known as the steady-state **thermal energy balance** or steady-state **heat balance**. By comparing the expression between the square brackets in equation (1.122) with the right-hand side of equation (1.121), one may easily appreciate the bracketed expression really is a heat effect.

The above derivation of the steady-state thermal balance, together with the discussion about enthalpy changes, illustrates how carefully the production term in a balance should be treated when, as in the case of thermal energy, the quantity concerned *does not remain conserved.*

Summary

A thermal energy balance can be drawn up, analogous to the overall energy balance. With the help of the first law of thermodynamics, the production of internal energy can be divided into a reversible work part and an irreversible part, the dissipation. As a result, the steady-state internal energy balance runs as

$$0 = \phi_m \left[u_{\text{in}} - u_{\text{out}} - \int_{\text{in}}^{\text{out}} p \cdot d\!\left(\frac{1}{\rho}\right) \right] + \phi_q + \phi_m\, e_{\text{diss}}$$

It ties in nicely with the overall energy balance, with the dissipation, as a producer of thermal energy, featuring prominently.

1.3.4 The mechanical energy balance and the Bernoulli equation

Now that the overall energy balance and the steady-state thermal energy balance have been formulated, constructing the steady-state **mechanical energy balance** is simple, given that

$$E_{\text{overall}} - E_{\text{thermal}} = E_{\text{mechanical}} \qquad (1.123)$$

Following on from this, subtracting equation (1.122) from the steady-state version of equation (1.91) while decomposing $d(p/\rho)$ into two terms as done in equation (1.120) leads to the steady-state mechanical energy balance

$$0 = \phi_m \cdot \left[\frac{1}{2} v_{\text{in}}^2 - \frac{1}{2} v_{\text{out}}^2 + g(z_{\text{in}} - z_{\text{out}}) + \int_{\text{out}}^{\text{in}} \left(\frac{1}{\rho}\right) dp \right] + \phi_w - \phi_m\, e_{\text{diss}} \qquad (1.124)$$

This equation is usually written slightly differently, with indices 1 and 2 instead of 'in' and 'out', respectively:

$$0 = \phi_m \left[\frac{1}{2} \left(v_1^2 - v_2^2\right) + g(z_1 - z_2) + \int_2^1 \left(\frac{1}{\rho}\right) dp \right] + \phi_w - \phi_m\, e_{\text{diss}} \qquad (1.125)$$

Notice that the production term has a minus sign here. This corresponds to the fact that dissipation destroys mechanical energy (both ϕ_m and e_{diss} are always positive, of course).

It should be noted here that the mechanical energy balance can be derived much more efficiently from the momentum balances (see further on in this textbook) by scalar multiplication with the velocity vector $\vec{v}$. The thermal energy balance is then arrived at by subtracting the mechanical energy balance from the overall energy balance.

It has been decided for this textbook to use the other method: on the one hand to avoid the route via momentum balances and the associated mathematical manipulations and on the other to familiarise the student with the idea that a balance equation can be drawn up for every quantity (conserved or not). Careful thought should be given to the physics of the production term; sometimes, it is useful or even necessary to model this production term. (This is the case, for example, when describing and simulating turbulent flows.) For **conserved quantities** (such as mass and total energy) the production term is of course zero.

The Bernoulli equation

There now follows a number of examples of the use of the mechanical energy balance in the exceptional events that no work is performed on or by the system ($\phi_w = 0$) and that the dissipation is negligible ($e_{\text{diss}} \approx 0$). In such cases, and providing that ρ is constant, the steady-state mechanical energy balance, equation (1.125), simplifies to what is known as the **Bernoulli equation**:

$$\frac{1}{2}v_1^2 + gz_1 + \frac{p_1}{\rho} = \frac{1}{2}v_2^2 + gz_2 + \frac{p_2}{\rho} \qquad (1.126)$$

or:

$$\frac{1}{2}v^2 + gz + \frac{p}{\rho} = \text{constant} \qquad (1.127)$$

In fact, this equation applies just along a flow line; the constant in equation (1.127) will have different values for different flow lines. As a matter of fact, fluid parcels while following flow lines cannot change their mechanical energy under the pertinent conditions.[2] The only possible transformation is from one form of mechanical energy into another – in line with the conditions for which equation (1.127) is valid. The terms in equation (1.127) all have the dimension of energy per unit of mass.

Depending on the system of interest, equation (1.127) is sometimes rewritten in terms of (pressure) heads, where all the terms have the dimension of length:

$$\frac{p}{\rho g} + z + \frac{v^2}{2g} = \text{constant} \qquad (1.128)$$

Working in terms of heads is illustrated in Figure 1.26. Applying the Bernoulli equation successively to the points a, b, and c (all three on a single horizontal line) teaches us that

$$\frac{p}{\rho g} = \frac{p'}{\rho g} + \frac{v^2}{2g} = \frac{p}{\rho g} \qquad (1.129)$$

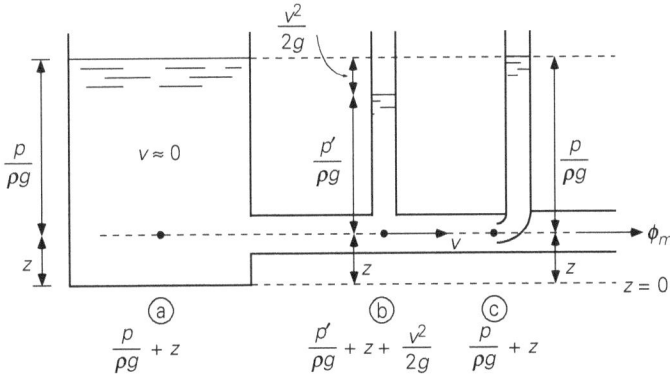

Figure 1.26

---

2 For a more detailed discussion, you are referred to, for example, Bird, R.B., W.E. Stewart & E.N. Lightfoot, Transport Phenomena, Wiley, 2nd ed., 2002.

This is shown in the diagram: the pressure p' at point b is lower than the pressure p at point a or point c. Indeed, at point b, the liquid has velocity, that is, kinetic energy, and therefore pressure p' is less than pressure p at point a. At point c – exactly in the bent opening of the standpipe – the liquid is motionless again and therefore pressure is equal to that at point a again. Point c will therefore be able to 'bear' a greater column of liquid than will point b; put another way, the kinetic energy present in point b is transformed in point c entirely into pressure energy. Point c is a so-called **stagnation point**.

The two heads in the middle part of equation (1.129) have their own name: *(static)* **pressure head** and **velocity head**, respectively. The 'missing' pressure at point b (see Figure 1.26) measuring $\frac{1}{2}\rho v^2$ – in line with the middle part of equation (1.129) – is known as the **dynamic pressure**, while p is named **static pressure** and ρgz is denoted by the term **hydrostatic pressure**.

Pressure in the curved opening of the standpipe at point c is known as **stagnation pressure** and is the sum of the local static pressure and the local dynamic pressure. This terminology returns in the treatment of the flow around immersed objects and the pertinent forces, in Section 2.3.1.

The Pitot tube

The aforementioned configuration can, in principle, be used to measure the velocity of the liquid. The most important precondition is that dissipation is indeed negligible. The **Pitot tube** (see Figure 1.27) shows how the principle of the configuration in Figure 1.26 takes shape in a compact device.

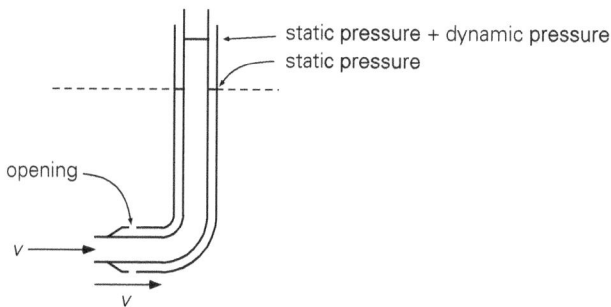

static pressure + dynamic pressure
static pressure
opening
v
v

Figure 1.27

The device consists of two bent concentric tubes, the innermost of which measures the local stagnation pressure, being the sum of static pressure and dynamic pressure, and the outermost tube only the local static pressure. A 'long way' before the tube, the pressure is p, after all, and the velocity v; so the following applies (at a single height, along a flow line):

$$p + \frac{1}{2}\rho v^2 = \text{constant} \qquad\qquad (1.130)$$

For the innermost tube, the flow is moving towards its opening. In order to prevent it (in a steady-state situation) from flowing in, a force in the opposite direction needs to be applied to the fluid. This is done by a higher pressure at the entrance – the local stagnation pressure. For the outermost tube, the fluid is flowing past the small opening at the side and does not try to flow into the outer tube. Hence, no force is needed to prevent fluid from entering. Consequently, no pressure difference is needed. [On the contrary, any pressure difference across the small hole would result in a flow in or out of the outer tube.] Thus, the difference between the two measured pressures is the dynamic pressure, $\frac{1}{2}\rho v^2$, which means it is a direct measure for the velocity of the liquid.

The Venturi tube
Another measuring instrument where, due to the nature of its construction, energy dissipation is negligible ($e_{\text{diss}} \approx 0$), is the Venturi tube. In the flow channel (see Figure 1.28), a gradual narrowing (apex angle < 25°) is followed by an even more gradual widening (apex angle < 8°) in order to comply with $e_{\text{diss}} \approx 0$. As long as the two criteria for the apex angle are met, then it can be relied on with certainty that the flow lines (along which the Bernoulli equation applies) will follow the shape of the geometry and that no eddies or energy dissipation will occur.

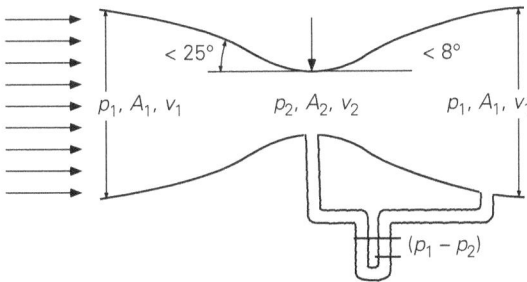

Figure 1.28

The mass flow rate ϕ_m, which runs through the flow channel in a steady-state situation, can now be derived from a mass balance, the Bernoulli equation, and the difference in pressure between point 1 a long way before (or a long way after) the throat (the narrowest passage) and point 2 in the throat. Denote the cross-sectional areas at points 1 and 2 by A_1 and A_2, respectively. From this, the mass balance for this steady-state condition can be obtained:

$$\phi_m = \rho\, v_1 A_1 = \rho\, v_2 A_2 \tag{1.131}$$

and from the Bernoulli equation (providing the liquid can be regarded as incompressible, that is, ρ = constant):

$$\frac{1}{2}v_1^2 + \frac{p_1}{\rho} = \frac{1}{2}v_2^2 + \frac{p_2}{\rho} \tag{1.132}$$

Combining the equations (1.131) and (1.132) gives the expression for the mass flow rate:

$$\phi_m = \frac{A_2}{\sqrt{1 - \left(\frac{A_2^2}{A_1^2}\right)}} \sqrt{2\rho(p_1 - p_2)} \tag{1.133}$$

Notice that thanks to the delicate design of this device, the possibility for reversibly (i.e. with hardly noticeable friction losses) transforming forms of mechanical energy is exercised – pressure energy becomes kinetic energy, which again becomes pressure energy.

Blood vessel defects

The Bernoulli equation may provide an elementary explanation of the risks associated with defects in blood vessels, viz. an aneurysm and a stenosis.

An (aortic) aneurysm is a bulge in a blood vessel, caused by a weakness of the arterial wall. As a result, velocity decreases and therefore, according to the Bernoulli equation, pressure increases, potentially causing the bulge to widen further and the blood vessel to rupture.

The term stenosis stands for an abnormal narrowing of a blood vessel due to the buildup of inflammatory substances and/or cholesterol deposits, called plaque. As a result, the velocity of the blood increases and (lateral) pressure decreases, both effects increasing the risks of deposits releasing into the blood circulation. In addition, some turbulence may be created, notably downstream of a stenosis. Particularly stenosis of a carotid artery (in the neck), which transports oxygenated blood from the heart to the brain, can lead to (symptoms of) a stroke.

Example 1.14. Flue gases through a chimney.
The flue gases from a furnace are fed to the base of a high chimney that is built on the ground. The diameter of the chimney is constant. The flue gases are at a constant temperature of 227 °C, and the outside air is 20 °C. Pressure p_1 at the bottom of the chimney is Δp_m below the atmospheric pressure p_a of 1 bar (measured at ground level). The heat losses from the flue gas in the chimney are negligible. In addition, there is no dissipation of any note.
On the basis of this information, and for $\Delta p_m = 250$ Pa, how high is the chimney?

Solution
Before this question can be answered, an investigation should first be carried out to find out why the flue gas actually flows out of the chimney. After all, it surely comes into the chimney at a pressure lower than the ambient pressure of 1 bar? At the top of the chimney, however, the air pressure has also fallen below 1 bar. As a matter of fact, pressure in the ambient air drops if the column of air being 'carried' gets smaller. The ambient pressure at height H of the chimney exit is therefore not p_0, but $p_0 - \rho_a gH$ (with ρ_a being the density of air at 20 °C = 1.2 kg/m³) and should be lower than the pressure $p_0 - \Delta p_m$ at the base of the chimney.

From equation (1.125), with $\phi_w = 0$ and given the constant diameter of the chimney, it follows for the chimney that

$$0 = \int_0^H \frac{1}{\rho_g}\, dp + gH \tag{1.134}$$

The pressure drop from bottom to top is very small – check this afterwards with equation (1.135) – and it is given that heat losses to the ambient air are minor. As a result, pressure and temperature of the flue gases hardly change if at all, and so the density ρ_g of the flue gases can also be regarded as constant (0.65 kg/m³ at 227 °C). This means equation (1.134) simplifies to a Bernoulli equation:

$$0 = \frac{1}{\rho_g}[(p_0 - \rho_a gH) - (p_0 - \Delta p_m)] + gH \tag{1.135}$$

$$\to H = \frac{\Delta p_m}{\left(\rho_a - \rho_g\right)g} \tag{1.136}$$

Substituting the data produces: $H = 46.3$ m.

A few trivial rules for designing a chimney follow from equation (1.136): a greater Δp_m requires a higher chimney (when ρ_a and ρ_g are equal); hotter flue gases (lower ρ_g) mean a lower chimney will suffice (due to a better 'draught').

Summary
In this section, the steady-state mechanical energy balance has been constructed by taking the difference from the steady-state overall energy balance and the steady-state thermal energy balance:

$$0 = \phi_m\left[\frac{1}{2}\left(v_1^2 - v_2^2\right) + g(z_1 - z_2) + \int_2^1 \left(\frac{1}{\rho}\right)dp\right] + \phi_w - \phi_m e_{diss}$$

The production term in it is always negative: dissipation destroys mechanical energy.

A number of cases were then examined in which no work was performed and the energy dissipation was negligible, while also ρ = constant. The steady-state mechanical energy balance then changes into the Bernoulli equation:

$$\frac{1}{2}v^2 + gz + \frac{p}{\rho} = \text{constant}$$

The latter shows conservation of mechanical energy: it is possible for pressure energy, kinetic energy, and gravitational energy to be transformed into each of the other types, but the sum of the three remains constant.

Finally, the Pitot tube and Venturi tube were dealt with, as were the terms pressure head, static head, velocity head, static pressure, hydrostatic pressure, dynamic pressure, and stagnation pressure.

1.4 The momentum balance

After the mass and energy balances we are left with the momentum balance. The quantity for which a balance now has to be drawn up is momentum $\vec{P}$ which usually is conceived as the product of the moving mass m (in kg) times velocity $\vec{v}$ (in m/s). The unit in which $\vec{P}$ is expressed is then Ns, analogous to Nm (or J) for energy E. As the notation shows, momentum involves an extra complication in relation to mass and energy. Like a force, momentum is a **vector** quantity, which means that three components have to be specified in order to establish the momentum. This is unlike mass or energy which are established with just one value (mass and energy are referred to as a scalar).

For momentum, therefore, three balances are needed. As a matter of fact, when using orthogonal axis systems like the much-used Cartesian coordinate system (x,y,z), the three momentum components $(p_x, p_y, p_z$ – denoting momentum concentrations per unit of mass, in Ns/kg, of per unit of volume, in Ns/m^3) only occur separately in these balances (note that p – with subscript! – does not refer to pressure here). In other words, the balance for momentum in the x-direction, p_x, contains no terms that include p_y or p_z.

First, we introduce the momentum balance in any direction x. The general form of the x-momentum balance is (for control volume V in Figure 1.29):

$$\frac{dP_x}{dt} = \text{flow of } x\text{-momentum in} - \text{flow of } x\text{-momentum out} + \\ + \text{production of } x\text{-momentum} \tag{1.137}$$

All terms of this equation have the unit Ns/s. The left-hand term represents the change over time of the overall amount of x-momentum in the control volume. The overall amount of x-momentum[3] in the control volume can be written as $P_x = M \cdot v_x$, in Ns.

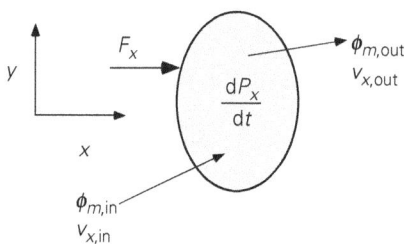

Figure 1.29

In this case, the flows in and out are the result of a mass flow rate that carries x-momentum with it. They therefore have the form of $\phi_m \cdot \{x$-momentum per unit of

---

3 Velocity v_x is actually a mean velocity defined as

$$\langle v_x \rangle = \frac{\int \rho v_x dV}{\int \rho dV}.$$

mass}. From classical mechanics, it follows that the x-momentum of a mass m is equal to mv_x. From this, it follows that the **x-momentum concentration** (per unit of mass) is equal to $(mv_x)/m = v_x$. *The velocity in the direction of x can therefore be taken as the concentration (per unit of mass) of x-momentum*, in Ns/kg. (From now on, the symbol p will no longer be used for momentum, only for pressure.) This means that, again entirely analogous to equations (1.9) and (1.10) and to energy flows in and out in equation (1.89), the flows of x-momentum in and out become:

$$\phi_{m,\text{in}} \cdot v_{x,\text{in}} \quad \text{respectively} \quad \phi_{m,\text{out}} \cdot v_{x,\text{out}}$$

By analogy, the concentration of x-momentum per unit of volume is given by ρv_x, in Ns/m^3.

All that remains now is the production of x-momentum. Here, too, classical mechanics can be invoked. In fact, Newton's second law

$$\frac{d}{dt}(mv_x) = \sum Fx \tag{1.138}$$

is already a simple form of the x-momentum balance that we are looking for. The left-hand side is the change to the x-momentum of the body. This change is the result of the x-components of the forces that are working on the body. Therefore, the right-hand side of equation (1.138), the sum of the forces in the x-direction, represents the momentum-production term. A force can therefore be regarded as a **producer of momentum** (Ns) per unit of time (s).

Classical mechanics further tells us that there are two types of force:

– **body forces**, which work on the mass in the control volume; the best-known exponent of this is gravity,
– forces that affect the surface of the control volume; these include pressure forces and frictional force (or the shear stress – see next chapter).

The momentum balances diverges from the mass and energy balance on this point. In the case of the latter two balances, production always takes place internally, while with the momentum balance it is possible for the production of momentum to occur on the surfaces. (It will be demonstrated below that this difference can be partly eliminated by introducing certain momentum flows that replace the frictional forces on the surface of the control volume.)

After all the above considerations, the general form of the x-momentum balance is now:

$$\frac{d}{dt}(Mv_x) = \phi_{m,\text{in}} \cdot v_{x,\text{in}} - \phi_{m,\text{out}} \cdot v_{x,\text{out}} + \sum F_x \tag{1.139}$$

It is useful to remember that the mass flow rates do not necessarily have to be in the direction of x (see Figure 1.29). If the situation is steady and the transport terms in equation (1.139) compensate each other (or are equal to zero), the x-momentum balance reduces to a **force balance**:

$$0 = \sum F_x \tag{1.140}$$

Similar balances for the y-momentum and z-momentum can of course be written down as well.

Example 1.15. The conveyor belt.
Sand from a hopper lands on a horizontal conveyor belt moving at a velocity of $v_b = 1.0$ m/s. The sand falls vertically at a mass flow rate of 225 kg/s. At first, the conveyor belt is empty and is gradually filled with sand (but does not fill up yet). Calculate the force needed to move the belt, assuming that it moves over its bearings without any friction.

Solution
The first question to be answered is this: why must a force be exercised on the belt if it is moving without any friction? The answer is that the sand that is falling continuously on the band has to be brought at the right horizontal velocity, that is, must receive momentum in the horizontal direction (the x-direction), for which a force in the horizontal direction is required! In order to calculate this force, a momentum balance has to be drawn up for a volume that covers all the sand that is lying on the belt (see Figure 1.30). Be sure to make the control volume so long at the right-hand side that no sand can leave the control volume in the time interval dt.

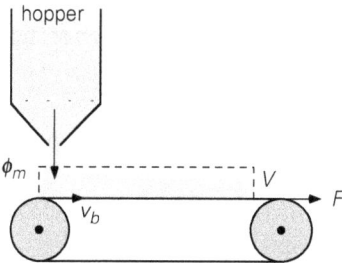

Figure 1.30

The balance for x-momentum runs as:

$$\frac{d}{dt}(M_z v_b) = \phi_{px,\text{in}} - \phi_{px,\text{out}} + \sum F_x \tag{1.141}$$

M_z is the mass of sand that is lying on the belt. The balance becomes simple because there is no x-momentum flowing in or out of the control volume. The sum of the forces is F, the force that moves the belt. The unsteady-state term can be simplified with the help of a mass balance for the same control volume

$$\frac{d}{dt} M_z = \phi_{m,\text{in}} - 0 = \phi_{m,\text{in}} \tag{1.142}$$

By combining the equations (1.141) and (1.142), and conceiving v_b as a constant, it follows that

$$F = v_b \cdot \phi_m = 225 \ N \tag{1.143}$$

Note we use and substitute the numerical values for belt velocity and sand flow rate only at the very end of the solution: this is a general rule! This way, we can see and check (the units inclusively) how the variable of interest depends on the independent variables, in this case how the force depends on belt velocity and sand flow rate.

It should also be pointed out that the same result is obtained if the sand has reached the end of the conveyor belt and is dropping down from the belt. Under these circumstances, the control volume should extend to the end of the belt; as a result, the quantity of sand on the belt in the control volume remains constant as now the sand is leaving the control volume while it carries x-momentum.

Example 1.16. Filling a container.
This example once more deals with the link between momentum flow and force as a producer of momentum.

In a factory, containers are filled with liquid. This is an automated process, in which the weight of the containers is determined with the help of a weighing scale. To give an idea: it takes about 20 s to fill a container and the velocity v of the vertical liquid jet that lands in the container is constant and amounts to 10 m/s.

What percentage of overweight must the scale show before the container contains the desired amount of liquid and the filling process can be stopped?

Solution
The control volume is an entire container. Although the liquid in a container may not be stagnant due to jet mixing during filling, on average the liquid in the container may be assumed to have virtually no vertical momentum; as a result, the left-hand side (the d/dt-term) of the balance for vertical momentum is zero and the balance then is (see Figure 1.31):

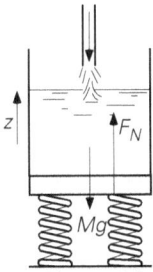

Figure 1.31

$$0 = -(\rho A v) \cdot v - 0 - M \cdot g + F_N \qquad (1.144)$$

where A is the surface of the diameter of the entrance and F_N is the force that the weighing scale exercises on the container. Equation (1.144) says that the weighing scale not only must carry the container (compensation of $M \cdot g$) but also has to eliminate the momentum flow (the thrust) of the incoming liquid (negative production, or consumption, of vertical momentum). Mass M is a function of time, which can easily be determined from a mass balance for the tank:

$$\frac{dM}{dt} = \rho A v \qquad (1.145)$$

Integrating this equation with the initial condition $M = 0$ at $t = 0$ results in the following expression for the mass at filling time τ:

$$M(\tau) = \rho A v \tau \qquad (1.146)$$

For the overweight percentage, it then follows that

$$\frac{F_N - M \cdot g}{M \cdot g} = \frac{\rho A v^2}{\rho A v g \tau} = \frac{v}{g \tau} = 5\% \tag{1.147}$$

Here again we use and substitute the numerical values of the parameters just at the very end of the solution.

Example 1.17. A hanging droplet.
Very slowly, individual water droplets are formed at the end of a vertically positioned glass capillary with an outer diameter of 2 mm. The flow through the capillary is negligibly small.
  If the surface tension of water is 70×10^{-3} N/m, use a force balance to calculate the diameter of the droplets that fall from the capillary.

Solution
A momentum balance for the control volume (the droplet) will result in just a force balance because of the very low velocities: both the increase in mass (the d/dt-term in the left-hand side of the momentum balance) and the inflow of momentum into the droplet during a time interval dt are insignificant. This simplification may be denoted as a quasi-steady-state approach.
  The forces that dominate are gravitational force F_g and the surface tension force F_σ of the liquid that is in contact with the rim of the capillary; see Figure 1.32. The buoyancy of the air on the drop is ignored here as the density of air is minor in comparison with that of the liquid. The two dominant forces have opposite direction and sign. This means that the force balance becomes:

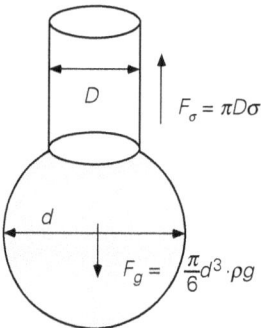

Figure 1.32

$$\sum F = 0 = F_\sigma + (-F_g) = \pi D \sigma - \frac{\pi}{6} d^3 \rho g \tag{1.148}$$

so that

$$d = \sqrt[3]{\frac{6 D \sigma}{\rho g}} = 4.4 \text{mm} \tag{1.149}$$

Comments about **surface tension**:
The molecules in the bulk of a liquid encounter forces from the surrounding molecules which cancel each other out. This is not the case with molecules next to air; a resulting force operates here, which is aimed perpendicularly at the surface of the liquid. A quantity of liquid on which no external forces are operating becomes spherical in shape; the surface energy has reached its minimum value at this point. It is as if the liquid were covered by an elastic film.

If the size of the surface of the liquid is increased, the number of molecules at the periphery also increases, and work is performed against the forces of attraction of the molecules in the liquid. The work (or energy), in joules, that is needed to increase the surface by 1 m^2, is represented by **surface tension** σ, expressed in J/m^2. Alternatively, the force required to increase a surface over a distance of b is σb (N) so that σ is also expressed in N/m.

At the rim of the capillary is a local and complicated force balance at the water-air-glass three-phase contact line. The liquid molecules in the boundary layer can be attracted more easily by a solid (**adhesion**) than by the liquid itself (**cohesion**), as a result of which, for example, the liquid rises in a capillary made from the solid in question.

In this example, the **adhesion force** is equal to the surface tension times the 'length' πD of the contact line.

Summary
Momentum, in Ns, is a vector. There are therefore three momentum balances, one for each component. It is recommended to think in terms of momentum concentrations: v, in Ns/kg, or ρv, in Ns/m^3. The three momentum balances have the familiar basic form with a d/dt term, in and out terms, and production terms:

$$\frac{d}{dt}(Mv_x) = \phi_{m,in} \cdot v_{x,in} - \phi_{m,out} \cdot v_{x,out} + \sum F_x$$

There are no terms involving y- or z-momentum in the x-momentum balance of course.

Forces can be seen as producers of momentum, in Ns/s. Here, too, there are no forces in the directions of y or z in the x-momentum balance.

Sometimes, a momentum balance simplifies to a force balance, for example, when in a quasi-steady-state approach all other terms of the momentum balance compensate.

1.5 Examples of combined use of mass, energy, and momentum balances

Example 1.18. Flow through *a* **bend**.
Water flows through a 90° bend in a well-insulated tube that has been positioned horizontally: Figure 1.33 shows a top view. The tube has a constant diameter A, and the water is assumed to be incompressible. The situation is a steady state. Draw up the mass balance, the energy balance, and the momentum balances for the whole bend.

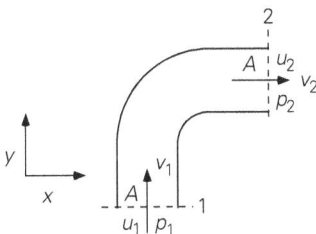

Figure 1.33

Solution

The situation is steady, so the left-hand side in all four balances is equal to 0. (Note: there are two relevant momentum balances!).

– *Mass balance*:

$$0 = \rho v_1 A - \rho v_2 A \longrightarrow v_1 = v_2 \tag{1.150}$$

Clearly the magnitude of the velocity is unchanged, but the direction has changed, of course.

– *Energy balance*:

$$0 = \rho v_1 A \cdot \left(e_1 + \frac{p_1}{\rho} \right) - \rho v_2 A \cdot \left(e_2 + \frac{p_2}{\rho} \right) \tag{1.151}$$

Thanks to equation (1.150), it then follows immediately that

$$u_1 + \frac{1}{2} v_1^2 + g z_1 + \frac{p_1}{\rho} = u_2 + \frac{1}{2} v_2^2 + g z_2 + \frac{p_2}{\rho} \tag{1.152}$$

Again because of equation (1.150) and with $z_1 = z_2$, it follows that

$$\frac{p_1 - p_2}{\rho} = u_2 - u_1 \tag{1.153}$$

This result implies that the pressure drop can only be predicted if the extent of the energy dissipation is known and vice versa.

– *Momentum balance in the y-direction*:

$$0 = \rho v_1 A \cdot v_1 - \rho v_2 A \cdot 0 + \sum F_y \tag{1.154}$$

The second term of the right-hand side states that no momentum is carried in the direction of y by the outgoing flow. The forces operating on the liquid in the y-direction are the pressure force $+ p_1 A$ exerted by the outside world in the positive y-direction on the liquid already inside the bend, and the reaction force of the wall to the liquid $+ F_{y,w \to f}$ (in either the positive or the negative y-direction: that follows from the sign); therefore:

$$F_{y,w \to f} = -\rho v_1^2 A - p_1 A \tag{1.155}$$

The wall must therefore supply a force in the negative y-direction and absorb a force $F_{y,f \to w}$ $(= -F_{y,w \to f})$ in the positive y-direction (action = reaction, Newton's third law). This force $F_{y,f \to w}$ stands for the sum of the pressure force on the liquid and the reaction force that arises when the momentum of the inflowing liquid is entirely absorbed by the wall.

– *Momentum balance in the x-direction*:

$$0 = \rho v_1 A \cdot 0 - \rho v_2 A \cdot v_2 + \sum F_x \tag{1.156}$$

The first term of the right-hand side states that no x-momentum is carried by the incoming flow.

The forces working on the liquid in the x-direction are the 'pressure force' $-p_2 A$ exerted by the outside world on the liquid still within the bend and acting in the negative x-direction, and the reaction force of the wall to the liquid $+ F_{x,w \to f}$ (in either the positive or the negative x-direction: that follows from the sign); therefore:

$$F_{x,w \to f} = \rho v_2^2 A + p_2 A \tag{1.157}$$

The wall must therefore supply a force in the positive x-direction and absorb a force $F_{x,f\to w}$ ($= -F_{x,w\to f}$) in the negative x-direction (again, action = reaction, Newton's third law).

A nice illustration of the combined effect of $F_{x,f\to w}$ and $F_{y,f\to w}$ is a garden sprinkler revolving about a vertical axis where the liquid causes the centre piece to rotate in the reverse direction, or the wagging motion of a loose garden hose from which water is spurting.

It has been assumed in the aforementioned that the velocity distributions at the entrance and exit are uniform (no differences in average velocity). In that case, the inflowing momentum in the direction of x is indeed $\rho A v_1 \cdot v_1$. If this is not the case, then the incoming momentum flow rate in the y-direction must be written as:

$$\rho \int v_1^2 dA \quad \text{or, for short} \quad \rho A \langle v_1^2 \rangle \tag{1.158}$$

where the brackets represent a cross-sectional average.

Example 1.19. The fire hose.
A jet of water comes out of the round hole (diameter $D_2 = 4$ cm) in a flange situated at the end of a horizontally held fire hose (internal diameter $D_1 = 10$ cm). The velocity of the water in the round hole is 15 m/s. The jet of water enters the outside air unimpeded (see Figure 1.34). The situation is steady.

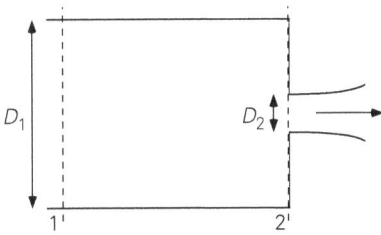

Figure 1.34

Determine the force F, in size and direction, that the flange exercises on the water. Assume that energy dissipation is negligible. (The contraction of the water jet just after the outflow opening may also be disregarded.)

Solution
In order to determine F, it will be necessary to use the mass balance, the mechanical energy balance, and the momentum balance for the horizontal direction. All three balances have the same control volume: the whole tube section between the planes 1 and 2 in Figure 1.34. Plane 1 has to be chosen 'far' enough upstream of the flange; plane 2 runs precisely along the inside of the end flange. The calculation of F is as follows:
The mass balance for the control volume is

$$0 = \rho A_1 v_1 - \rho A_2 v_2 \rightarrow v_1 = \frac{A_2}{A_1} v_2 = \left(\frac{D_2}{D_1}\right)^2 v_2 = 2.4 \text{m/s} \tag{1.159}$$

The mechanical energy balance for the same control volume is

$$0 = \rho A_1 v_1 \cdot \left(\frac{1}{2} v_1^2 + \frac{p_1}{\rho}\right) - \rho A_2 v_2 \cdot \left(\frac{1}{2} v_2^2 + \frac{p_2}{\rho}\right) \tag{1.160}$$

This results in a pressure drop between 1 and 2: the increase in kinetic energy is derived from the drop in pressure energy. Using the mass balance produces the following:

$$p_1 - p_2 = \frac{1}{2}\rho v_2^2 - \frac{1}{2}\rho v_1^2 = 1.1 \cdot 10^5 \text{ Pa} \tag{1.161}$$

which actually is the Bernoulli equation (remember: steady state; no work, no heat exchange; constant ρ; no energy dissipation). Because the pressure in the freely outflowing jet of water is equal to the pressure of the outside air ($p_2 = 1$ bar), the pressure in the hose is $p_1 = 2.1 \cdot 10^5$ Pa.

These pressures also occur in the momentum balance in the flow direction:

$$0 = \rho A_1 v_1 \cdot v_1 - \rho A_2 v_2 \cdot v_2 + p_1 A_1 - p_2 A_2 + F \tag{1.162}$$

All the quantities in this equation are now known from the previous steps, except F; entering the data (as the last step of the solution!) produces: $F = -1{,}286$ N. The force that the flange exercises on the water therefore operates in the opposite direction to that of the flow.

Notice that the pressure occurs in two equations: both in the mechanical energy balance and in the momentum balance(s). In general: which equation can or should be used for which purpose depends entirely on the nature of the problem under consideration.

Example 1.20. Flow round an immersed obstacle.
Two litres of water per second flow through a horizontal pipe with a constant diameter $D = 5$ cm. There is an obstacle in the pipe. This obstacle forms an (extra) element of resistance for the flow; conversely, the flow exercises a force F on the obstacle. The situation is steady. With the help of a pressure sensor, the pressure drop is measured for the (long) part of the pipe in which the obstacle is located. The friction on the wall of this part of the pipe may be ignored in relation to the flow resistance of the obstacle. The difference in pressure is 980 Pa.

First determine the force F that the water is exercising on the obstacle, and how much energy is dissipated per second.

Solution
To this end, draw up a balance for the momentum in the direction of the flow for the section of pipe over which the pressure drop is measured (see Figure 1.35). Note that the planes 1 and 2 should be located far enough from the obstacle to make sure that the flow in both planes should be free of effects associated with the presence of the obstacle. The momentum balance then is

$$0 = \phi_v \cdot \rho v_1 - \phi_v \cdot \rho v_2 + p_1 \cdot \frac{\pi}{4} D^2 - p_2 \cdot \frac{\pi}{4} D^2 - F \tag{1.163}$$

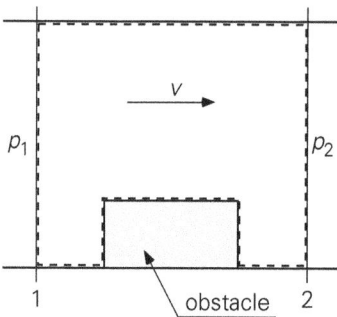

Figure 1.35

Because of $v_1 = v_2$ (because of the mass balance) and $p_1 - p_2 = 980$ Pa (measured), as the last step of the solution equation (1.163) results in:

$$F = \frac{\pi}{4}D^2 \cdot (p_1 - p_2) = 1.92 \text{ N} \qquad (1.164)$$

As a result of friction along the surface of the obstacle and in the wake behind the obstacle (in eddies), mechanical energy is dissipated (transformed into heat). With the help of the mechanical energy balance, this energy dissipation is

$$0 = -\phi_m \frac{p_2 - p_1}{\rho} - \phi_m e_{fr} \qquad (1.165)$$

From this, it is easy to see how much mechanical energy is destroyed per unit of time:

$$\phi_m e_{fr} = 1.96 \text{ J/s} \qquad (1.166)$$

It was possible to calculate the energy dissipation in this example because the pressure drop over the obstacle was measured and therefore known. This will not generally be the case however, and it will be necessary to use a model for e_{fr} so that **pressure drop calculations** for pipeline systems are possible. This will be dealt with in Chapter 5.

Example 1.21. Outflow from a vessel through a hole.
A large vessel is full of water up to height z_0. A small hole with a diameter D, which is much smaller than the vessel diameter, is located at height z in the wall of the vessel below the water line. Water obviously pours out through the hole. Assume that the hole has a sharp edge so that – due to the absence of wall – friction plays a subordinate role only.
    How does the flow rate through the hole depend on the height of the liquid, z_0?

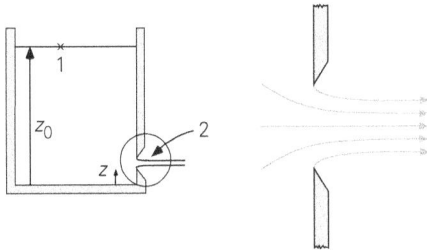

Figure 1.36

Solution
Application of the Bernoulli equation between points 1 and 2 (see Figure 1.36) – where both pressures are equal to the ambient pressure p_0 – produces for the velocity v in the hole (whereby $v = v_2$ and $v_1 \ll v_2$)

$$v = \sqrt{2g(z_0 - z)} \qquad (1.167)$$

And therefore the following applies to the flow rate through the hole:

$$\phi_v = \frac{\pi}{4}D^2 \cdot v = \frac{\pi}{4}D^2 \cdot \sqrt{2g(z_0 - z)} \qquad (1.168)$$

In practice, the flow rate is less than calculated using equation (1.168), as contraction of the discharging liquid jet occurs: the inertia of the liquid that from all sides flows radially towards the discharge hole does not allow the liquid to instantaneously give up this radial velocity (momentum) and to utilise the full cross-sectional area of the discharge opening. Only some distance downstream, outside the hole, do the flow lines become parallel and only at this downstream position does the pressure p_2 in the liquid jet become equal to the ambient pressure p_0 – this position therefore proves to be the point 2 presumed in view of equation (1.167). At this position, the cross-sectional area of the liquid jet is therefore smaller than that of the discharge opening.

This contraction effect is usually taken into account by including a **discharge coefficient** C_d on the right-hand side of equation (1.168), with $C_d < 1$ actually denoting the ratio of the cross-sectional area of the liquid jet at the position where the flow lines have become parallel, to that of the discharge opening:

$$\phi_V = C_d \cdot \frac{\pi}{4} D^2 \cdot \sqrt{2g(z_0 - z)} \tag{1.169}$$

The above phenomenon is generally denoted by the term **vena contracta**. Usually, the (sometimes very) small effect of friction losses is also included in the above discharge coefficient, but with sharp entrance edges of the discharge opening the role of friction may safely be ignored. The lowest possible value of C_d is 0.6 and relates to turbulent flow (at a high value of the Reynolds number: see Chapter 2) through a sharp-edged hole. The value of C_d increases as the entrance edges of the discharge opening are more rounded-off and the pertinent contraction of the discharging liquid jet decreases.

Appendix 1A

Many balance equations result in a first-order linear differential equation of the form

$$\frac{dy}{dx} = \lambda y \tag{A.1}$$

This differential equation is first order and linear because both the derivative of y and y itself only occur to the power of 1. The equation states that dy/dx, that is, the change of y by x, has a linear dependency on y itself. Notice that λ can be both positive and negative.

Because differentiating y versus x leads to a statement that gives the function y itself, $e^{\lambda x}$, which is also written as $\exp(\lambda x)$, is a solution to this differential equation. This means the general solution to the above differential equation is $y = C_1 e^{\lambda x}$, where C_1 is a constant yet to be determined – that is, that C_1 does not depend on x and therefore gives zero in the case of differentiating versus x. For a specific situation, C_1 is obtained with the help of the boundary (or initial) condition that $y = y_0$ when $x = 0$, resulting in the one solution, $y = y_0 e^{\lambda x}$. Wherever dy/dx depends on y itself, the solution to y is always an exponential function of x.

Another way of solving the above differential equation is the **separation of variables** method. This is particularly useful when the right-hand side – as is often the case – is a more complicated statement that is still a linear function of y. With the separation of variables, the dependent variable y and the independent variable x are separated: every y goes to the left-hand side (as a result of division), and every x goes to the right-hand side (through multiplication); equation (A.1) then becomes

$$\frac{dy}{y} = \lambda dx \tag{A.2}$$

Integrating this equation gives

$$\int \frac{dy}{y} = \ln y = \lambda x + C_2 \tag{A.3}$$

This result can be modified into an explicit outcome for y:

$$y = \exp(\lambda x + C_2) = C_2' \exp(\lambda x) \tag{A.4}$$

where the integration constant C_2 – and therefore also C_2' – have to be found by substituting the boundary or initial condition $y = y_0$ at $x = 0$ – which is often written as $y\,(x = 0) = y_0$ – into equation (A.4) because the solution of equation (A.4) also applies to $x = 0$; $C_2' = y_0$ then follows. Remember that every integration stage produces an integration constant for which a boundary or initial condition is required in order to arrive at a specific solution.

With transport phenomena, it is often necessary to work with a more complicated version of differential equation (A.1). This can be shown in general as

$$\frac{dy}{dx} = \alpha y + \beta \tag{A.5}$$

where α and β are constants. **Separation of variables** is a very suitable method for solving this differential equation. This separation step results in

$$\frac{dy}{\alpha y + \beta} = dx \tag{A.6}$$

The next step is to ensure that the same statement is made after the 'd' in the numerator on the left-hand side as in the denominator on the left-hand side in order that the integration of the left-hand side produces $\ln(\alpha y + \beta)$. To that end, the left and right-hand side of equation (A.6) are first multiplied by α:

$$\frac{\alpha dy}{\alpha y + \beta} = \alpha dx \tag{A.7}$$

Then – providing that α and β are constants, that is, that they do not depend on y or x – the left-hand side is modified in stages:

$$\frac{\alpha dy}{\alpha y + \beta} = \frac{d\alpha y}{\alpha y + \beta} = \frac{d(\alpha y + \beta)}{\alpha y + \beta} \tag{A.8}$$

where it is handy to remember that the derivative of $(\alpha y + \beta)$ is precisely the α by which equation (A.6) was multiplied. Substituting equation (A.8) into equation (A.7) and integrating results in

$$\ln(\alpha y + \beta) = \alpha x + C_3 \tag{A.9}$$

From this it follows:

$$ay + \beta = \exp(ax + C_3) = C_3{}' \exp(ax) \tag{A.10}$$

Then $C_3{}'$ is obtained by entering $x = 0$, $y = y_0$ into equation (A.10):

$$C_3{}' = ay_0 + \beta \tag{A.11}$$

This ultimately gives us the explicit solution for y:

$$y = \frac{-\beta}{a} + \frac{ay_0 + \beta}{a} \exp(ax) \tag{A.12}$$

Appendix 1B

The field of transport phenomena sometimes involves **inhomogeneous linear differential equations**. They have the following form

$$\frac{dy}{dx} = ay + b(x) \tag{B.1}$$

This type of equation forms an extension of the differential equations of the type at (A.5), because the second term on the right-hand side is now a function of the independent variable x. There are several, closely related methods of solving this type of inhomogeneous differential equation (B.1).

First of all, the route via the homogeneous differential equation: the first step is to solve the homogenous variant of equation (B.1), therefore

$$\frac{dy}{dx} = ay \tag{B.2}$$

The solution to this is

$$y = K_1 \exp(ax) \tag{B.3}$$

Note that K_1 may not be determined with the help of an initial or boundary condition because every initial or boundary condition is part of the **overall** solution to equation (B.1) and not of a partial solution. Equation (B.3) is not of course sufficient as a solution to equation (B.1): when substituting the solution of equation (B.3) into the left-hand side <u>and</u> right-hand side of equation (B.1), we would find $0 = b(x)$. Alongside solution (B.3), another term therefore has to be drawn up – one that does meet the requirements of equation (B.1). Because the form of this extra term will strongly depend on $b(x)$, there is no generally valid form for the term, so a **particular solution** will have to be found via **trial and error**. This is a disadvantage of this method. Once the particular solution has been found, it has to be added to the (partial) solution of equation (B.3), after which the integration constant follows through the application of the initial or boundary condition.

If $b(x)$ is an exponential function of x – say, $b(x) \propto \exp(\gamma x)$ – then the particular solution can be found by trying a solution of the form $k + h(x)\cdot\exp(\gamma x)$. It will then hopefully be possible to find k and $h(x)$ by substituting this trial function into equation (B.1). If not, another trial function should be looked for. The result should be an equation in which a number of terms occur that contain $\exp(\gamma x)$ as well as several terms without $\exp(\gamma x)$: k is found from the terms without $\exp(\gamma x)$ – this is because they remain when every term with $\exp(\gamma x)$ disappears for $x = 0$ or for $x \to \infty$, depending on the sign of γ – after which $h(x)$ follows from the rest of the equation.

If $b(x)$ is a linear function of x, then the particular solution can be found by looking at the end situation (for $x \to \infty$) if at least dy/dx has then become very small.

A second way of solving differential equation (B.1) is through **variation of constants** where on the basis of solution (B.3)

$$y = K(x)\exp(ax) \tag{B.4}$$

is tested to see whether it can meet the requirements of equation (B.1). Substituting equation (B.4) into equation (B.1) then gives

$$\frac{dy}{dx} = \frac{dK(x)}{dx}\exp(ax) + aK(x)\exp(ax)$$
$$= ay + b(x) = aK(x)\exp(ax) + b(x) \tag{B.5}$$

from which it follows that

$$\frac{dK(x)}{dx}\exp(ax) = b(x) \tag{B.6}$$

and therefore

$$K(x) = \int b(x)\exp(-ax)dx \tag{B.7}$$

by which

$$y = \exp(ax)\int b(x)\exp(-ax)dx \tag{B.8}$$

Note that the integration constant that results from the integration on the right-hand side of equation (B.8) should also be multiplied by $\exp(ax)$. The integration constant that results from integration on the right-hand side is again dealt with by the initial or boundary condition.

A third method for solving differential equation (B.1) is with an **integrating factor**. In this method, the differential equation is first rewritten as

$$\frac{dy}{dx} - ay = b(x) \tag{B.9}$$

The solution being sought $y(x)$ is then multiplied by a function $F(x)$, to which the following must apply:

$$\frac{d}{dx}[F(x)y(x)] = F(x)\frac{dy(x)}{dx} + \frac{dF(x)}{dx}y(x)$$

$$= F(x)\left[\frac{dy}{dx} - ay\right] \tag{B.10}$$

From this, it follows that $F(x)$ must satisfy

$$\frac{dF(x)}{dx} = -aF(x) \tag{B.11}$$

for which $F(x) = \exp(-ax)$ is a solution. Substituting this result into equation (B.10) and combining it with equation (B.9) gives

$$\frac{d}{dx}[\exp(-ax)y(x)] = \exp(-ax)\left[\frac{dy}{dx} - ay\right] =$$

$$= b(x)\exp(-ax) \tag{B.12}$$

from which it follows, through integration, that

$$\exp(-ax)\,y(x) = \int b(x)\exp(-ax)dx \tag{B.13}$$

When $b(x)$ is also an exponential function in particular, the integration of the right-hand side is very straightforward. Thanks to equation (B.13), the final solution to equation (B.1) is

$$y(x) = \exp(ax)\int b(x)\exp(-ax)dx \tag{B.14}$$

Note that the integration constant that results from the integration on the right-hand side of equation (B.14) should also be multiplied by $\exp(ax)$. The integration constant then follows through application of the initial or boundary condition.

2 Mechanisms, non-dimensional numbers, and forces

2.1 Molecular transport

2.1.1 Moving molecules

So far, we have looked primarily at flows of mass, energy, and momentum resulting from 'collective' behaviour. Mass of a certain species, energy of any type, or momentum flowed into (or out of) the control volume because it was carried, as it were, 'under the arm' of a mass flow rate that flowed into (or out of) the control volume. It could be said that this involved the molecules going 'together'. The form of the expressions for the inflows and outflows is always a product of a mass (or volume) flow rate and a concentration in that flow rate (per unit of mass or volume). Such flows are forms of **convective transport**.

However, individual molecules are also able to realize, or contribute to, the (net) transport of mass, energy, and momentum. At a molecular scale, molecules in a liquid or gas move around somewhat chaotically, criss-crossing and colliding with each other as a result of their thermal motion. Each molecule transports its own mass, momentum, and kinetic energy. Because they collide, they are able to transfer their momentum and energy wholly or partly to other molecules. In time, net transport over larger distances is possible. The mobility of individual molecules in a solid is, of course, much less, but here too, due to the fact they vibrate they are perfectly able to pass, or conduct, heat by colliding with neighbouring molecules.

Notwithstanding the fact that this **molecular transport** is caused by the movements of individual molecules, it is useful to talk in terms of 'mean quantities' (continuum quantities) like density, (mass) concentration, and temperature. This section describes the aforementioned transport in a phenomenological manner.

Diffusion

The following experiment serves to give us a greater understanding of the mechanism of molecular transport. A container is divided into two equal parts by a partition, in which there is a small hole (see Figure 2.1). The right-hand side of the tank is filled completely with distilled water, and the left-hand side with saltwater. The pressures on both sides are equal, which means no convective flow occurs from one compartment to the other. However, after a long period of time, both sides will contain equal levels of saltwater. This equilibrium situation will never return to the first situation in which one side contained distilled and the other side saltwater. What is the explanation for this? A close-up of the hole (see Figure 2.2) makes this clear.

https://doi.org/10.1515/9783111246574-002

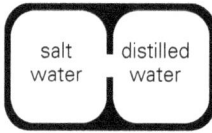

Figure 2.1

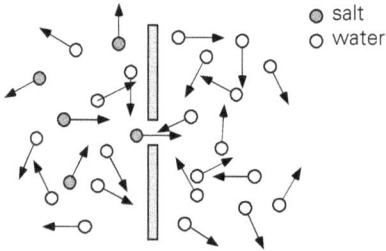

Figure 2.2

A veritable bombardment of molecules takes place from both sides onto the partition wall and thus also onto the imaginary plane of the hole: as a result, as many molecules find their way through the hole from left to right as they do from right to left, maintaining the zero pressure difference between the two compartments. However, if we look at the salt molecules,[4] things are different. The number of salt molecules approaching the hole from either side depends on the number density (concentration) of salt molecules on that side. At first, salt molecules only travel through the hole from left to right – there are no salt molecules on the right, after all. In time, the number of salt molecules in the right compartment increases and so does the number of salt molecules that is able to return to the left. The **net** transport from left to right gradually decreases until the concentrations of salt are equal on both sides.

The process described above is known as **molecular diffusion** and originates in macroscopic differences in concentration. It is to be expected that the resulting net mass transport rate ϕ_m simply depends on this difference in concentration: the greater the difference, the stronger the net transport rate, and the greater the distance between the places of high and low concentration, the weaker the transport rate (as it may take the molecules longer to cover a larger distance).

Conduction

Molecular heat transport goes by analogy. The driving 'force' here is a difference in temperature. Temperature is actually the resultant of the kinetic energy of a very large number of atoms. An example in this context is that of a copper rod, the left side of

---

4 It is actually incorrect to talk of salt molecules, as the dissolution of salt always involves dissociation into ions. This is a typical example of the use of an 'engineering point of view': the 'removal' of the physical reality by the introduction of all kinds of simplifications to the phenomenon under review in order to highlight and describe the essence of the phenomenon under consideration.

which has a higher temperature than the right. This means on average that the molecules on the left have a greater kinetic energy than those on the right. In a solid, neighbouring molecules collide with each other all the time due to their vibrations about their average position. The effect of these collisions is that the fast-vibrating molecules transfer their kinetic energy to the more slowly vibrating ones. This is why the temperature on the left of the rod decreases slightly, while the temperature on the right increases. This continues until the temperature is equal on both sides and of course the net transport of energy (heat) is zero. The conclusion is that a difference in temperature gives rise to an energy flow – that is, transport of kinetic energy of molecules.

Looking at this phenomenon from a macroscopic perspective, there is simply transport, or 'flow', of heat. This mechanism of heat transport is not just confined to solids: the mechanism occurs within gases and liquids as well. In the case of the latter two, energy is passed on between neighbouring molecules not just via collisions but also because of the criss-crossing movements of the molecules. 'Warm' molecules diffuse to a 'cold' location where they share their kinetic energy, while 'cold' molecules move away from that cold location and pick up some 'heat'. The term used in the case of heat transport resulting from the activity of individual molecules is **conduction**. The net heat transport rate due to conduction is denoted by ϕ_q.

Momentum

Finally, it is also possible for momentum to be transported by individual molecules.

Here, the molecules carry their own momentum. This effect can be easily understood by looking at the individual molecules in a **laminar** (= layered) flow. Conceptually, a flow of this type consists of adjacent layers (*in Latin: laminae*, plural of **lamina**) where the velocity varies slightly from one layer to another (see Figure 2.3). For the sake of simplicity, it is assumed here that the substance involved is pure (i.e. that all the molecules are the same). Molecules in layer 1 have, on average, superimposed a velocity $v_{x,1}$ on top of their criss-cross movement. For layer 2, of course, this is $v_{x,2}$ on average.

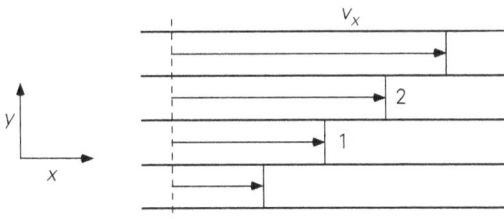

Figure 2.3

A molecule (with molecule mass m) that 'diffuses' from layer 1 to layer 2 will bring, on average, momentum $mv_{x,1}$ to it. Here, too, it is useful to conceive $v_{x,1}$ as an x-momentum concentration (in Ns/kg), as explained in Section 1.4. As a result, $mv_{x,1}$ will then stand for the x-momentum of that molecule. A molecule from layer 2 will, on average, bring momentum $mv_{x,2}$ to layer 1. In the case of this diffusion, (about) as many molecules will go from layer 1 to layer 2 as they will vice versa. This will cause the overall momentum in layer 2 to fall: molecules richer in x-momentum will be replaced by more poor ones. For layer 1, it is the very opposite that is true. In addition, upon arriving in a new layer, molecules will collide with their new neighbours and exchange x-momentum, thus reducing the velocity differences between the **laminae**.

Viewed from a macroscopic perspective, a net momentum flow is involved once again; note that *x-momentum is now transported in the y-direction!* (see Figure 2.4). This net x-momentum transport, denoted[5] by ϕ_{px} is in the negative y-direction in this case from layers with a higher to those with a lower x-momentum concentration, as dv_x/dy is positive.

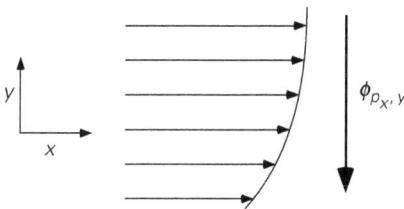

Figure 2.4

Summary
The following applies to molecular transport, based on the thermal movement of individual molecules:
   difference in concentration → net mass transport rate ϕ_m of a particular species
   difference in temperature → net energy transport rate ϕ_q (or heat 'flow')
   difference in velocity → net momentum transport rate ϕ_p

In the first case we speak of diffusion, and in the second of (heat) conduction. In the third case, with laminar flow, we have to specify which momentum is transported by the individual molecules in which direction: in a laminar flow in the x-direction, the velocity gradient dv_x/dy results in a net x-momentum transport in the y-direction. When dv_x/dy is positive, the molecular transport is in the negative y-direction, while for negative dv_x/dy values the transport takes place in the positive y-direction.

2.1.2 Fick, Fourier, Newton

With a view to arriving at practical descriptions of all three molecular transport mechanisms discussed above, simple equations have been drawn up – which are

---

5 In this elementary explanation, compliant with the use of the symbol P_x in Section 1.4, we use the symbols p_x, p_y, p_z to denote momentum concentrations in the respective directions. Further on, we will switch to the usual v_x, v_y, v_z for momentum concentrations or velocities.

valid for a large category of substances – that link the driving forces, the above 'differences', to the transport rates. As mentioned in the discussion on mass diffusion, a greater difference in concentration Δc causes a greater flow (rate) while a greater distance L, over which the difference occurs, leads to a lesser flow (rate). It appears that the $\Delta c/L$ ratio is decisive. This proves to be true not only at finite distances but also locally.

To put it more accurately, the derivative of concentration c to position x, that is, the local dc/dx, is decisive for the local mass flow rate. Of course, the mass is transported from a location with a high concentration to one with a low concen-tration, which is why the flow ϕ_m is a function of $(-dc/dx)$. A similar reasoning may be applied to both heat conduction and molecular momentum transport as a function of temperature gradient and velocity gradient, respectively.

It is easy to understand that, as with any flow or transport, the size of molecular transport is also directly proportional to the size of the plane through which the transport occurs. For this reason, it is useful to introduce the notion of a **flux**. A flux is a flow rate per unit of surface area (where the plane is perpendicular to the direction of transport). Therefore, if mass flow or transport rate ϕ_m passes through surface A (with A perpendicular to the direction of ϕ_m), then the flux is

$$\phi''_m = \frac{\phi_m}{A} \tag{2.1}$$

The superscript " indicates that it concerns[6] a flux, that is, a flow rate per m².

The equations for the molecular transport of a species, of heat, and of momentum are given in terms of their fluxes. In the simplest approach, each flux is assumed to be directly proportional to the pertinent gradient. The three equations, then, are:

mass flux	$\phi''_{m,x} = -ID\dfrac{dc}{dx}$	Fick's law	(2.2)
heat flux	$\phi''_{q,x} = -\lambda\dfrac{dT}{dx}$	Fourier's law	(2.3)
y-momentum flux	$\phi''_{py,x} = -\mu\dfrac{dv_y}{dx}$	Newton's (viscosity) law	(2.4)

All three 'flows' are moving in the direction of x. Of course, Newton's viscosity law also implies that an x-momentum flux in the y-direction is possible. The minus sign in these laws means that the transport always runs from high to low concentration; the direction of the flux (in the positive or negative x-direction) always depends of course on the sign of the pertinent gradient (value less than or greater than zero).

---

6 Later, when dealing with ϕ, we will also use a single ' to indicate reference to a flow per unit of length – for example, ϕ_m'.

The proportionality coefficient ID is called the **diffusion coefficient**; the SI unit for ID is m^2/s. The coefficient λ denotes the **thermal conductivity coefficient** and its SI unit is J/msK. The coefficient μ is the **dynamic viscosity**, the SI unit of which is Ns/m^2; the latter unit shows that μ stands for the momentum (in Ns) that in a laminar flow per unit of area is transported across the interface of two **laminae**.

Of the three laws, Fourier's law is the most generally applicable. Fick's law applies only to binary systems (these are system of just two components) and is also subject to other restrictions. This will be dealt with in detail in Chapter 4. Newton's viscosity law too is of limited validity. Liquids and gases that obey this law are referred to as **Newtonian fluids**. Liquids that fall into this category are more the exception than the rule. Liquids not obeying Newton's viscosity law are called **non-Newtonian liquids**. This will also be covered more extensively at a later stage (in Chapter 5).

In line with our earlier recommendation of working in terms of concentrations of species, energy, and momentum, we would prefer to formulate the laws as follows:

difference in **mass** concentration → **mass** flow rate

difference in **energy** concentration → **energy** flow rate

difference in **y-momentum** concentration → **y-momentum** flow rate

Fick's law of diffusion already meets this requirement. The other two can be rewritten for the special case of substances with a **constant** ρc_p or **constant** ρ, respectively:

$$\text{Fourier:} \quad \phi''_{q,x} = -\lambda \frac{dT}{dx} = -\frac{\lambda}{\rho c_p} \frac{d}{dx}(\rho c_p T) = -a \frac{d}{dx}(\rho c_p T) \tag{2.5}$$

If ρc_p = constant

$$\text{Newton:} \quad \phi''_{p_y,x} = -\mu \frac{dv_y}{dx} = -\frac{\mu}{\rho} \frac{d}{dx}(\rho v_y) = -\upsilon \frac{d}{dx}(\rho v_y) \tag{2.6}$$

if ρ = constant

The combination $\rho c_p T$ is the concentration of thermal energy (unit: J/m^3). The coefficient $a \equiv \lambda/\rho c_p$ is called the **thermal diffusivity coefficient** and has the same unit as ID: m^2/s. The combination ρv_y is the y-momentum concentration (in Ns/m^3); and its associated coefficient $\upsilon \equiv \mu/\rho$ is the **kinematic viscosity** (again, the unit is m^2/s). Note that equations (2.2), (2.5), and (2.6) are completely analogous, with all three of the respective coefficients ID, a and υ having the same SI unit m^2/s. This analogy between molecular mass transport, heat transport, and momentum transport was already a key feature in the technique of drawing up balances and in describing convective transport in Chapter 1.

Note further that the numerical values of the dynamic viscosity μ and the kinematic viscosity υ may differ by several orders of magnitude; for water, for example, $\mu = 10^{-3}$ Ns/m^2 and $\upsilon = 10^{-6}$ m^2/s (at $T = 20$ °C and $p = 1$ bar). It is therefore important to keep these two viscosities firmly apart.

Example 2.1. Heat **conduction** in a copper rod.

A copper rod (with a constant diameter and length L) is kept at a constant temperature at both ends: T_0 at $x = 0$ and T_L at $x = L$ with $T_0 > T_L$. The rod is also thermally insulated from the surroundings.

In a steady-state situation, what is the temperature profile in the rod? How great is the heat flow through the rod, in that case?

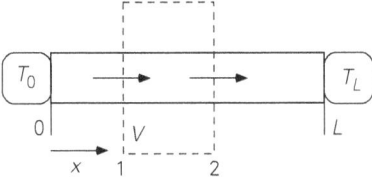

Figure 2.5

Solution
As the question is about the temperature profile *in* the rod, there is no point in selecting the whole rod as the control volume and drawing up a **macro-heat balance**: this would at best link the mean temperature of the rod to heat flows in and out. Instead, you should select a small but representative control volume **inside** the rod, as illustrated in Figure 2.5, and draw up a **micro-heat balance** for this small(er) control volume.

Because the rod is insulated, apart from the two ends, the heat transport will only occur lengthwise in the rod (from one end towards the other). The control volume can therefore cover the whole cross-sectional area of the rod – after all, nothing occurs transversely. This therefore concerns just a one-dimensional transport problem (without heat production).

In a steady-state situation, the heat balance for the control volume V in Figure 2.5 is therefore

$$0 = \phi_q\big|_{x=x_1} - \phi_q\big|_{x=x_2} \tag{2.7}$$

where, due to the given $T_0 > T_L$, the first term on the right-hand side stands for the 'flow in' and the second term for the 'flow out'. Because of the constant diameter of the rod

$$0 = \phi_q''\big|_{x=x_1} - \phi_q''\big|_{x=x_2} \tag{2.8}$$

As only heat conduction is involved, the fluxes can be written as follows:

$$\phi_q''\big|_{x=x_1} = -\lambda \frac{dT}{dx}\bigg|_{x=x_1} \quad \text{and} \quad \phi_q''\big|_{x=x_2} = -\lambda \frac{dT}{dx}\bigg|_{x=x_2} \tag{2.9}$$

Substituting this into equation (2.8) shows that the two heat fluxes through the planes 1 and 2 are equal, implying that – providing that λ is constant – the temperature gradient does not depend on position:

$$\frac{dT}{dx} = C_1 = \text{constant} \tag{2.10}$$

Solving this very simple differential equation gives, with the help of boundary conditions $x = 0 \to T = T_0$ and $x = L \to T = T_L$

$$T(x) = \frac{T_L - T_0}{L}x + T_0 \tag{2.11}$$

The temperature profile is therefore linear, which means the heat flow through the rod is

$$\phi_q = -\lambda A \frac{dT}{dx} = \lambda(T_0 - T_L) \cdot \frac{A}{L} \tag{2.12}$$

where A is the cross-sectional area of the rod.

Remember how earlier, in Section 1.2.2, also a small control volume **inside** the reactor was selected, the purpose of which was to arrive at a concentration profile along the reactor. There, we opted for a thin slice dx because of the chemical reaction, with the view of obtaining dc/dx and the pertinent differential equation. In the above case, this is not necessary. The above case does illustrate again the use of a **micro-balance** to derive a profile; in this case a temperature profile.

Example 2.2. Evaporation of naphthalene from a tube.
A long, narrow tube (length 25 cm, internal diameter $D = 1$ cm) is partly filled with solid naphthalene. The top of the tube is exposed to the air (see Figure 2.6). The length of the tube above the naphthalene is 20 cm. The ambient temperature is 20 °C.

How long does it take for a 1 mm layer of the naphthalene to evaporate?
Naphthalene details:

diffusion coefficient in air	$ID = 7 \cdot 10^{-6}$ m^2/s
density of solid naphthalene	$\rho = 1{,}150$ kg/m^3
molar mass	$m = 106$ kg/kmol
vapour pressure at 20 °C	$p^* = 0.05$ mm Hg

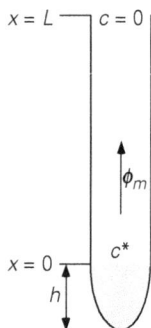

Figure 2.6

Solution
Drawing up a mass balance for the solid naphthalene in the tube produces:

$$\frac{dM}{dt} = \rho A \frac{dh}{dt} = -\phi_{m,\text{out}} = -A \cdot \phi''_{m,\text{out}} \tag{2.13}$$

A is the cross-sectional area of the tube. In order to progress further, it is necessary to determine the flux of naphthalene vapour. Naphthalene disappears only as a result of diffusion. At the interface between the solid naphthalene and air, the naphthalene concentration c^* in the gas phase may be estimated with the help of the ideal gas law from the local vapour pressure p^*:

$$c^* = m \frac{p^*}{RT} \tag{2.14}$$

where p^* is the equilibrium vapour pressure above the solid naphthalene that depends on temperature. Because the top of the tube is exposed to the surrounding air, there is hardly any naphthalene present there: $c(L) \approx 0$. There is therefore clearly a driving concentration difference between $x = 0$ and $x = L$, resulting in diffusive transport.

Given that solid naphthalene diffuses away very slowly because of its extremely low vapour pressure at room temperature – the driving force for diffusion is very small – the diffusion in the tube can be regarded as quasi-steady: it is as if the situation is steady at any given moment (this has to be checked afterwards). Over a longer period of time, of course, this assumption becomes increasingly pointless.

In a derivation entirely analogous to that leading to equation (2.11), with the heat flux $\phi_q'' = -\lambda\, dT/dx$ replaced by the diffusive mass flux $\phi_m'' = -\text{ID}\, dc/dx$, the following result is obtained:

$$c(x) = \frac{c(L) - c(0)}{L}\, x + c(0) = c^* \left(1 - \frac{x}{L}\right) \tag{2.15}$$

From this, it follows for the flux at $x = 0$ (at the surface of the naphthalene), that:

$$\phi_m'' = -\text{ID}\frac{dc}{dx} = \text{ID}\frac{c^*}{L} \tag{2.16}$$

Combining equations (2.13), (2.14), and (2.16) now produces

$$\frac{dh}{dt} = -\frac{\text{ID}\ \eta p^*}{\rho\ RTL} \tag{2.17}$$

Solving this differential equation gives: $\Delta h = 1$ mm if

$$\Delta t = \frac{\rho\ RTL}{\text{ID}\ \eta p^*}\ \Delta h = 1.1 \cdot 10^8\ s \approx 3.6\ \text{years} \tag{2.18}$$

Notice that the earlier assumption about the diffusion rate was therefore correct. In fact, it is not entirely appropriate to use Fick's law in this situation: it is only because the vapour pressure of naphthalene is so low that Fick's law produces a satisfactorily accurate result. This situation will be dealt with again in Section 4.3.

Convective transport is generally a much more effective way of transporting a particular quantity than is molecular (or diffusive) transport. Although the velocity of a liquid package is much lower than that of a molecule – a typical liquid velocity is 1 m/s, while the random velocity of a molecule in water at room temperature is of the order of 600 m/s – molecules move collectively in the case of convective transport. When it comes to molecular transport, the molecules behave as genuine individuals that each go their own way, and their movement is not in any particular direction. Finally, with convective transport, it is the quantity itself (e.g. thermal energy) that counts, while with molecular transport it is all about gradients in this quantity.

Summary
For a large category of substances, molecular transport can be described with simple phenomenological laws. The transport of mass, thermal energy (heat), and momentum is described by Fick's, Fourier's, and Newton's law, respectively:

$$\phi_{m,x}'' = -\text{ID}\frac{dc}{dx};\ \phi_{q,x}'' = -\lambda\frac{dT}{dx};\ \phi_{py,x}'' = -\mu\frac{dv_y}{dx}$$

All three laws link flow, or rather flux, to the driving gradients via a proportionality constant: the diffusion coefficient ID, the thermal conductivity coefficient λ, and the dynamic viscosity μ, respectively. Fourier's law is the most generally applicable, while the other two are subject to stricter requirements. Providing that ρ and $\rho\, c_p$ are constant, all three laws can be written as

$$\text{flux of} \left\{ \begin{array}{c} \text{mass} \\ \text{thermal energy} \\ y\text{-momentum} \end{array} \right\} \propto -\frac{d}{dx} \left\{ \begin{array}{c} \text{mass concentration} \\ \text{thermal energy concentration} \\ y\text{-momentum concentration} \end{array} \right\}$$

for molecular transport in the x-direction. In this formulation, illustrating the firm analogy of molecular mass, heat, and momentum transport, the proportionality constants are the diffusion coefficient ID, the thermal diffusivity a, and the kinematic viscosity υ, respectively, all in m²/s.

The minus sign in all these laws means that the transport always runs from high to low concentrations; the direction of the flux depends on the concentration gradient sign.

2.1.3 Transport coefficients

In the above naphthalene example, the rather high value for time Δt is attributable to a significant degree to the fact that the diffusion coefficient ID has such a low value. In order to be able to make a quick estimate of the size of the various flows in practice, it is useful to know the order of magnitude of the various transport coefficients (these are ID, λ, a, μ, and υ). The following table could come in handy for this purpose:

	ID (m²/s)	υ (m²/s)	a (m²/s)	λ (W/m · K)
Gas	10^{-5}	10^{-5}	10^{-5}	10^{-2}
Liquid	10^{-8} to 10^{-9}	10^{-5} to 10^{-6}	10^{-7}	10^{-1}
Solid	10^{-11} to 10^{-15}			1–500

Remember that these transport coefficients are physical properties which in general depend on the atomic or molecular composition of a substance and on pressure and temperature.

The above estimates can be made for **gases** with the help of the **kinetic gas theory**; the kinetic gas theory can also be used to determine how, for gases, the transport coefficients depend on pressure and temperature. For the diffusion coefficient, for example, the following simplified model can be set up.

Imagine that the molecules of a gas may be conceived as hard spheres. These spheres have a mean velocity v_m and travel on average a distance ℓ until they collide (ℓ being the average free path length). Suppose now that there are two types of molecule – A and B – both of about the same size and weight. Molecules B form the background with molecules A diffusing in between them, n_A denoting the number of molecules A per unit of volume. If we focus on a plane at position x (see Figure 2.7), let us assume that the number of A molecules to the left of the plane (n_{A-}) is slightly smaller than to the right of the plane (n_{A+}). The question now is how great the flow of A molecules through the plane is?

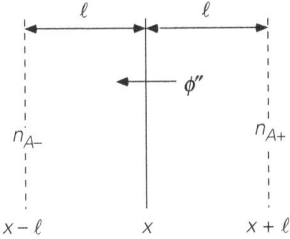

ℓ ℓ

ϕ''

n_{A-} n_{A+}

$x-\ell$ x $x+\ell$ **Figure 2.7**

The movements of the molecules are random. For the sake of simplicity, we can say that 1/3 of all molecules move in the x-direction (and 1/3 in the y-direction and 1/3 in the z-direction). Of this, half, that is 1/6 of the total, move in the positive x-direction. During a short interval of time τ, therefore, 1/6 of all molecules located on the left-hand side within a distance $v_m \cdot \tau$ of plane x, pass this plane. The same applies to the molecules to the right of x. Net, this gives a flux (= number of particles per unit of surface and per unit of time) of:

$$\phi'' = \frac{\frac{1}{6} n_{A-} \cdot v_m \tau - \frac{1}{6} n_{A+} \cdot v_m \tau}{\tau} \tag{2.19}$$

An estimate of the two particle densities can be obtained by attributing density n_{A-} at the plane $x - \ell$, that is, one mean free path length removed from plane x. The particles located here just pass by plane x without colliding. Similarly, put n_{A+} at plane $x + \ell$. This means that the following applies to the particle flux:

$$\phi''_n = \frac{1}{6} v_m \cdot [n_A(x-\ell) - n_A(x+\ell)] \tag{2.20}$$

Further,

$$n_A(x \pm \ell) \approx n_A(x) \pm \frac{dn_A}{dx} \cdot \ell \tag{2.21}$$

So equation (2.20) becomes

$$\phi''_n = -\frac{1}{3} v_m \cdot \ell \cdot \frac{dn_A}{dx} \tag{2.22}$$

From this, it is easy to make a mass flux by multiplying it by molecule mass m:

$$\phi''_m = -\frac{1}{3} v_m \cdot \ell \cdot \frac{dc_A}{dx} \tag{2.23}$$

Comparing equations (2.2) and (2.23) gives for the diffusion coefficient:

$$ID = \frac{1}{3} v_m \cdot \ell \tag{2.24}$$

(This result can also be obtained through a much more precise method.) Equation (2.24) expresses that in an ideal gas a diffusion coefficient (in m^2/s) may be described in terms of the product of a molecular velocity (in m/s) and the mean free path length (in m).

The mean velocity v_m of a molecule can be determined from

$$\frac{1}{2} m v_m^2 = \frac{3}{2} k_B T \tag{2.25}$$

where k_B is the Boltzmann constant, and therefore it is found that

$$v_m = \sqrt{\frac{3RT}{\mathfrak{m}}} \tag{2.26}$$

Here, R is the gas constant and $\mathfrak{m}$ the molar mass.

The free path length ℓ is inversely proportional to the particle density n, and the frontal surface area of the particles is $\pi\sigma^2$, where σ is the diameter of the hard spheres. Rewriting this with the help of Avogadro's number N_{avo} gives:

$$\ell \propto \frac{1}{n\sigma^2} = \frac{mN_{avo}}{mn \cdot N_{avo}\sigma^2} = \frac{\mathfrak{m}}{\rho} \cdot \frac{1}{N_{avo}\sigma^2} \tag{2.27}$$

For the magnitude of ID and the variation of ID with T, p, and ρ, it then follows that (for ideal gases, at least!):

$$ID \propto \frac{\sqrt{3RT\mathfrak{m}}}{\rho} \propto \frac{\sqrt{T}}{\rho} \tag{2.28}$$

It is also possible to estimate the other transport coefficients in ideal gases with the help of the same simple model.[7] After all, heat conduction involves the same molecular motions that now carry, on average, $\rho c_p T$. In the case of molecular momentum

---

7 The above analysis is based on the assumption that molecules are hard, inelastic spheres. A more complex analysis that takes a degree of elastic behaviour by molecules into account would lead to

$$\lambda \propto T^{0.70 \text{ à } 0.75}, \ \mu \propto T^{0.65 \text{ à } 0.70}, \ ID \propto T^{0.70 \text{ à } 0.75}/\rho$$

These results, confirmed by experiment, apply to real gases when the pressures and temperatures are not too extremely high.

transport, the molecules carry ρv on average. Here we assume that the exchange of energy and momentum during the collisions of the molecules is fast with respect to the molecular motions which then are rate determining for the transport of energy and momentum. The exact same as for ID therefore applies to a and υ. Accordingly, we can write down immediately:

$$\lambda \propto \left(\frac{1}{3}v_m.\ell\right) \cdot \rho c_p \propto \sqrt{RT\eta} \cdot c_p \propto \sqrt{T} \tag{2.29}$$

$$\mu \propto \left(\frac{1}{3}v_m.\ell\right) \cdot \rho \propto \sqrt{RT\eta} \propto \sqrt{T} \tag{2.30}$$

It therefore appears that λ and μ are independent of pressure.

For **liquids** it can be deduced for the case of molecules A diffusing amidst molecules B, that the following relation applies to the diffusion coefficient

$$ID_{AB} \approx \frac{k_B T}{3\pi\mu_B\sigma_A} \tag{2.31}$$

with k_B is the Boltzmann constant, μ_B the viscosity of liquid B, and σ_A the diameter of molecule A. This equation is known as the Stokes-Einstein equation.

The viscosity of liquids depends to a major degree on temperature as already reported by Frenkel (1926) and Andrade (1930); to a good approximation, liquid viscosities decrease with temperature according to:

$$\log \mu \sim \frac{1}{T} \tag{2.32}$$

Summary

We have looked in this section at the order of magnitude of the transport coefficients. The table in this section gives a good overview of this.

Using simple kinetic gas theory it has been found that, for ideal gases, the diffusion coefficient may be conceived as the product of a molecular velocity and a free path length. As a result of this model, the transport coefficients for ideal gases depend on temperature and pressure (density) as

$$ID \propto \frac{\sqrt{T}}{\rho}, \ \lambda \propto \sqrt{T}, \ \mu \propto \sqrt{T}$$

For liquids, the transport coefficients are a function of temperature only: the diffusion coefficient is proportional to temperature while dynamic viscosity obeys to

$$\log \mu \sim \frac{1}{T}$$

2.1.4 Shear stress: an alternative description of molecular momentum transport

In Section 1.4, dealing with the momentum balance, it was stated that forces are producers of momentum. This idea can be used for developing an alternative view to molecular momentum transport. Let us again look at a control volume located in a neat, laminar (= layered) flow (see Figure 2.8).

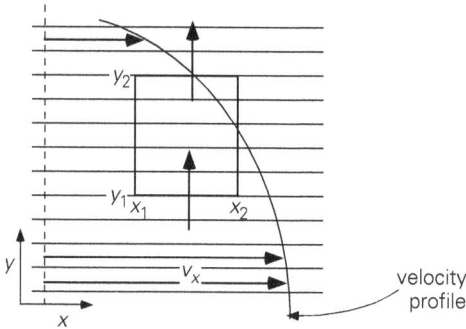

Figure 2.8

The flow is steady and is moving in the x-direction, where v_x is a function of the y-coordinate: $v_x = v_x(y)$. The convective x-momentum flows through planes $x = x_1$ and $x = x_2$ are of the same magnitude (this can be easily shown with a mass balance!) and cancel each other out. Applying Newton's viscosity law to this situation implies that a molecular transport of x-momentum $\phi_{px,y}|_{y=y_1}$ is entering the control volume through plane $y = y_1$, due to the velocity gradient dv_x/dy in plane $y = y_1$, while similarly a transport $\phi_{px,y}|_{y=y_2}$ is leaving through plane $y = y_2$. Therefore, the x-momentum balance looks as follows in a steady-state situation:

$$0 = \phi_{px,y}|_{y=y_1} - \phi_{px,y}|_{y=y_2} + \sum F_x \tag{2.33}$$

in which $\sum F_x$ stands for all forces acting in the x-direction such as pressure forces driving the flow.

It is also possible, however, to look at this case from another perspective, given that adjacent layers interact due to molecular cross-overs: a faster moving layer will drag a slower one, and a slower moving layer will retard a faster one. This interaction may be denoted by the terms 'friction' or 'friction forces'. Such a frictional force is nothing more than a different interpretation (or model) of the net molecular momentum transport rate $\phi_{px,y}$. Along the plane $y = y_1$, a mutual frictional force acts between the layer directly under plane $y = y_1$ and the layer directly above it – the mutuality is because of Newton's Third Law of classical mechanics ('Action = – Reaction').

The frictional force is of course proportional to the size of the area A of the plane on which it acts. It is therefore better to look at the force per unit of area – just as equations (2.2)–(2.4) relate to fluxes. Because the force acts parallel to the plane (and not normal to it), the force per unit of area is not referred to as a pressure, but as a **shear stress**, for which the symbol τ is used. In short:

$$\tau_{yx} \equiv \phi''_{px,y} \tag{2.34}$$

There is one very important convention associated with the term **shear stress**: the choice of the direction in which this force generally goes. The convention is:

$\tau_{yx}|_{y=y_1}$ is the shear stress that acts in the x-direction in (or along) the plane $y = y_1$ and which the liquid layer with a y-coordinate of just less than y_1 exerts on the layer with a y-coordinate of just greater than y_1.

Note that – for just the one-dimensional flows we are considering here – the first subscript of τ_{yx} refers to the direction normal to that of the flow; the second subscript relates to the x-direction in which either τ_{yx} or $-\tau_{yx}$ acts.

By definition, this shear stress appears in a force balance with a **plus** sign. Whether τ_{yx} acts in the positive or the negative x-direction, depending on the sign of τ_{yx} and – due to Newton's viscosity law – on the sign of dv_x/dy, is not important when drawing up a force balance, but will be the result of the derivation that follows.

Therefore, for the situation considered here (Figure 2.8), it is possible to write $\tau_{yx}|_{y=y_1} \cdot A$ instead of $\phi_{px,y}|_{y=y_1}$. The momentum flux $\phi''_{px,y}|_{y=y_2}$ can also be replaced by a shear stress: the layer with $y < y_2$ exerts a shear stress $\tau_{yx}|_{y=y_2}$ on the layer just above it where $y > y_2$.

Again thanks to Newton's Third Law, the layer just above plane $y = y_2$ exerts a shear stress $-\tau_{yx}|_{y=y_2}$ on the layer below it, which forms part of the control volume. The molecular 'flow out' of x-momentum $-\phi_{px,y}|_{y=y_2}$ in equation (2.33) can therefore be replaced by the force $-\tau_{yx}|_{y=y_2} \cdot A$.

On the basis of the above alternative perspective (model) and the introduction of the concept of **shear stress**, the **momentum** balance of equation (2.33) can also be written as a **force** balance:

$$0 = A \, \tau_{yx}|_{y=y_1} - A \, \tau_{yx}|_{y=y_2} + \sum \text{other forces in the } x\text{-direction} \tag{2.35}$$

Equations (2.33) and (2.35) are therefore completely identical and differ only through the model associated with them:

Equation (2.33) → net molecular x-momentum transport as a result of individual molecule movements

Equation (2.35) → shear stress as a result of layers exerting and feeling mutual friction

Combining equations (2.34) and (2.4) results in an expression relating shear stress to a velocity gradient with a minus sign on the right-hand side. Many textbooks present a form of Newton's viscosity law without this minus sign. Our minus sign is due to the analogy between the molecular transport of species, heat and momentum.

! **Summary**

There are two ways of looking at molecular momentum transport:

1. as a kind of diffusion of momentum: in terms of individual molecular motions, leading to fluxes such as $\phi''_{px,y}$
2. as a frictional force per unit of area: a shear stress τ_{yx} that is exerted by two layers on each other as they move past each other at different velocities.

Due to these two interpretations, the **momentum** balance for a simple laminar flow in steady-state conditions can also be regarded as a **force** balance.

There is an important convention with regard to the signs used for the shear stress: the force exerted by the layer with the small coordinate on the layer with the large coordinate is given a '+' sign in the balance.

2.2 Dimensional analysis

2.2.1 Non-dimensional numbers

After convective transport of mass, energy, and momentum were dealt with in detail in Chapter 1, molecular transport of mass, heat, and momentum were covered in Section 2.1. In practice, many combined situations are present in fluids, in which both convective and molecular transport occur, though not necessarily in the same direction. In many cases, it is useful and usual to characterise processes or situations with the help of **non-dimensional numbers** that indicate the relationship of two such fluxes. An example could be the convective flux of a quantity (such as mass) in relation to the molecular flux of the same quantity, but also the ratio of the transport coefficients for two molecular fluxes is a commonly used non-dimensional number.

To start with the latter possibility: equations (2.2), (2.5), and (2.6) give the molecular fluxes as a function of the gradients in the respective concentrations (providing that ρ and ρc_p are constant). The transport coefficients ID, a, and υ all have the same dimension (i.e. m^2/s) and their values express how effectively the individual molecules of the relevant substance in the state of matter under consideration diffuse and transport heat and momentum, respectively, given a certain value for the concentration gradients.

Consider the υ/a ratio for an ideal gas. Thanks to equations (2.29) and (2.30), it is clear to see that molecules in an ideal gas transport heat as easily as they do momentum: $\upsilon/a = 1$. But for liquids, it appears that $\upsilon/a = 10–100$: liquid molecules obviously transport momentum more easily than they do heat. The υ/a ratio is referred to as the **Prandtl** number or Pr for short:

$$\mathrm{Pr} = \frac{\upsilon}{a} \tag{2.36}$$

Similarly, the **Schmidt** (Sc) and **Lewis** (Le) numbers are defined thus

$$\mathrm{Sc} = \frac{\upsilon}{\mathrm{ID}} \tag{2.37}$$

and

$$\mathrm{Le} = \frac{a}{\mathrm{ID}} = \frac{\mathrm{Sc}}{\mathrm{Pr}} \tag{2.38}$$

For ideal gases, all three numbers have the value of one: after all, the mobility of the molecules in an ideal gas is the limiting factor for molecular transport rather than the effectiveness of exchange of energy or momentum between the colliding molecules. For liquids, however, Pr generally has values of between 10 and 100, while Sc is often of the order of 10^3.

As already discussed, however, for a certain quantity, the magnitude of the convective transport in a certain situation can be compared with its molecular transport to decide if one of the two may be ignored with respect to the other. If this occurs with a particular species in a multi-species system, the **Péclet** number is obtained:

$$\mathrm{Pe} = \frac{vL}{\mathrm{ID}} \tag{2.39}$$

After all, when considering simultaneous convective and molecular (or diffusive) transport of a species C in the x-direction, the convective transport is equal to $A{\cdot}v{\cdot}c$ (in which v is the velocity, A is the cross-sectional area of the flow channel, and c is the concentration of C), while the molecular transport of C is given by $- A\mathrm{ID}dc/dx$:

$$\mathrm{Pe} = \frac{\text{convective transport}}{\text{diffusive transport}} = \frac{Avc}{-A\,\mathrm{ID}\dfrac{dc}{dx}} \tag{2.40}$$

If the concentration gradient (including the minus sign) may be approximated by c/L – where L is the typical length scale or dimension over which the concentration varies in the case of interest: see equation (2.16) for example – expression (2.39) is obtained.

Example 2.3. Purging.
A liquid containing a dissolved aggressive species B (concentration c_{B0}) is flowing through a tube. However, the pressure sensor that is being used to measure the pressure is affected by corrosion due to contact with species B unless the concentration c_B is less than $c_{B0}/1{,}000$. This can be resolved by mounting the sensor not in the wall of the (main) tube, but in a small branch away from the main tube (see Figure 2.9).

Nonetheless, B may still reach the sensor as a result of diffusion. By creating a low liquid flow (with velocity v) through the branch towards the main tube – this method is known as purging – it is possible to keep the concentration of B low and to limit the corrosion. The distance between the sensor and the junction of the branch and the main tube is denoted by L. The question is this: how great must v be in order for $c_B(L) < c_{B0}/1{,}000$?

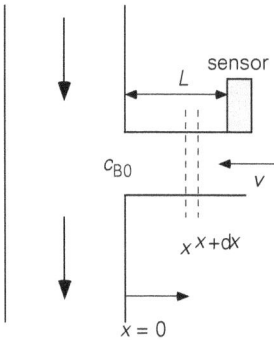

Figure 2.9

Solution

The method that always works is that of the recipe from Chapter 1: draw up a mass balance for species B! In this case this should again be a micro-balance, for example, over a thin slice dx somewhere between $x = 0$ and $x = L$ – on the analogy of the plug flow reactor of Section 1.2.2 – as we are interested in the decay of the concentration along the branch. From such a micro-mass balance it then can be derived how c_B depends on x – from which it can be determined how great the distance L must be chosen to the effect that $c_B(L) < c_{B0}/1{,}000$.

Such a micro-balance for a steady-state situation contains four terms: a convective and a diffusive term at x and a convective and a diffusive term at $x + dx$, where it should be realised that the velocity v is in the negative x-direction:

$$0 = -v A\, c_B|_x - A\,\mathrm{ID}\frac{dc_B}{dx}\Big|_x +$$

$$-\left\{ -v A\, c_B|_{x+dx} \right\} - \left\{ -A\,\mathrm{ID}\frac{dc_B}{dx}\Big|_{x+dx} \right\} \tag{2.41}$$

from which the second-order differential equation

$$0 = v\frac{dc_B}{dx} + \mathrm{ID}\frac{d^2 c_B}{dx^2} \tag{2.42}$$

follows. Integrating this equation once gives

$$v \cdot c_B + \mathrm{ID}\,\frac{dc_B}{dx} = K_1 \tag{2.43}$$

The integration constant K_1 follows from the (boundary) condition that in the steady-state situation no net transport of B occurs over a randomly selected plane x between $x = 0$ and $x = L$:

$$-A \cdot \mathrm{ID}\frac{dc_B}{dx} + A \cdot (-v) \cdot c_B = 0 \tag{2.44}$$

After all, in this case it is also possible to immediately start with equation (2.44), irrespective of drawing up a micro-balance over a thin slice dx. As a matter of fact, equation (2.44) expresses that in any plane, at position x, the diffusive transport to the right (the first term of the left-hand side) compensates the convective transport to the left (the second term of the left-hand side, the minus sign preceding v indicating

the direction). The method of drawing up a balance for a thin slice dx, however, is the recommended procedure: it always works.

The solution to the differential equation (2.44) is obtained via the method of separation of variables
(see Appendix 1A), with the boundary condition: $x = 0 \rightarrow c_B = c_{B0}$:

$$\frac{c_B(x)}{c_{B0}} = \exp\left(-\frac{vx}{\mathrm{ID}}\right) \tag{2.45}$$

The requirement for v therefore amounts to: $v > (\mathrm{ID}/L) \cdot \ln 1{,}000$. In the event that $L = 10$ cm and ID = 10^{-6} m^2/s,
this requirement is $v > 7 \cdot 10^{-5}$ m/s!

For $x = L$, equation (2.45) can also be written in terms of the non-dimensional Péclet
number Pe $\equiv v\,L/\mathrm{ID}$:

$$\frac{c_B(L)}{c_{B0}} = \exp(-\mathrm{Pe}) \tag{2.46}$$

This is a good illustration of the fact that non-dimensional numbers are not just a con-
ceptual nicety, but in many cases follow directly from an exact analysis (based on a
balance!).

Summary
Various non-dimensional numbers have been introduced which all represent the ratio of two transport
properties. The Prandtl (Pr = v/a), Schmidt (Sc = v/ID), and Lewis (Le = a/ID) numbers, for example, represent the ratio of two molecular transport coefficients.

Another non-dimensional number introduced is the Péclet (Pe = vL/ID) number which shows how, in
a certain situation, convective and molecular fluxes of a particular species relate to each other. These
non-dimensional numbers very often show up when – starting from some macro- or micro-balance – a
transport problem is solved.

2.2.2 The Reynolds number and the transition from laminar to turbulent flow

A similar thing as in Section 2.2.1 for mass transport applies to the convective and molecular transport of momentum. We previously discussed the laminar flow $v_x(y)$: the flow
runs in orderly, parallel (or in a cylindrical tube, coaxial) layers. For this situation and in
the case of Newtonian fluids it is easy to estimate what the relationship is between convective x-momentum transport in the x-direction and molecular x-momentum transport
in the y-direction:

$$\text{molecular:} \quad \phi''_{px,y} = -\mu\frac{dv_x}{dy} \approx \mu\,\frac{v_x}{D} \tag{2.47}$$

$$\text{convective:} \quad \phi''_{px,x} = v_x \cdot \rho v_x \tag{2.48}$$

In the case of the molecular transport estimate in equation (2.47), a typical dimension D has been introduced for the flow. This means that the gradient dv_x/dy can be replaced (or estimated) by v_x/D – in fact, the choice of D should be such that the estimate of the gradient is reasonable. The ratio of the convective momentum flux and the molecular momentum flux is known as the **Reynolds number** (Re):

$$\text{Re} = \frac{\text{convective transport}}{\text{molecular transport}} = \frac{v_x \cdot \rho v_x}{\mu v_x / D} = \frac{\rho v_x D}{\mu} \tag{2.49}$$

The subscript x is generally omitted from this definition of Re. Note that the concept and the definition of Re apply to both liquids and gases, that is, to any fluid.

The momentum flux $v \cdot \rho v$ in the numerator of the Reynolds number is often denoted by the term **inertia**: the property whereby a fluid keeps moving straight ahead in the event of no force being exerted on it: see also the discussion on momentum and forces in Section 1.4. The denominator of Re can be regarded as shear stress, that is, as friction or deceleration. This gives us:

$$\text{Re} \approx \frac{\text{inertia}}{\text{friction}}$$

The Reynolds number can also be interpreted in terms of forces. The numerator $v \cdot \rho v$ can be read as ρv^2: a quantity that – apart from a factor of $\frac{1}{2}$ – we already have encountered in the context of the Bernoulli equation (see Section 1.3.4) as pressure energy and as **dynamic pressure**. In this perspective, Re may be conceived as the ratio of dynamic pressure and shear stress.

Note that in this typical example, the convective x-momentum transport and the molecular x-momentum transport occur in different directions. In other words, it concerns two competing transport mechanisms: convective transport thanks to – usually – a difference in pressure, and molecular (or diffusive) transport thanks to viscosity. The molecular transport mechanism aims to re-distribute the momentum as evenly as possible over the field of flow: after all, the molecules try and transport momentum from locations with a high momentum concentration (or velocity) to locations with a lower momentum concentration, even in a direction that is normal to the direction of the flow (the convective transport). This is of no interest to the inertia of the fluid: it wishes only to move straight ahead. The Reynolds number therefore characterises a flow: it is a measure for how great the differences in velocity still can be in a direction normal to the convective momentum transport.

If Re is sufficiently small (i.e. below a certain critical value), then the flow is **laminar**, as presupposed in the above example. The momentum concentration in a fluid element in some of the layers is in that case not so great that the shear stress is incapable of keeping this element or layer in check: the velocities are only relatively low. However, if the Re number becomes too great, this will no longer be so. It will no longer be possible for the neat layered structure to be maintained. The differences in momentum concentration of adjacent fluid layers may then become so great that the molecules can no longer handle them or, in other words: the shear stresses become too large. The result is that the fluid layers stumble and tumble over each other, as it were. The flow now turns **turbulent**. All kinds of **eddies** are then formed in the flow, and thanks to these eddies the momentum is distributed very effectively across the flow field.

These eddies derive their energy from the main flow and in turn pass it on to smaller eddies. In a completely developed turbulent flow, the kinetic energy in the eddies is passed on to ever-smaller eddies without any notable dissipation. After all, the viscosity (→friction) is far too weak to resist the eddies. It is only with eddies with very small dimensions that the Reynolds number on the scale of such an eddy is low enough for the viscosity to be capable of exercising its dissipating effect. Ultimately, therefore, the mechanical energy is dissipated by friction in heat. To the human eye and mind, the dynamics of turbulent flow appears to be stochastic; but in reality it is chaotic. After all, the interactions between the fluid elements or packages are still governed by the laws of physics, but the effects of these laws in turbulent flows are many times more complex than in the case of laminar flow.

Flow through a cylindrical tube is **laminar** as long as the Reynolds number relating to the tube diameter D is less than around 2,000. The transition to turbulent flow occurs when Re is somewhere between 2,000 and 2,500, depending on exactly how the experiment is carried out (the transition to turbulent is very sensitive to small disruptions, such as vibrations from pumps: these enhance the instability of the laminar flow). In other geometries, such as right-angled ventilation channels, and for the flow around immersed bodies, other critical values of the Reynolds number apply to the laminar-turbulent transition point.

Summary

The Reynolds number

$$Re = \frac{\rho v D}{\mu}$$

represents in a certain situation the ratio of a convective momentum flux (inertia) to a molecular momentum flux (friction) – in different directions, incidentally. This is most clear when Re is written as

$$Re = \frac{v \cdot \rho v}{\mu v / D}$$

Thus, the Reynolds number characterises the state of the flow. Various inter-pretations of the Reynolds number have been discussed.

Whenever viscosity (or friction) is no longer capable of suppressing flow disturbances or instabilities, the flow sooner or later turns turbulent. Whether a flow is laminar or turbulent can be deduced from the value of the Reynolds number.

The critical value of Re beyond which the flow is or becomes turbulent varies from geometry to geometry; for a cylindrical pipe, the critical value is in the 2,000–2,500 range.

2.2.3 Dimensional analysis: the concept

Several non-dimensional numbers have been introduced above, in which transport terms are compared with each other. The numerator and denominator of these non-dimensional numbers must of course have the same dimension in order for their ratio to be non-dimensional. It is in fact no different to the various terms in the balances in Chapter 1: every term in a balance has to have the same dimension (hint: this makes it an excellent checking mechanism!).

However, there is another way of using **dimensions**. It is possible to determine beforehand how the various quantities and physical properties that govern a process can be merged into a smaller number of non-dimensional groups. In the final solution to the problem under review, it will only be these groups that occur as variables. In general, this will mean far fewer variables than there were at the beginning. The technique for achieving this is known as **dimensional analysis**. How the technique works and the dangers that lurk in the technique of dimensional analysis will be covered in detail below.[8]

The problem in Example 2.3 can also be dealt with using dimensional analysis. The starting point of every dimensional analysis is the difficult question: what does $c_B(L)$ actually depend on? Actually, it is about an analysis to find the independent variables leading up to, in this case, the dependent variable $c_B(L)$; in most cases, one should be able to change, or control, the independent variables and, in this indirect way, to attain a different value of the variable of interest.

As seen in Example 2.3, the answer is that it depends on c_{B0}, L, v, and ID:

$$c_B(L) = f(c_{B0}, L, v, \text{ID}) \tag{2.50}$$

The dimensional analysis technique works as follows: write the relationship between the dependent variable $c_B(L)$ and the independent variables c_{B0}, L, v, and ID with the help of unknown exponents α, β, γ, δ, and a **non-dimensional** coefficient or constant k as

$$c_B(L) = k \cdot c_{B0}^{\alpha} \cdot L^{\beta} \cdot v^{\gamma} \cdot \text{ID}^{\delta} \tag{2.51}$$

Make sure that the right-hand side of equation (2.51) only contains independent variables that you are able to choose separately from each other.

This equation (2.51) has to be dimensionally correct and consistent: the dimensions on the right-hand side must be equal to those on the left-hand side. This regulates the powers α, β, γ, and δ. The SI units (or dimensions) of the five variables c_B, c_{B0}, L, v, and ID are known of course, although the variables have to be expressed in their basic units of kg, m, s, and K. In this case:

---

8 We prefer the 'exponent method' over the method of 'repeating variables'.

$$c_B, c_{B0}: \quad \text{kg m}^{-3}$$
$$L: \quad \text{m}$$
$$v: \quad \text{m} \cdot \text{s}^{-1}$$
$$ID: \quad \text{m}^2 \cdot \text{s}^{-1}$$

Filling in the units into equation (2.51) means the various basic units occur as follows:

$$\text{kg:} \quad 1 = \alpha$$
$$\text{m:} \quad -3 = -3\alpha + \beta + \gamma + 2\delta$$
$$\text{s:} \quad 0 = -\gamma - \delta$$

(remember that coefficient k is non-dimensional). This is a system of three equations with four unknowns. Three of the four unknowns in this system can be expressed in the fourth, for example

$$\alpha = 1, \quad \delta = -\gamma, \quad \beta = \gamma$$

where the latter result is obtained by substituting the earlier results for α and δ into the only equation containing β. With this result, equation (2.51) is reduced to

$$c_B(L) = k \cdot c_{B0} \cdot \left(\frac{vL}{ID}\right)^{\gamma} \tag{2.52}$$

The first conclusion that can be drawn from equation (2.52) is that the **non-dimensional** concentration $c_B(L)/c_{B0}$ is only a function of the **non-dimensional** group vL/ID which is recognised as the Peclet number Pe, known from Example 2.3. Formulated mathematically

$$\frac{c_B(L)}{c_{B0}} = f(\text{Pe}) \tag{2.53}$$

It is not possible to say anything about function f on the basis of this dimensional analysis. For that, it is necessary to find an exact solution to the problem: in this case, it is known from Example 2.3 that f is an exponential function.

In general, it is not the case that $c_B(L)$ can be calculated on the basis of equation (2.52) with values for k and γ which apply to the entire domain of the Pe number. In practice, k and γ can be determined empirically for a limited Pe range only. In the case of Example 2.3, this amounts to determining a tangent on the curve of equation (2.46) about a certain value of Pe: viewed this way, it is clear that k and γ are not constant in the entire domain of possible values for Pe.

Nonetheless, a very useful conclusion can be drawn from equation (2.52), that is, on the basis of the dimensional analysis: as long as the Peclet number is constant, the non-dimensional concentration does not change. The values of the individual parameters within the Pe number do not matter – only the value of the Pe number matters.

This has important implications. Even if a problem cannot be solved analytically and experiments are not possible (too dangerous, too costly, installation not yet built), it is possible with the help of experiments in a small-scale model, for example, to make a forecast about the process or phenomenon on a large scale, namely by letting the dominant non-dimensional number be of equal value in both situations (small-scale model and large-scale installation).

With a view to experimenting at a different scale or under different conditions there is another illustration of the power of dimensional analysis: for example, in the 'purging' case of Example 2.3 it is possible to replace the aggressive species B with a harmless species, X. Suppose then that the diffusion coefficient of X is eight times greater than that of B: $ID_X = 8ID_B$. And that in the case of substance X experimentally it is found that the velocity needed to get $c_X(L) = c_{X0}/1{,}000$ is $v_X = 4 \cdot 10^{-4}$ m/s. The result of the dimensional analysis can then be used to determine what v_B should be in order to ensure that $c_B(L) = c_{B0}/1{,}000$. This is because the following applies:

$$\frac{c_X(L)}{c_{X0}} = f(\text{Pe}) \quad \text{and} \quad \frac{c_B(L)}{c_{B0}} = f(\text{Pe}) \tag{2.54}$$

In both cases, the function f is the same, and as here, too:

$$\frac{c_X(L)}{c_{X0}} = \frac{c_B(L)}{c_{B0}} \tag{2.55}$$

should therefore be

$$f(\text{Pe}_X) = f(\text{Pe}_B) \rightarrow \text{Pe}_X = \text{Pe}_B \tag{2.56}$$

or, put another way

$$\frac{v_X L}{ID_X} = \frac{v_B L}{ID_B} \rightarrow vB = \frac{ID_B}{ID_X} \cdot v_X = 5 \cdot 10^{-5}\,\text{m/s} \tag{2.57}$$

! **Summary**

Dimensional analysis is a technique for finding relations between the various quantities and physical properties that determine a problem, without having to solve the problem exactly or analytically. However, the knowledge that is acquired is relatively approximate. A dimensional analysis condenses the variables into non-dimensional groups that usually represent the ratio between two mechanisms, fluxes, or physical properties.

With the help of dimensional analysis, it is possible to compare different problems that relate to similar situations. In such situations, the non-dimensional groups should be kept constant. The rationale behind this approach is that the exact solution includes only the non-dimensional numbers that are found.

2.2.4 Dimensional analysis: technique and the Buckingham-π theorem

For carrying out a dimensional analysis, some understanding of the problem under consideration is very important. This does not necessarily mean that already from the very beginning it should be understood *how* quantity A depends on a number of other quantities. Rather, the starting point is an analysis of *which independent variables* quantity A may depend on. This requires some physical insight as to causal relations between quantities.

As an example, physical insight regarding the mobility of molecules in an ideal gas – as embodied in the model leading to equation (2.28) – tells us that in ideal gases the diffusion coefficient ID depends among other things on temperature and density which can be varied independently of each other indeed: $ID = f(T, \rho)$. Pressure should not be added here, as – due to the ideal gas law – pressure, density, and temperature are mutually dependent: the variables at the right-hand side should be independent of each other. In addition, it can be said a priori that in a dimensional analysis $ID = f(T, \rho)$ or, alternatively, $ID = f(T, p)$ will never lead to a decent result, as unlike ID both $f(T, \rho)$ and $f(T, p)$ contain the basic units K and kg. The gas constant R and the molar mass η of the species considered should be incorporated!

Further remarks and suggestions with respect to the selection of independent variables will be made in the context of the examples of Section 2.2.5.

The **general recipe for carrying out a dimensional analysis** is as follows:

1. Select the quantity that is to be examined: A
2. Determine the independent physical variables $B_1, \ldots, B_{n-1}$ that determine A according to $A = f(B_1, \ldots, B_{n-1})$. In total, there will now be n variables involved: A, $B_1, \ldots, B_{n-1}$.
3. Write the relation in the following form

$$A = k\, B_1^{\beta_1} \cdot B_2^{\beta_2} \ldots B_{n-1}^{\beta_{n-1}} \tag{2.58}$$

where $\beta_1, \ldots, \beta_{n-1}$ are unknown exponents (real numbers, not necessarily integers) and k a non-dimensional coefficient.

4. Express in both the left-hand side of the latter relation and the right-hand side all dimensions in the basic dimensions or all (SI) units in the basic units. The basic dimensions (with their SI unit in brackets) are length (m), mass (kg), time (s), and temperature (K). As an example: for energy $\rightarrow J = N \cdot m = kg \cdot m^2 \cdot s^{-2}$. In the remainder of this textbook we will work in SI units.

5. Make sure that this equation is dimensionally sound, which means that left and right in relation (2.58) the basic units should occur just as often. You do this by counting the basic units left and right and by then writing down with the help of the coefficients $\beta_1, \ldots, \beta_{n-1}$ for each basic unit an algebraic equation such that left and right this basic unit occurs just as often.

6. Solve the algebraic equations for $\beta_1, \ldots, \beta_{n-1}$ from stage 5. Whenever the number of unknowns $\beta_1, \ldots, \beta_{n-1}$ is greater than the number of algebraic equations (given the number of basic units), then express one or more of these β's in terms of the other ones (just as in the **purging** example).

7. Substitute the values found in stage 6 into equation (2.58) and combine $A, B_1, \ldots$, B_{n-1} to **non-dimensional** groups or **numbers**.

8. It is usual to work towards the commonly known non-dimensional numbers, such as Pr, Sc, Le, Pe, and Re which were introduced earlier in this chapter. This may be done by mutually multiplying or dividing the non-dimensional numbers found in stage 7. When transport terms such as fluxes and physical properties, which occur in these common non-dimensional numbers, play a role with respect to a quantity A, these common non-dimensional numbers appear virtually automatically.

The number of non-dimensional groups that is found depends, among other things, on the number of independent variables that are considered. As to their number, it is hard to present a guideline; there should in any case be a sufficient number. The analysis at the start as to which independent variables B may play a role is of crucial importance. As said, in the examples of Section 2.2.5 ample attention will be paid to this issue.

There is a general rule, the **Buckingham π theorem**, that can be used to calculate in advance the number of non-dimensional groups (p) from the total number of variables (n) and the number of basic units (m) involved:

$$p = n - m \qquad (2.59)$$

The number of non-dimensional (p) that is found therefore directly depends on the number of independent variables ($n - 1$). This once more illustrates how important it is to identify the proper independent variables at the start of the analysis.

The proof of equation (2.59) is simple and is as follows: Equation (2.58) contains n variables (or parameters) $\{A, B_1, \ldots, B_{n-1}\}$ and $n - 1$ unknown powers $\{\beta_1, \ldots, \beta_{n-1}\}$. The number of different basic units that occur is m. In general, there are then m linear equations for $n - 1$ unknown powers. This means that there must be $\{(n-1) - m\}$ unnamed powers and that m powers can be expressed in these 'unnamed powers'. If we now suppose that exactly every power $\{\beta_1, \ldots, \beta_{n-1}\}$ can be calculated, then there would be precisely one non-dimensional group. Therefore, $\{(n-1) - m\}$ unnamed powers produce

$$p = 1 + \{(n-1) - m\} = n - m \qquad (2.60)$$

non-dimensional groups.

The Buckingham π theorem is a useful tool for quickly seeing how many non-dimensional groups a particular combination of variables will produce. Additionally, it is possible to make a quick check – after carrying out the dimensional analysis – whether the number of groups are in keeping with the theorem. The Buckingham π theorem also illustrates how due to a dimensional analysis the number of independent variables can be reduced by exploiting the dimensions of the variables and by focusing on non-dimensional numbers. This is another powerful advantage of applying a dimensional analysis.

Example 2.4. Purging II.
In Example 2.3, it was found that

$$c_B(L) = f(c_{B0}, L, v, ID) \tag{2.61}$$

So, there are five quantities, three basic units → two non-dimensional groups according to the Buckingham π theorem. This is in line with the result of the above dimensional analysis.

Summary

An eight-stage recipe has been given for performing a dimensional analysis in which it is very important to carefully select the independent variables at the start of the analysis. A sound physical understanding of the problem is then very helpful. It is recommended to arrive at the commonly known non-dimensional numbers as much as possible.

The Buckingham π theorem is the number of non-dimensional groups or numbers (p) that follows from a dimensional analysis with n parameters and m basic units:

$$p = n - m$$

An important advantage of dimensional analysis is that for the problem at hand the number of variables is strongly reduced, viz. from n to p.

2.2.5 Examples of dimensional analysis

Example 2.5. Pulling a wire.
A smooth wire (diameter d) is pulled through a tank that is filled with a liquid. The wire is wetted over a length L. Consider the force F needed to pull the wire at a constant velocity v through the liquid.

Which variables does force F depend on? By means of which non-dimensional groups can this process be described?

Solution
The answer to the first question is the diameter d and length L of the wire, the velocity v, and the density ρ and viscosity μ of the liquid. The reasoning behind this is that as a result of the movement of the wire ($\rightarrow v$) the liquid will also flow. In other words, the liquid will gain momentum, which is related to ρ. Finally, the transfer of momentum has to do with the frictional force ($\rightarrow \mu$) between the surface of the wire ($\rightarrow d$ and L) and the adjacent liquid. The effect of temperature on density and viscosity may be ignored as being a secondary effect; in addition, the condition that the variables have to be mutually independent excludes this dependence being taken into account. As long as the velocity of the wire is constant, the density of the material of the wire is irrelevant.

The answer to the second question can be found with the help of dimensional analysis. We will use the aforementioned recipe:
1) quantity: force: $\qquad F \quad \rightarrow \text{kg} \cdot \text{m/s}^2$

2) independent physical variables: $d, L \;\rightarrow$ m

$$v \;\rightarrow \text{m/s}$$
$$\rho \;\rightarrow \text{kg/m}^3$$
$$\mu \;\rightarrow \text{kg/m} \cdot \text{s}$$

3) the relation $F = k\, d^\alpha \; L^\beta \; v^\gamma \; \rho^\delta \; \mu^\varepsilon$

4) in basic units: $\dfrac{\text{kgm}}{\text{s}^2} = m^\alpha \, m^\beta \left(\dfrac{m}{s}\right)^\gamma \left(\dfrac{\text{kg}}{\text{m}^3}\right)^\delta \left(\dfrac{\text{kg}}{\text{m} \cdot \text{s}}\right)^\varepsilon$

5) for the various basic units, it follows that:

$$\text{kg: } 1 = \delta + \varepsilon$$

$$\text{m: } 1 = \alpha + \beta + \gamma - 3\delta - \varepsilon$$

$$\text{s: } -2 = -\gamma - \varepsilon$$

6) These are three equations with five unknowns. Of these, two can be 'chosen at will' in order that they can be used to express the other three. The choice of these two 'free options' is determined by two considerations. The first of these is this: in the three equations, is it possible to see what a sensible choice would be? If so, make that choice. If not, ask the question of whether it is possible to predict which groups will occur on the basis of the stages 2 and 3 of the recipe and base your choice on this (a matter of experience).

In this example, at stage 2, the variables d and L can be seen which together form a non-dimensional group; so don't choose β and α together as this would result in a conflict, but just β, for instance. Stage 2 also includes the variables ρ, v, d (or L), and μ, which together form a Reynolds number. Choose ε in this case, for example. Solving the equation now produces

$$\delta = 1 - \varepsilon$$

$$\gamma = 2 - \varepsilon$$

$$\alpha = 2 - \beta - \varepsilon$$

7) Substituting the above values for α, γ, and δ into the equation of stage 3 gives

$$F = k\, d^{2-\beta-\varepsilon} \; L^\beta \; v^{2-\varepsilon} \; \rho^{1-\varepsilon} \; \mu^\varepsilon$$

Combining now produces the non-dimensional groups

$$\frac{F}{\rho v^2 d^2} = k \left(\frac{L}{d}\right)^\beta \cdot \text{Re}^{-\varepsilon} \tag{2.62}$$

8) Incidentally, if β and δ were to be chosen at stage 6 rather than β and ε, then the following would have resulted, instead of equation (2.62):

$$\frac{F}{d v \mu} = k \left(\frac{L}{d}\right)^\beta \cdot \text{Re}^\delta \tag{2.63}$$

This result is also correct and can be derived from equation (2.62) by multi-plying left and right by Re; clearly, $\delta = 1 - \varepsilon$.

This is generally applicable with dimensional analysis: the groups that are found can be multiplied with each other in order to make a new group, and one of the old groups may be replaced by the new one. Of course, the non-dimensional groups describing the process or phenomenon should be mutually independent, just like the dimensional variables of step 2 (see also Section 2.2.4).

Dimensional analysis is a very powerful technique for reducing the number of variables used for describing a problem. However, dimensional analysis is not without danger: the selection of independent variables in stage 2 of the recipe determines the outcome. The selected variables must actually completely cover the relevant physical processes – otherwise, dimensional analysis may result in senseless, non-physical results. If, for instance, heat transfer is to be analysed, and then in stage 2 the temperature difference driving the heat transfer should be taken as a variable rather than the two individual temperatures; in the latter case, one would arrive at the ratio of these temperatures (in K or in °C?) but such a ratio does not govern heat transport (see further Chapter 3). Now, the need of selecting a proper set of variables will be demonstrated in the next example:

Example 2.6. A stirred tank.
A polymerisation reaction occurs in a stirred tank, during which the liquid viscosity μ increases. The impeller is linked to an engine, which ensures the impeller rotates at a constant number of impeller revolutions per unit of time, N. How does the power P, which the engine supplies to the stirrer, depend on N and μ and other relevant variables?

First attempt
Suppose that P is only dependent on μ and N:

$$P = k N^\alpha \mu^\beta$$

so that the basic units involved should obey to:

$$\frac{kg\,m^2}{s^3} = \left(\frac{1}{s}\right)^\alpha \left(\frac{kg}{m \cdot s}\right)^\beta$$

The equations as to the individual basic units run as:

$$
\begin{aligned}
kg: \quad & 1 = \beta \\
m: \quad & 2 = -\beta \\
s: \quad & -3 = -\alpha - \beta
\end{aligned}
$$

The first and second equations **conflict**! This attempt, therefore, is, erroneous: there are not enough physics in the equation for P, which anyhow should also be scale-sensitive.

Second attempt
P depends on μ, N, and the diameter of the stirrer, D.

$$P = k N^\alpha \mu^\beta D^\gamma$$

so that

$$\frac{kg\,m^2}{s^3} = \left(\frac{1}{s}\right)^\alpha \left(\frac{kg}{m \cdot s}\right)^\beta m^\gamma$$

Then, the equations are:

$$
\left.
\begin{aligned}
kg: \quad & 1 = \beta \\
m: \quad & 2 = -\beta + \gamma \\
s: \quad & -3 = -\alpha - \beta
\end{aligned}
\right\} \quad \rightarrow \alpha = 2,\ \beta = 1,\ \gamma = 3
$$

Therefore, $P = f(N, \mu, D)$ leads to a single non-dimensional group which is a constant:

$$\frac{P}{N^2 \mu D^3} = \text{constant} \tag{2.64}$$

In the example, therefore, when $N =$ constant and $D =$ constant: $P \propto \mu$

Third attempt

N, μ, and D were chosen in the second attempt, and by combining $N \cdot D$, so was implicitly a velocity. These are three of the four variables that determine the Re. For that reason, let us also include ρ. It is likely to be of importance as far as the power is concerned whether the liquid being stirred is 'light' or 'heavy'. Try

$$P = k \; N^\alpha \; \mu^\beta \; D^\gamma \; \rho^\delta$$

Equations:

$$\text{kg:} \quad 1 = \beta + \delta$$
$$\text{m:} \quad 2 = -\beta + \gamma - 3\delta$$
$$\text{s:} \quad -3 = -\alpha - \beta$$

This gives:

$$\alpha = 3 - \beta$$
$$\gamma = 5 - 2\beta$$
$$\delta = 1 - \beta$$

It therefore follows that

$$P = k \rho N^3 D^5 \left(\frac{\rho N D^2}{\mu}\right)^{-\beta} \tag{2.65}$$

The latter group is the predicted **Reynolds** number. The velocity at the edge or at the tip of the stirrer can be taken as the typical velocity, to which $v = \pi D \cdot N$ applies in any case. It therefore follows that

$$\rho \frac{ND^2}{\mu} = \frac{1}{\pi} \rho \frac{(\pi ND)D}{\mu} \propto \frac{\rho v D}{\mu} = \text{Re} \tag{2.66}$$

In a dimensional analysis, the factor $\frac{1}{\pi}$ is of no importance. With the help of the non-dimensional group $\text{Po} = P/(\rho N^3 D^5)$ – which is known as the **Power number** – equation (2.65) can therefore be written in non-dimensional form as

$$\text{Po} \equiv \frac{P}{\rho N^3 D^5} = k \left(\frac{\rho N D^2}{\mu}\right)^{-\beta} = k \; \text{Re}^{-\beta} \tag{2.67}$$

Notice that for $\beta = 1$, density drops out of equation (2.67), and the result obtained is the same as that of attempt 2, where no effect of the density and therefore of inertia was included. In the $\beta = 1$ regime, the viscous effects, which are in the denominator of the Reynolds number, evidently predominate over the inertia forces which are found in the numerator of the Reynolds number: see equation (2.49). Clearly, $\beta = 1$ is associated with low Reynolds number values which correspond with the laminar flow regime. For $\beta = 0$, the viscosity drops from equation (2.64); $\beta = 0$ is therefore associated with turbulent flow conditions where viscosity does not play a role (at the macro-scale). Experimentally, the diagram in Figure 2.10 has indeed been found.

Fourth attempt

$P = f(N, \mu, \rho)$: the length scale has been omitted from the problem. This produces

$$P \propto \left(\frac{N\mu^5}{\rho^3}\right)^{1/2} \tag{2.68}$$

It will be clear that this result does not tally with the experiments. The reason for this is that the physics of the problem *can* not be properly described through proposition $P = f(N, \mu, \rho)$: according to this description, it does not matter as far as the power is concerned whether the tank in question is large or small, while the density cannot occur in the denominator either, from a physics point of view!

Next, an examination could be carried out into how the power of the engine depends on the geometry of the stirrer (such as the height and thickness of the stirring blade or the degree of curvature of the

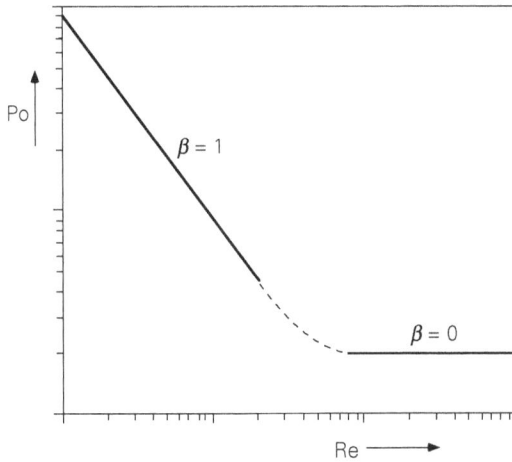

Figure 2.10

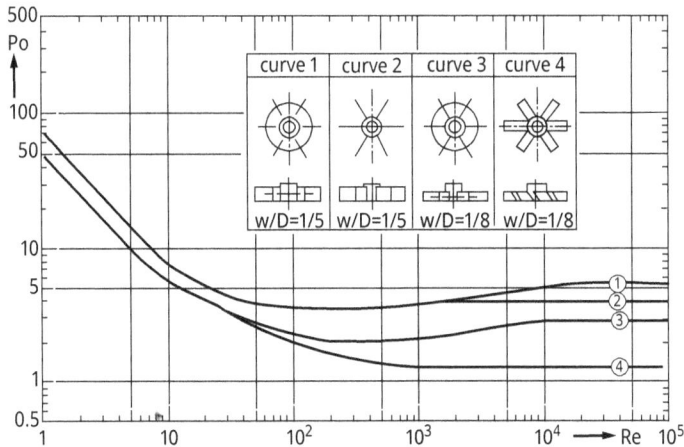

Figure 2.11

stirring blade) and the position of the stirrer in the container. According to the Buckingham π theorem, the addition of a variable keeps resulting in a new non-dimensional number – in geometric numbers in this case. Figure 2.11 shows how the Power number of so-called turbine stirrers also depends on the size of the disk and on the ratio of the height, w, of the vertical or tilted impeller blades mounted onto the disk and the impeller diameter.

The above example shows that incorporating both μ and ρ as independent variables results in finding the Reynolds number. If from the outset it is clear, however, that under the conditions of interest the flow is laminar, that is, viscosity dominated, then ρ can be ignored in the analysis (providing gravity can be ignored as well) and Re will not show up. Similarly, if from the outset it is clear that in a specific case the flow is turbulent, or inertia dominated, viscosity can be ignored and Re will not show up either. Just in the general case, with no a priori information about the type of flow, the Reynolds number will be one of the non-dimensional numbers found, indicating the variable of interest may be (very) different in laminar and turbulent flows. This observation not only applies to stirred vessels but also to any case where convection plays a role.

In vertical laminar flows (and some non-laminar flow cases), gravity may play a dominant role as the force driving the flow. In such cases, ρ might be better combined with the gravitational acceleration g to the new variable $\gamma = \rho g$ (in kg/m^2s^2), denoted as specific weight, e.g. occurring in equation (5.97) – to restrict the number of variables and the number of non-dimensional numbers and to prevent Re from appearing.

Example 2.7. Slot coating.[9]
Figure 2.12 shows a cross-sectional view through the die of a specific coating machine. The very viscous Newtonian liquid (usually a polymer) is supplied via the vertical channel and is then entrained by the lower plate or belt (web) that moves to the right with velocity U. The result is that that this plate or belt gets a coating of thickness δ_∞ which is smaller than the gap height δ.

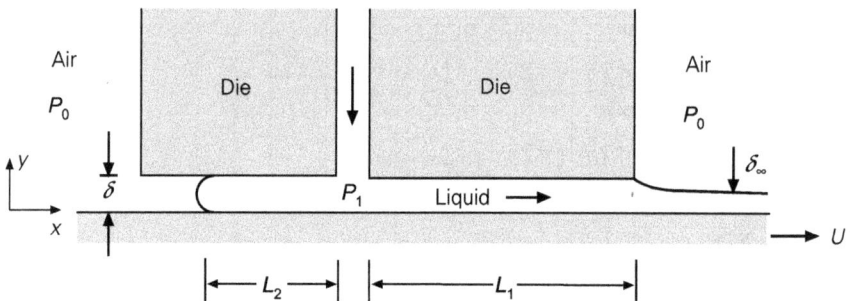

Figure 2.12

---

9 This problem has been derived from a problem in W.H. Deen, *Analysis of Transport Phenomena*, Oxford University Press, 2nd Ed., 2011.

Consider a steady-state one-dimensional flow in the horizontal x-direction. As pressure P_1 is higher than the atmospheric ambient pressure P_0, the viscous liquid will not only flow to the right – in the same positive x-direction as the web is dragging it – but also to the left – in the 'wrong' negative x-direction opposing the drag exerted by the web in the positive x-direction.

The question is to find out by a dimensional analysis how in a steady-state operation the distance L_2 to which the liquid is capable of reaching to the left, depends on, among other things, the web velocity U.

Solution

In a dimensional analysis, an inventory has to made of all, or at least several, highly relevant independent variables that have an effect on the variable of interest, in this case L_2. The most relevant parameters are:
– the web velocity U;
– the pressure difference $\Delta P = P_1 - P_0$ rather than the individual pressures P_1 and P_0: see the remark in the paragraph preceding Example 2.6;
– the channel height δ: a larger δ would leave more room for the flow in the 'wrong' direction; and
– the viscosity of the liquid: not only because the overall flow is viscous but also owing to the viscous drag or shear stress exerted on the liquid by the moving web.

These four independent variables together determine L_2. That makes five variables in total, leading (according to Buckingham's theorem) to two non-dimensional numbers (given the three basic units involved).

Following the recipe of the dimensional analysis may result in the finding that the ratio L_2/δ depends on the non-dimensional group $\delta \Delta P/\mu U$ that stands for the ratio of ΔP to $\mu U/\delta$, where the latter – given Newton's viscosity law – is a measure of the viscous drag force (per unit of area) or the shear stress exerted by the moving web on the viscous liquid. Of course, L_2 increases with increasing ΔP and h and decreases with increasing μ and U.

This example will be revisited in Chapter 5 to derive an analytical expression for L_2 which is fully in line with the above result of the dimensional analysis. Note that the fluid density does not play a role in this coating process – as the flow is viscous.

Conclusion

A significant difficulty when applying dimensional analysis is that there is almost always something that emerges from the analysis, regardless of the choice of parameters. However, making the wrong choice of independent variables that 'stretch' the problem will result in failure: anyone putting insufficient physics into the analysis phase or not inserting enough common sense into the analysis will eventually find themselves in conflict with the laws of physics. Great care should therefore be taken when selecting quantities and physical properties.

The conclusion, then, is this: dimensional analysis is very useful, but needs careful consideration.

2.3 Flow around immersed objects and net forces

2.3.1 Flow field and flow resistance

Any object that moves through a liquid or a gas experiences – and exerts – a force: the fluid tries to resist the motion of the object, and reversely the object tries to entrain fluid. Mutatis mutandis; this is also the case if the object is at a fixed position and the fluid flows around it. The 'relative motion' of the immersed object with respect to the fluid is what matters and evokes a particular flow field and a related pressure field around the immersed object.

The analysis will remain restricted to rectilinear (relative) motions of (more or less) spherical and cylindrical particles through fluids which overall either are stagnant or move at a uniform velocity. The focus is on the interaction force, or **drag force**, exerted by a moving fluid on a stationary object or particle in the direction of the fluid flow; in case the particle itself moves, this drag force (exerted by the fluid) opposes the particle motion and is directed in the direction opposite to the particle motion since with respect to the particle the fluid moves in the opposite direction. Only steady-state situations will be considered here – that is, particle velocity and/or background velocity of the fluid being taken constant. Lateral forces, such as the lift force on an airplane wing or on a red blood cell in the laminar flow in a vein, as well particle rotations are left out of consideration.

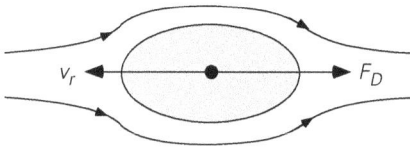

Figure 2.13

With the help of dimensional analysis (!), it is possible to gain an understanding of the groups that determine the **drag force** F_D on the object. F_D will depend on the relative velocity, v_r, of the object with respect to the fluid (see Figure 2.13), i.e. the velocity of the object minus the undisturbed fluid velocity far from the object. It is this relative velocity that determines the flow field and the pressure field around the object. Obviously, the viscosity, μ, and the density, ρ_f, of the fluid will play a role, and the dimensions of the object will determine the magnitude of the force as well. For this, we take a typical 'diameter' D, therefore

$$F_D = F(v_r, \mu, \rho_f, D) \tag{2.69}$$

The result of the dimensional analysis is

$$\frac{F_D}{\rho_f v_r^2 D^2} = k \cdot \mathrm{Re}^\alpha \tag{2.70}$$

Therefore, the following applies to the drag force F_D:

$$F_D = f(\mathrm{Re}) \cdot D^2 \cdot \rho_f v_r^2 \tag{2.71}$$

The latter factor on the right-hand side of this equation stands for the inertia of the flow, although we will see in Section 2.3.2 that it is usually rewritten as $\frac{1}{2}\rho_f v_r^2$ for the purpose of calculating the drag force. We can also recognise it as a **dynamic pres-**

sure; see the discussion about equations (1.129) and (1.130). At the nose of the immersed object, that is, at the so-called **stagnation point**, the fluid is 'stopped' locally; as a result, in line with the Bernoulli equation, all kinetic energy arriving at that position is converted into pressure energy, rendering pressure highest there, called the **stagnation pressure** p_{stagn}:

$$p_{stagn} = p + \frac{1}{2}\rho_f v_r^2 \qquad (2.72)$$

where the first term on the right-hand side denotes the local static pressure. The idea here is that the flow imposes a pressure drop on the object that can be estimated with the help of the dynamic pressure at the nose.

The second factor on the right-hand side of equation (2.71) has the dimension of a surface. This is hardly surprising: after all, the drag force acts on the surface of the object. The function of the Reynolds number on the right-hand side of equation (2.71) accounts for the 'details' of the area of flow around the object, where of course it matters whether the flow around the object is laminar or turbulent.

In order to elucidate all this, two contributions to the overall **drag force** F_D must be distinguished namely the form drag and the friction drag (or skin friction).

The term **form drag** can be understood from the example below. Consider a flat plate (surface area A) that is perpendicular to the flow of a fluid (see Figure 2.14). The fluid approaching the plate will be slowed down and will build up a **dynamic pressure** of the order of magnitude of $\frac{1}{2}\rho_f v_r^2$. However, behind the plate, in its wake, the liquid contains numerous eddies. As a result, this stagnation pressure is absent here: after all, there is kinetic energy in the eddies in the wake, which is deducted from the pressure energy according to the Bernoulli equation, and in addition substantial energy dissipation takes place. Consequently, there is a pressure drop over the plate. This results in a force on the plate in the direction of the flow:

$$F \propto A \cdot \Delta p = A \cdot \frac{1}{2}\rho_f v_r^2 \qquad (2.73)$$

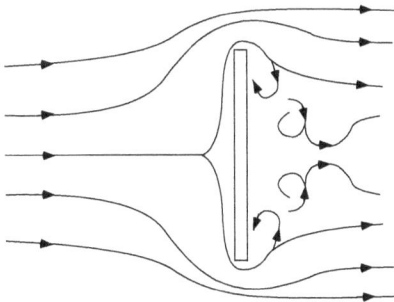

Figure 2.14

Friction drag (or **skin friction**) occurs in particular if the flow passes parallel to the surface of the object. The flat plate is a good example of this, too. Now, though, it concerns a motionless plate parallel to the flow. On the plate, the velocity of the liquid at the surface of the plate is equal to that of the plate (according to the usual so-called no-slip condition) – zero, in other words. In the **boundary layer** (see Figure 2.15), the velocity builds up in the direction of y until value v_r is reached at the edge of the boundary layer. This profile (which depends on both x and y) can be calculated with the help of the x and y-momentum balance, the mass balance, and boundary layer theory. From this, we can determine, from Newton's viscosity law, what is important to us – the shear stress between the liquid and the plate. The eventual result (not derived here) is

$$\tau_w = 0.644 \left(\frac{\rho_f v_r x}{\mu} \right)^{-\frac{1}{2}} \cdot \frac{1}{2} \rho_f v_r^2 \quad \text{providing that } \mathrm{Re}_x < 3 \cdot 10^5 \tag{2.74}$$

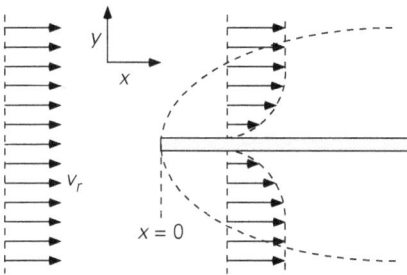

Figure 2.15

The overall force on the plate (length L in the direction of x, breadth B) can now be determined by adding up the contribution of the bottom and the top of the plate:

$$F_{\text{friction}} = 2 \int_0^L \tau_w \cdot B \, dx = 2.66 \left(\frac{\rho_f v_r L}{\mu} \right)^{-\frac{1}{2}} \cdot BL \cdot \frac{1}{2} \rho_f v_r^2 \tag{2.75}$$

With the help of the above illustrations of **form drag** and **friction drag** in extreme situations, it is easy to understand that the overall drag force on a random object in a flow is determined by the result of all the local pressure and shear stress values to which every part of the surface of the object is subjected. In principle, therefore, the overall drag force can be calculated by taking the component in the flow direction of all those local forces and integrating them across the entire surface of the object – as written explicitly for a flat parallel plate exposed to a flow in the first part of equation (2.75). The **form drag** is then the resultant of integrating all the pressure forces on the object, while the **friction drag** results from integrating all the skin friction contributions.

We should also keep in mind that the velocity field around the object (from which all the local values of the shear stress are calculated) and the pressure field are generally time-dependent and three-dimensional. The current state of technology makes a numerical calculation of the overall interaction force as described above possible for an increasing number of flow conditions, but in general, very simplified calculations are still widely used.

Summary
Immersed objects that move through and with respect to the surrounding fluid experience a counteracting force – the drag force. This force attempts to have the body move at the same velocity as the
liquid. There are two distinctive effects: form drag and friction drag which have been explained extensively. Together, they form the drag force.
    By means of a dimensional analysis, an expression has been derived for this drag force:

$$F_D = f(\text{Re}) \cdot D^2 \cdot \rho_f v_r^2$$

In this context, the concept of stagnation pressure was discussed again, along with the contribution of
shear stresses acting between object and fluid. Of course, the drag force strongly depends on the
exact way the fluid flows around the object and therefore on the particle Reynolds number.

2.3.2 Drag force and drag coefficient

Only in very exceptional cases can the drag force be calculated analytically. For example, this is the case with creeping flow around a sphere; that is, the flow with low
Reynolds number values (Re < 1) round a sphere, if the fluid is incompressible – that
is, if it has a constant density. For such low Reynolds numbers, the flow around the
sphere is symmetrical. However, this does not mean that there is no form drag because there is loss of pressure over the sphere as a result of dissipation. The analytical
solution leads to an expression for F_D that is known as Stokes' law:

$$F_D = -6\pi\mu R\, v_r \quad \text{providing that Re} < 1 \text{ (or, for more precise results, Re} < 0.1) \qquad (2.76)$$

with R being the radius of the sphere. Splitting this into form and friction drag gives

$$F_{\text{form}} = -2\pi\mu R v_r \qquad (2.77)$$

$$F_{\text{friction}} = -4\pi\mu R v_r \qquad (2.78)$$

In the engineering world, the overall drag force on an object is generally modelled on
the basis of the following empirical equation:

$$F_D = C_D(\text{Re}) \cdot A_\perp \cdot \frac{1}{2}\rho_l v_r^2 \qquad (2.79)$$

in which $A_\perp$ is the largest (projected) cross-sectional area of the object normal to the
direction of flow or motion. In using equation (2.79) a uniform approach velocity of the
fluid or a stagnant fluid is tacitly assumed (see the introduction of Section 2.3.1). Equation (2.79) shows, unsurprisingly, much similarity with equations (2.70), (2.73), and
(2.75). Note that the case of a plate being parallel to a fluid flow is not covered by equation (2.79) because $A_\perp$ approaches zero in that case.

    According to Newton's Third Law, it is not just the medium that exerts a drag force
on a particle, as modelled by equation (2.79), but such a particle also exerts a reaction
force on the medium: the force in equation (2.79) is given a minus sign. Note that a minus

sign also means that the force works in the opposite direction. A particle that moves through a medium at velocity v_r experiences a counteracting force from this medium. This is apparent from the following reformulation of equation (2.79) in vector notation:

$$\vec{F}_D = -C_D \cdot A_\perp \cdot \frac{1}{2}\rho_l|\vec{v}_r|\vec{v}_r \tag{2.80}$$

in which $|\vec{v}_r|$ stands for the modulus of the velocity, that is, stands for the magnitude of the velocity (regardless of the direction). The minus sign expresses that $\vec{v}_r$ and $\vec{F}_D$ by definition have opposite directions (see also Figure 2.13): after all, $\vec{v}_r$ stands for the velocity of the object with respect to the fluid, while $\vec{F}_D$ represents the force of the fluid on the object.

Coefficient C_D is called the **drag coefficient** and is therefore a function of Re, as follows from the dimensional analysis. Apart from liquid properties and the relative velocity, Re includes a typical dimension (size) of the object. For a sphere or a cylinder in a cross flow, the characteristic size is the diameter. The shape of the object also shows up in the precise form that function C_D (Re) has. Figure 2.16 shows C_D for several types of object.

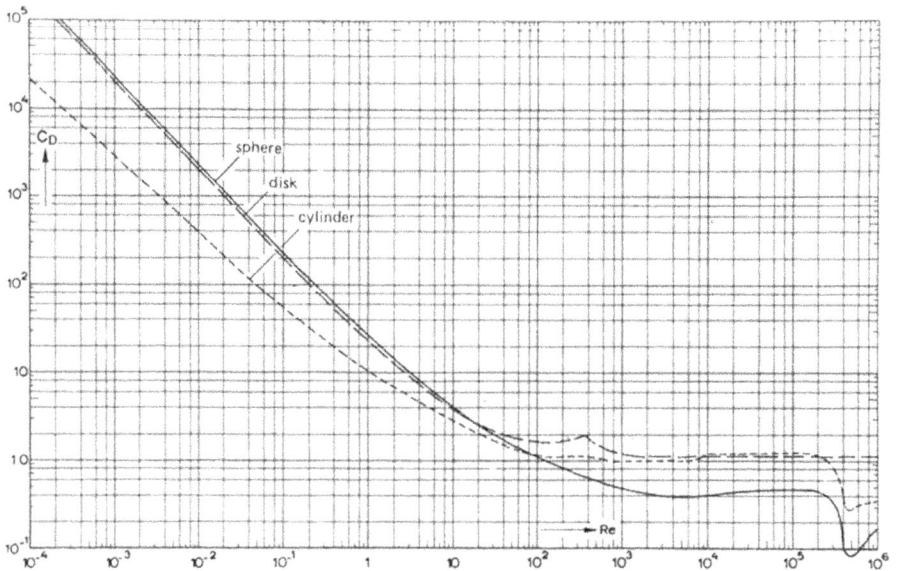

Figure 2.16

Stokes' law can also be written in terms of a drag coefficient. Combining equations (2.76) and (2.79) produces (with diameter $D = 2\,R$):

$$C_D = \frac{24\mu}{\rho v_r D} = \frac{24}{\text{Re}} \quad \text{(providing that Re} < 1 \text{ or even better Re} < 0.1) \tag{2.81}$$

If the Re number becomes any greater, eddies form on the rear side of the sphere. A wake is then formed behind the sphere, when the boundary layer separates from the object as a result of the inertia of the fluid that prefers to carry on moving straight ahead. In the case of greater Re numbers, the eddies become bigger, and the point at which the boundary layer separates shifts further forward. Where the Re numbers are greater still, the wake becomes irregular and turbulent. Eddies are then carried by the flow and new ones are produced behind the sphere. Finally, with very high Re numbers, the boundary layer becomes turbulent and the point where it separates, shifts once again to the rear of the sphere. This is all illustrated in Figure 2.17.

It is also possible to give an analogous description for the flow around cylinders in a cross flow. Another notable phenomenon occurs here, for $10^2 < \mathrm{Re} < 5 \cdot 10^3$. The eddies behind the cylinder – there are two, in the form of cylinders – then become unstable. The cylinder now sheds off eddies to the left and right, alternately. A so-called **Kármán vortex street** now occurs behind the cylinder (named after **Von Kármán**). Such a street is schematically shown in Figure 2.18.

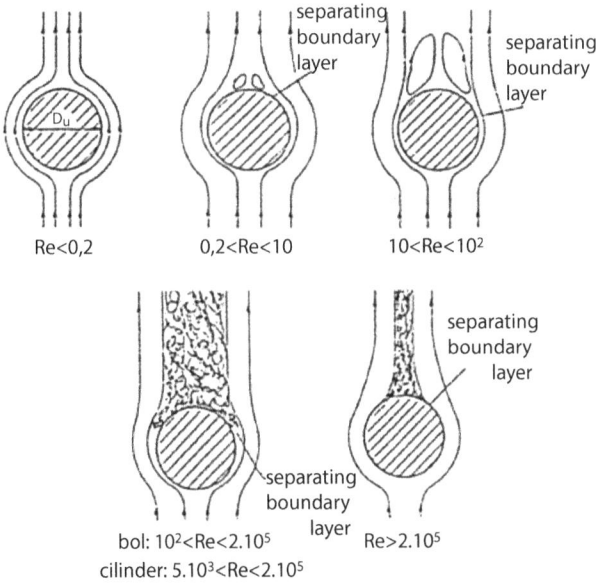

Re<0,2 0,2<Re<10 10<Re<10²

separating boundary layer

separating boundary layer

bol: 10²<Re<2.10⁵
cilinder: 5.10³<Re<2.10⁵

separating boundary layer Re>2.10⁵

separating boundary layer

Figure 2.17

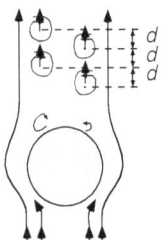

$10^2 < \mathrm{Re} < 5 \cdot 10^3$

Figure 2.18

! **Summary**

Generally, the drag force exerted by a flow on an immersed object is modelled (in a steady state situation) with the stagnation pressure and the projection of the surface of the body on a plane that is perpendicular to the relative velocity of the body and liquid:

$$F_D = C_D(\text{Re}) \cdot A_\perp \cdot \frac{1}{2}\rho_l v_r^2$$

The coefficient C_D in this expression is called the drag coefficient and is a function of the particle Reynolds number and of the form of the body.

For creeping flow round a spherical particle (for preferably Re smaller than 0.1) Stokes' law applies:

$$F_D = 6\pi \mu R v_r$$

which is in agreement with the previous equation for $C_D = 24/\text{Re}$.

Fluid flow around immersed spheres and cylinders in cross-flow has been discussed qualitatively. One should distinguish between laminar and turbulent flow. An important aspect is the separation of the boundary layer from the object and the exact position where this happens.

2.3.3 Force balance and terminal velocity

The above concepts relating to the drag force that an object is subjected to in a flow field can be used for calculating (relative) particle motions. The behaviour of a particle is, after all, determined by the resultant of all the forces acting upon it. If the resultant does not equal zero, the particle is accelerating (Newton's Second Law); if the resultant is equal to zero, then the particle is motionless or it is carrying out a uniform movement (Newton's First Law).

The magnitude of the resultant is determined by a force balance. The drag force is one of the forces that features in such a balance and depends on the relative velocity v_r. If the result of the forces is zero, the relative velocity, generally denoted as the terminal velocity, can be derived directly from the force balance. In case the sum of the forces is not (yet) zero, the particle will (still) accelerate (or decelerate) until the drag force balances with the other two forces and the particle reaches its terminal velocity.

Consider a particle falling through a fluid at a constant velocity. It is known from classical mechanics that then the sum of the forces on the particle must be equal to zero. This also follows from a steady-state momentum balance over the particle, of course: no momentum flows convectively into or out of the particle and therefore the steady-state momentum balance for the particle as the control volume – see equations (1.139) and (1.140) – reduces to a force balance:

$$\sum F_z = 0 \tag{2.82}$$

For a particle with density ρ_p falling at its terminal velocity, this means the sum of gravity, buoyancy, and drag force is zero (see Figure 2.19):

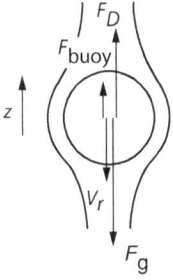

Figure 2.19

$$F_g + F_{\text{buoy}} + F_D = 0 \tag{2.83}$$

With the z-coordinate and the forces taken positive in the upward direction, the following applies to the buoyancy (according to Archimedes):

$$F_{\text{buoy}} = \frac{\pi}{6} D^3 \cdot \rho_f g \tag{2.84}$$

to the gravity:

$$F_g = -\frac{\pi}{6} D^3 \cdot \rho_p g \tag{2.85}$$

and to the drag force – see equation (2.79):

$$F_D = C_D \cdot A_\perp \cdot \frac{1}{2} \rho_f v_r^2 \tag{2.86}$$

where ρ_f denotes the density of the ambient fluid and $A_\perp = \frac{\pi}{4} D^2$ (and not πD^2, as that is the entire surface of the sphere, while the drag force is concerned with the projected area!). It should also be pointed out that the drag force is positive, directed upwards: as the particle falls, the surrounding fluid exerts a force in the opposite direction (in other words, the fluid flows upwards in relation to the particle and so the drag force is directed upwards).

Substituting the equations (2.84)–(2.86) into the balance of equation (2.83) produces

$$C_D \cdot \frac{\pi}{4} D^2 \cdot \frac{1}{2} \rho_f v_r^2 = \frac{\pi}{6} D^3 \cdot (\rho_p - \rho_f) g \tag{2.87}$$

from which follows

$$v_r = -\left(\frac{4}{3}gD \cdot \frac{\rho_p - \rho_f}{\rho_f} \cdot \frac{1}{C_D}\right)^{1/2} \tag{2.88}$$

where the minus sign accounts for the fact that the particle is moving downwards – that is, against the upward direction of z.

For calculating the terminal velocity from such a balance, it is necessary to know which value to use for C_D. Now C_D is a function of Re and therefore of the relative terminal velocity itself. It is possible to break out of this dilemma through **iteration**. This will be illustrated through two examples.

Example 2.8. A hailstone.
Consider the uniform free fall of a spherical hailstone (diameter $D = 4$ mm, density $\rho_p = 915$ kg/m³) through motionless air.
  How great is the terminal velocity v_r?

Solution
It is now possible to read in Figure 2.16 a value of C_D for an initial guess of Re. By using equation (2.88), a v_r can then be calculated with which a new Re is calculated. This is then used in turn to read a new C_D value and to calculate a second v_r. This process of iteration is repeated until v_r no longer changes (to any notable degree).

On the basis of commonplace experience, you might expect with the hailstone that v_r would be so great that Re >> 1 (this is rapidly the case, as the air viscosity is very low: $\mu_a = 1.8 \cdot 10^{-5}$ Ns/m²). On the basis of the initial guess that Re > 10^4 the iteration may proceed as follows:

$$\text{Re} > 10^4 \rightarrow C_D = 0.43: \text{ enter in equation (2.88)} \rightarrow v_r = 9.6 \text{ m/s} \rightarrow$$
$$\text{Re} = 2{,}570 \rightarrow C_D = 0.40 \, v_r = 10.0 \text{ m/s}$$
$$\text{Re} = 2{,}660 \rightarrow C_D = 0.40$$

Further iteration is no longer necessary: the terminal velocity is some 10 m/s. Note that the density of air is so small in relation to the density of the hail stone that an exact value for the air density is irrelevant and may be ignored completely without loss of accuracy. The use of the double logarithmic plot of Figure 2.16 implies anyhow that the terminal velocity of a particle can not be calculated with great accuracy – in many cases, this is not really a problem. We call this engineering accuracy.

Example 2.9. An air bubble in glue.
A spherical air bubble (density $\rho_a = 1.2$ kg/m³, with subscript a for air) is slowly rising in a bottle of glue at a constant slip velocity, v_s. The bubble has a diameter of 3 mm and rises 2.5 cm in 5 s (easy to measure!). The density of the glue is $\rho_f = 10^3$ kg/m³.
  Find the viscosity of the glue.

Solution
Again, the solution is derived from a force balance, see equation (2.83). With the positive z-axis vertically upwards, now the drag force is negative, as the air bubble moves upwards. Then equation (2.83) becomes:

$$-\frac{\pi}{6}D^3 \cdot \rho_a g + \frac{\pi}{6}D^3 \cdot \rho_l g - C_D \cdot \frac{\pi}{4}D^2 \cdot \frac{1}{2}\rho_l v_s^2 = 0 \qquad (2.89)$$

From this, it follows for the drag coefficient that

$$C_D = \frac{4}{3}\frac{gD}{v_s^2}\frac{\rho_l - \rho_a}{\rho_l} \qquad (2.90)$$

From the data it then follows that $C_D = 1{,}570$. Such a high C_D means that Re < 1 (see Figure 2.16). But then it is also true that Stokes' law can be used, equation (2.81), i.e. $C_D = 24/\mathrm{Re}$ (the bubble is spherical in shape) and so it follows that Re $= 24/C_D = 1.53 \cdot 10^{-2}$. With the data, the velocity of the bubble $v_s = 5 \cdot 10^{-3}$ m/s. It is then established that the viscosity of the glue is

$$\mu_l = \frac{\rho_l v_s D}{\mathrm{Re}} = 0.98 \ \mathrm{Ns/m^2} \qquad (2.91)$$

By using Stokes' law, a much more accurate value can be found for the viscosity; now, the uncertainty in the viscosity value may be just due to inaccuracies in glue density, bubble size and rise velocity.

Summary

The drag force is one of the forces acting on immersed particles moving through and with respect to a surrounding fluid, along with gravity and buoyancy. The drag force therefore plays an important role in the calculation of particle motions. Particle velocity is constant when the sum of the forces acting on the particle is zero.

An iterative calculation of the relative velocity of a particle is often indispensable since the drag coefficient depends on the particle Reynolds number.

3 Heat transport

3.1 Steady-state heat conduction

The fact that heat can be transported by **conduction** was introduced in Section 2.1. Heat conduction, or molecular transport of heat, refers to the ability of molecules to create a net heat transport without net mass transport. The important equation here is **Fourier's law** which sets out the link between heat flux and the 'driving' temperature gradient:

$$\phi_{q,x}'' = -\lambda \frac{dT}{dx} \tag{3.1}$$

According to this law, a gradient in the temperature in the direction of x causes a heat flux in the direction of x that always makes the heat flow from a place with a high temperature to one with a low temperature – which is why there is a minus sign on the right-hand side of equation (3.1). Throughout this chapter, the thermal conductivity coefficient λ will be assumed to be constant (i.e. independent of temperature) for the sake of simplicity; the extension to a temperature-dependent λ is trivial.

We will first deal with thermal conductivity in steady-state conditions for various elementary one-dimensional configurations. At the end of this Section 3.1, we will introduce numerical techniques in use for more complex, more-dimensional geometries.

3.1.1 Heat conduction in Cartesian coordinates

The first situation concerns a flat slab (see Figure 3.1) with thickness D and is made from a solid with thermal conductivity coefficient λ. The left-hand side of the slab is kept at a constant temperature T_1, and the right-hand side at a constant temperature T_2. In a steady-state situation, the heat balance for a thin slice somewhere in the slab between x_1 and x_2 is

$$0 = \phi_{q,\text{in}} - \phi_{q,\text{out}} = \phi_q|_{x_1} - \phi_q|_{x_2} \tag{3.2}$$

In other words:

$$\phi_q\Big|_{x_1} = \phi_q\Big|_{x_2} \tag{3.3}$$

Given that this equation applies to any x_1 and x_2 in the slice, the heat flow is independent of x and is therefore a constant. This also applies to the heat flux, given that the surface area A of the slice through which the heat flow passes is also independent of x.

https://doi.org/10.1515/9783111246574-003

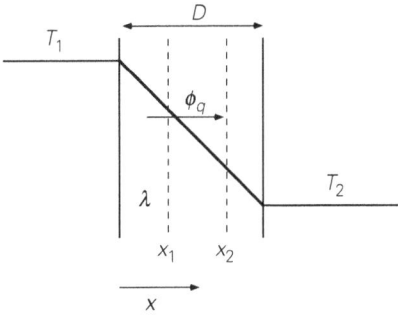

Figure 3.1

From

$$\phi_q'' = -\lambda \frac{dT}{dx} = \text{constant}$$

it follows that

$$\frac{dT}{dx} = -\frac{\phi_q''}{\lambda} = \text{constant} \tag{3.4}$$

Here, ϕ_q'' is a constant that has yet to be calculated. Integrating equation (3.4) produces a (second) integration constant. For this reason, two boundary conditions are needed in order to be able to give a complete description of the heat transport problem. With $x = 0 \rightarrow T = T_1$ and $x = D \rightarrow T = T_2$ as the boundary conditions, this gives the following temperature profile across the slab:

$$\frac{T_1 - T(x)}{T_1 - T_2} = \frac{x}{D} \tag{3.5}$$

while for the heat flux, the following applies:

$$\phi_q'' = \frac{\lambda}{D}(T_1 - T_2) \tag{3.6}$$

Equation (3.5) shows that the temperature profile in the slab is a straight line.

Equation (3.5) can also be obtained, incidentally, by drawing up an energy balance for a thin slice between x and $x + dx$. The balance is

$$0 = A \cdot \left(-\lambda \frac{dT}{dx}\bigg|_x\right) - A \cdot \left(-\lambda \frac{dT}{dx}\bigg|_{x+dx}\right) \tag{3.7}$$

Equation (3.7) can be simplified with the help of the following rule:

$$f(x + dx) = f(x) + \frac{df}{dx} \cdot dx \tag{3.8}$$

after which it becomes (after dividing by $A \cdot \lambda$, assuming that λ does not depend on place x via T)

$$0 = -\frac{dT}{dx}\bigg|_x + \left\{ \frac{dT}{dx}\bigg|_x + \frac{d}{dx}\left(\frac{dT}{dx}\right)dx \right\} \tag{3.9}$$

This result can be simplified to

$$\frac{d}{dx}\left(\frac{dT}{dx}\right) = \frac{d^2 T}{dx^2} = 0 \tag{3.10}$$

After integrating twice and applying the two boundary conditions, this produces equation (3.5) again.

Example 3.1: Heat production in a bar.
In part $L \leq x \leq 2L$ of a copper bar with length $2L$ and cross-sectional area A, a constant and uniform heat production q takes place, where q is expressed in W/m³. The one end of the bar ($x = 0$) is kept at a constant temperature T_0, while the remainder of the bar (i.e. the whole surface area including the other end $x = 2L$) is perfectly insulated.
 The question is to derive the temperature profile in the whole bar in this steady-state situation.

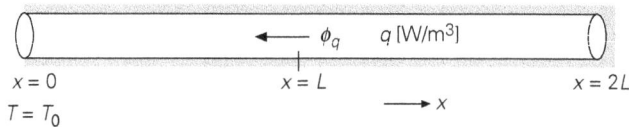

$x = 0$ $x = L$ $x = 2L$
$T = T_0$

Figure 3.2

Solution
A sketch of the problem is represented in Figure 3.2 in which the shading indicates that no heat losses take place at the pertinent surfaces. This keeps the heat transport one-dimensional, while the transport is in the negative x-direction. As the heat transport is due to conduction with $dT/dx > 0$, Fourier's law results in $\phi_q'' < 0$. A heat balance for the entire right-hand part of the bar for this steady-state situation learns that all heat produced in this right part is delivered through the plane at $x = L$ to the left part:

$$0 = 0 + \phi_q\big|_{x=L} + qAL \tag{3.11}$$

This shows that the heat flux ϕ_q'' must be negative, that is, is directed to the left and leaving the control volume, while the temperature gradient (at $x = L$) must be positive, meaning that the temperature increases to the right.
 This means that the size of the heat flux ϕ_q'' is equal to qL:

$$\phi_q''\big|_{x=L} = -\lambda \frac{dT}{dx}\bigg|_{x=L} = -qL \tag{3.12}$$

From equations (3.11) and (3.12), however, no conclusions can be drawn as to the temperature profiles in the left part or the right part of the bar. Exactly as with the concentration profile in a plug flow reactor where a species balance had to be drawn up for a thin slice dx inside the reactor (see Section 1.2.2), now a heat balance has to be drawn up for a slice dx from the bar. In doing this, the part $0 \leq x \leq L$, without heat supply should be treated separately from the part $L \leq x \leq 2L$ in which heat is produced: after all, a heat balance over a thin slice dx in the part $0 \leq x \leq L$ does not contain a production term, while a heat balance over a slice dx in the part $L \leq x \leq 2L$ does.

Drawing up a heat balance for a slice in the left part $0 \le x \le L$ results in the second-order differential equation (3.10) – the derivation starting at equation (3.7) perfectly fits this case. The two boundary conditions needed to arrive at the temperature profile in this part of the bar are $T = T_0$ or $x = 0$ (given in the question) and the expression for the temperature gradient at $x = L$ as given in equation (3.12). The result for the part $0 \le x \le L$ is

$$T = T_0 + \frac{qL}{\lambda} x \qquad (3.13)$$

Drawing up a heat balance for a slice dx in the part $L \le x \le 2L$ gives

$$0 = -\lambda A \frac{dT}{dx}\bigg|_x - \left(-\lambda A \frac{dT}{dx}\bigg|_{x+dx}\right) + qAdx \qquad (3.14)$$

from which now

$$\frac{d^2T}{dx^2} = -\frac{q}{\lambda} \qquad (3.15)$$

follows. Here, too, two boundary conditions are needed as differential equation (3.15) has to be integrated twice: perfect insulation implies $dT/dx = 0$ at $x = 2L$, while for the temperature at $x = L$, an expression can be derived with the help of equation (3.13). For the temperature profile in the part $L \le x \le 2L$, the following expression is then found:

$$T = T_0 + \frac{2qL}{\lambda} x - \frac{q}{2\lambda}\left(x^2 + L^2\right) \qquad (3.16)$$

Notice that for solving this problem, three different heat balances are needed.

3.1.2 Analogy with Ohm's law

From equation (3.6), it directly follows that the temperature difference $\Delta T \equiv T_1 - T_2$ over the slab is simply related to the heat flux as

$$\Delta T = \phi_q'' \frac{D}{\lambda} = \phi_q \frac{D}{\lambda A} \qquad (3.17)$$

Equation (3.17) shows that driving force ΔT results in a heat flux ϕ_q''. The equation has a resemblance to **Ohm's law**, known from the field of electricity:

$$\Delta V = I \cdot R = I \frac{\rho L}{A} \qquad (3.18)$$

where L is the length of the conducting wire, A the cross-sectional area, and ρ the specific resistance of the material of which the conducting wire is made.

This similarity gives us a slightly different view of thermal conductivity. As Ohm's law says: if an electric potential difference (a voltage) ΔV is created across a resistance R, then this **driving force** ΔV results in an electric current I, which is directly proportional to ΔV. The proportionality constant R describes the resistance of a piece of material to the transport of electricity. With this consideration in mind, ΔT can be viewed in equation (3.17) as the 'driving force' and heat transport ϕ_q as the 'flow'.

This also means that $D/\lambda A$ in equation (3.17) can be interpreted as a resistance, but now to the transport of heat. Remember that in equation (3.17), D is the path length to be covered for the heat transport, just as L in equation (3.18) is the path length that the electric current must cover. The role of the reciprocal thermal conductivity coefficient, that is, of $1/\lambda$, is therefore entirely comparable with that of specific resistance ρ. Note that both λ and ρ are physical properties of the compound involved. The quantity D/λ is therefore the specific resistance (i.e. the resistance per unit of surface area) to heat transport: the greater D/λ is, the smaller is the heat flux ϕ_q for a given driving force of ΔT.

The situation can be slightly complicated by placing two slabs of different thicknesses, D_1 and D_2, of different materials, with thermal conductivity coefficients λ_1 and λ_2 against each other. The left-hand side of slab 1 is kept at temperature T_1, and the right-hand side of slab 2 at T_2 (see Figure 3.3). Here, too, it is possible to derive the link in a steady-state situation between driving force ΔT over the two slabs and the heat flux through the two slabs. This situation presents itself when a layer of an insulating material is applied in order to reduce the heat transport from left to right (the heat loss).

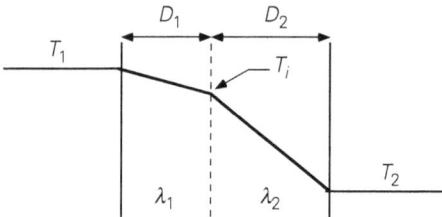

Figure 3.3

If the temperature on the interface between the two slabs is called T_i, the link between the driving force $T_1 - T_i$ over slab 1 and the associated flux $\phi_{q,1}''$ can be written down directly on the basis of equation (3.17):

$$T_1 - T_i = \phi_{q,1}'' \cdot \frac{D_1}{\lambda_1} \tag{3.19}$$

and, analogously, for the second slab:

$$T_1 - T_2 = \phi_{q,2}'' \cdot \frac{D_2}{\lambda_2} \tag{3.20}$$

For this case, $\phi_{q,1}''$ is equal to $\phi_{q,2}''$, as can be easily seen from a heat balance for a volume with material 1 on the one side, and material 2 on the other. Adding equations (3.19) and (3.20) causes the interim unknown temperature T_i to be omitted:

$$\Delta T = T_1 - T_2 = \phi_q'' \left(\frac{D_1}{\lambda_1} + \frac{D_2}{\lambda_2} \right) = \phi_q \left(\frac{D_1}{\lambda_1 A} + \frac{D_2}{\lambda_2 A} \right) \qquad (3.21)$$

The above situation concerning two slabs also has a well-known electric analogue – that is, two electrical resistances in series (see Figure 3.4), to which the following applies:

$$\Delta V = I \cdot (R_1 + R_2) \qquad (3.22)$$

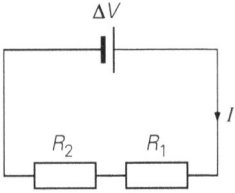

Figure 3.4

Equations (3.21) and (3.22) are entirely analogous: in equation (3.22) too, the individual resistances in a series circuit have to be added up in order to obtain the overall resistance. In which order the resistances are installed is irrelevant for both the heat transport and the electrical current.

3.1.3 Heat conduction in cylindrical coordinates

In each of the cases examined up to now, the flux through the material was independent of its position in the material. This is actually the exception, because in reality the flux is usually place-dependent. This can be illustrated on the basis of the link between the driving force and the flux in the case of a hollow cylinder. A typical example of such a situation is a pipe through which water flows at a temperature different from that of the medium outside. The result is radial heat transport through the pipe wall due to conduction.

Assume that the situation as shown in Figure 3.5 is steady. The inner radius of the hollow cylinder is R_1 and the outer radius is R_2. The inside surface is kept at temperature T_1, and the outer surface at T_2. In order to calculate the (local) heat flux through the cylindrical wall, Fourier's law says that the temperature profile in the wall has to be known.

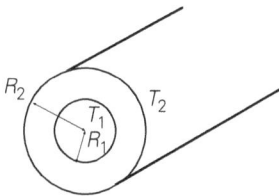

Figure 3.5

To this end, consider a ring (length L) in the cylindrical wall with coaxial radii r and $r + dr$ (see Figure 3.6). A heat balance for this ring, in steady-state conditions without heat production, is

$$0 = \phi_{q,\text{in}} - \phi_{q,\text{out}} \tag{3.23}$$

It is still the case that $\phi_{q,\text{in}} = \phi_{q,\text{out}} = \text{constant}$. However, this does not mean that the flux is also constant: after all, the geometry is now curved. The radial flux is now shown as follows:

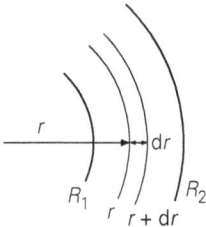

Figure 3.6

$$\phi_q'' = -\lambda \frac{dT}{dr} \tag{3.24}$$

Additionally, in this case, the following applies to the link between flow and flux:

$$\phi_q = A \cdot \phi_q'' = 2\pi r L \cdot \phi_q'' \tag{3.25}$$

Combining equations (3.23)–(3.25) produces the following differential equation:

$$-\lambda 2\pi r L \frac{dT}{dr} = C_1 \tag{3.26}$$

The general solution to equation (3.26) is

$$T(r) = -\frac{C_1}{2\pi\lambda L}\ln r + C_2 \tag{3.27}$$

Constants C_1 and C_2 can be determined with the help of boundary conditions: $r = R_1 \rightarrow T = T_1$ and $r = R_2 \rightarrow T = T_2$. This gives the following as the solution to the temperature profile in the cylindrical wall:

$$\frac{T - T_2}{T_1 - T_2} = \frac{\ln(r/R_2)}{\ln(R_1/R_2)} \tag{3.28}$$

Now that the temperature profile is known, it is also possible to calculate the flux. Differentiating equation (3.28) and substituting the temperature gradient into (3.24) – or using equation (3.26) – produces

$$\phi_q'' = -\lambda \frac{dT}{dr} = -(T_1 - T_2) \frac{\lambda}{\ln(R_1/R_2)} \frac{1}{r} \tag{3.29}$$

Flux ϕ_q'' is indeed now dependent on its place in the cylindrical wall: as r increases, so ϕ_q'' decreases. Equation (3.29) also provides the link between ΔT (= $T_1 - T_2$) and ϕ_q:

$$\Delta T = \frac{\ln(R_2/R_1)}{2\pi\lambda L} \phi_q \tag{3.30}$$

Here too, the 'flow' is proportional to the 'driving force', but in this case the link between geometry and 'resistance' is more complicated. When, however, R_2 does not differ too much from R_1, for example, when the thickness $\delta = R_2 - R_1$ of a pipe wall is pretty small with respect to R_1, owing to a series expansion – for $\varepsilon \ll 1$, one can write $\ln(1 + \varepsilon) \approx \varepsilon$ – equation (3.30) can be rewritten as

$$\Delta T = \frac{\delta}{\lambda} \frac{\phi_q}{2\pi R_1 L} \tag{3.31}$$

in which δ/λ is recognised as the resistance to heat transport of a flat layer, as in equation (3.17).

Once again, the same result can be obtained by further working out the steady-state heat balance (3.23). For this, the 'in' and 'out' flows should be expressed in terms of local temperature gradients:

$$0 = -[A]_r \cdot \lambda \frac{dT}{dr}\bigg|_r - \left(-[A]_{r+dr} \cdot \lambda \frac{dT}{dr}\bigg|_{r+dr} \right) \tag{3.32}$$

In this equation, $[A]_r$ is the surface area of the ring with radius r and $[A]_{r+dr}$ is the surface area of the ring with radius $\{r + dr\}$. Now, the surface area of a ring with radius r is equal to

$$[A]_r = 2\pi r \cdot L \tag{3.33}$$

Equation (3.32) can be simplified with the help of equation (3.33) by dividing by $2\pi\lambda L$ and by putting terms r and $r + dr$ with the local temperature gradient:

$$[A]_{r+dr} \frac{dT}{dr}\bigg|_{r+dr} = \left\{ A \frac{dT}{dr} \right\}\bigg|_{r+dr} = \left\{ 2\pi r L \frac{dT}{dr} \right\}\bigg|_{r+dr} \tag{3.34}$$

so that

$$0 = -\left\{ r \frac{dT}{dr} \right\}\bigg|_r - \left\{ -r \frac{dT}{dr} \right\}\bigg|_{r+dr} \tag{3.35}$$

Here, $r\,(dT/dr)$ should be regarded as one function, $f(r)$. Equation (3.35) will then contain the difference between $f(r + dr)$ and $f(r)$ when going back from $r + dr$ to r, which of course is $(df/dr) \cdot dr$ for small dr. Entering $r\,(dT/dr)$ for f again produces

$$\frac{d}{dr}\left(r\frac{dT}{dr}\right) = 0 \tag{3.36}$$

Solving equation (3.36), with the boundary conditions $r = R_1 \rightarrow T = T_1$ and $r = R_2 \rightarrow T = T_2$, produces equation (3.28) of course.

3.1.4 Heat conduction in spherical coordinates

A second example of a curved geometry in which the heat transport is independent of place, but the heat flux is not, is that of spherical geometry. Consider a sphere (with radius R) that is kept at temperature T_1. The sphere is surrounded by a stagnant medium such that the heat transport is due to conduction only. At a long distance from the sphere, the temperature of the medium is T_∞. What is the link between driving force and heat transport, in this case, in a steady-state situation?

In order to answer this question, of course, a heat balance will have to be drawn up, this time for a concentric spherical shell with radii of $r\,(>R)$ and $r + dr$. The heat balance is

$$0 = \phi_{q,\text{in}} - \phi_{q,\text{out}} \tag{3.37}$$

In this case, too, surface area A, through which the heat is transferred, is a function of r: $A = A(r)$. The link between the local heat transport rate and the local temperature gradient can therefore be written as follows:

$$\phi_q = A(r) \cdot \left(-\lambda\frac{dT}{dr}\right) \tag{3.38}$$

Using this in heat balance (3.37) produces

$$0 = -[A]_r \cdot \lambda\frac{dT}{dr}\bigg|_r - \left(-[A]_{r+dr} \cdot \lambda\frac{dT}{dr}\bigg|_{r+dr}\right) \tag{3.39}$$

Equation (3.39), analogously to the above explanation on the derivation of equation (3.36), can be written as the following differential equation:

$$0 = \lambda\frac{d}{dr}\left(A(r)\frac{dT}{dr}\right) \tag{3.40}$$

For $A(r)$, the following now applies:

$$A(r) = 4\pi r^2 \tag{3.41}$$

This means that equation (3.40) becomes

$$\frac{d}{dr}\left(r^2\frac{dT}{dr}\right) = 0 \tag{3.42}$$

The general solution to differential equation (3.42) is

$$T(r) = -\frac{C_1}{r} + C_2 \tag{3.43}$$

Substituting the boundary conditions $r = R \rightarrow T = T_1$ and $r \rightarrow \infty \rightarrow T = T_\infty$ produces

$$T(r) = (T_1 - T_\infty)\frac{R}{r} + T_\infty \tag{3.44}$$

Using this expression for finding the temperature gradient in equation (3.38) produces a relation between the driving force ΔT (= $T_1 - T_\infty$) and the heat transport ϕ_q, or the heat flux ϕ_q'', at the surface of the sphere (with diameter $D = 2R$):

$$\Delta T = \frac{1}{4\pi\lambda R}\phi_q = \frac{D}{2\lambda}\phi_q'' \tag{3.45}$$

Example 3.2. Radioactive sphere I.
In a sphere, 10 cm in diameter and consisting of uranium oxide, heat production q (in W/m^3) as a result of the radioactive decay takes place uniformly distributed over the volume. The surface of the sphere is kept at a low and constant temperature T_w.
  Calculate the temperature at the centre of the sphere for $q = 6$ MW/m^3, $T_w = 20$ °C, and $\lambda = 8$ W/mK.

Solution
The heat production inside the particle, along with the surface temperature being held constant at a low value, results in a radial temperature profile with the maximum temperature at the centre. Such a temperature profile cannot be found by means of a macro-balance for the whole sphere, but requires a look inside the sphere: inside, a spherical shell with thickness dr has to be selected as a micro control volume for a steady-state heat balance:

$$0 = -4\pi r^2\lambda\frac{dT}{dr}\Big|_r - \left(-4\pi r^2\lambda\frac{dT}{dr}\Big|_{r+dr}\right) + q4\pi r^2 dr \tag{3.46}$$

into which – compared to equation (3.39) – right away $4\pi r^2$ has been substituted for $A(r)$, and the subscripts r and $r + dr$ relate to both the area $4\pi r^2$ and the gradient dT/dr. The next step is dividing all terms of equation (3.46) by $4\pi\lambda dr$ and results in

$$\frac{d}{dr}\left(r^2\frac{dT}{dr}\right) = -\frac{q}{\lambda}r^2 \tag{3.47}$$

where it is extremely important to keep r^2 between the brackets behind the d/dr on the left-hand side, as when going from position r to position $r + dr$, the shell area increases. Integrating equation (3.47) once results in

$$r^2 \frac{dT}{dr} = -\frac{q}{3\lambda} r^3 + C_1 \qquad (3.48)$$

As at $r = 0$, the temperature is maximum, that is, $dT/dr = 0$, the integration constant C_1 is concluded to be zero. Then, it follows that

$$\frac{dT}{dr} = -\frac{q}{3\lambda} r \qquad (3.49)$$

Integrating once more and using the second boundary condition $T = T_w$ at $r = R$ eventually gives

$$T = T_w + \frac{q}{6\lambda} (R^2 - r^2) \qquad (3.50)$$

Substituting the given values for q, T_w, and λ results in the answer $T_0 = 1{,}270 \ °C$ at $r = 0$.

3.1.5 A numerical treatment for two-dimensional Cartesian

All the examples that we have looked at so far have been solved entirely analytically. This was possible because the geometries were simple. In many practical cases, finding an analytical solution is difficult, if not impossible. Fortunately, although many problems can be tackled perfectly well with the help of numerical techniques, and for some steady-state problems, the means are surprisingly easy. Solving problems numerically is now an essential part of modern transport phenomena. Consider, as an illustration (Figure 3.7a), a two-dimensional problem: what is the steady-state temperature distribution in a square where two sides are kept at constant temperature $T_0 = 0 \ °C$ and the other two sides at $T_1 = 50 \ °C$?

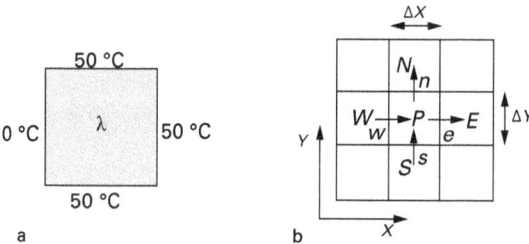

Figure 3.7

The problem can be excellently described in a Cartesian system. First, the square is divided up into $N \cdot M$ small rectangles with sizes Δx and Δy. These small rectangles are shown in Figure 3.7b. The centre rectangle is shown as P, with the four adjacent rectangles having the four directions of the compass, North, East, South, West, shortened to N, E, S, W. The 'in' and 'out' heat flows have also been drawn for rectangle P.

A steady-state heat balance for the rectangle with centre P is therefore

$$0 = -\lambda\frac{\partial T}{\partial x}\bigg|_{w}\Delta y - \left(-\lambda\frac{\partial T}{\partial x}\bigg|_{e}\right)\Delta y +$$

$$+ \left(-\lambda\frac{\partial T}{\partial y}\bigg|_{s}\right)\Delta x - \left(-\lambda\frac{\partial T}{\partial y}\bigg|_{n}\right)\Delta x \qquad (3.51)$$

where the partial derivative symbol ∂ is used (to express that T is a function of both x and y). The indices w, e, s, and n here refer to the interfaces of the rectangle around P (see Figure 3.7b). It is also the case that no heat production occurs in the rectangle around P. Formally, the balance applies only to an infinitesimally small rectangle. As far as realistic dimensions are concerned, this is an approximation. Each of the four gradients in the above equation can be calculated, such as the heat flow through the left-hand side of the rectangle around P:

$$\frac{\partial T}{\partial x}\bigg|_{w} \approx \frac{T_P - T_W}{\Delta x} \qquad (3.52)$$

If this is done for every term in equation (3.51), if $\Delta x = \Delta y$, and λ is taken as a constant, then heat balance (3.51) can be written as

$$0 = T_W + T_E + T_S + T_N - 4T_P \qquad (3.53)$$

In other words, the temperature of P is the mean of its four neighbours. This can be converted very easily into a computer code. Start with an estimate. Let us suppose that the entire square has a uniform temperature of T_0. Now, for each smaller square, take the mean of its neighbours and regard this temperature as an improved version of the original estimate. Of course, the original estimate changes because temperature T_1 applies on two of the sides. After this first iteration, the new temperature field can be compared with the old one. If the difference between both fields is still too great, we repeat the process until the solution has converged and we have a numerically obtained calculation of the temperature.

In principle, it is possible to arrive at any given point that is near the solution by dividing up the calculation field (the square) into a very large number of elements and by very clearly defining the convergence criterion. Below is a simple algorithm that can be used for the problem described above, with the help of Matlab, for example:

```
%    steady state long-term heating, algorithm for Matlab
%    square with constant temperature at sides:
%    bottom and left-hand side: 0 °C
%    top and right-hand side: 50 °C

clear all;

N=20;   %number of internal cells = N*N
```

```
Told=zeros(N+2,N+2);         %initialise first estimate
                             %make the temperature field N+2 at N+2 so that we
                             %can include boundary conditions via so-called virtual
                             %points. These are points situated outside the square.
for i=1:N+1                  %define the virtual points associated with boundary condition T=50 °C
   Told(i+1,N+2)=50;            %top side at 50 °C
   Told(N+2,i+1)=50;            %right-hand side at 50 °C
end
Tnew=Told;                   %initialise new field

eps=1;                       %initialise convergence parameter
epsconv=1e-3;                %select the convergence criterion

while eps>epsconv            %start iterating until convergence
   for i=2:N+1               %calculate new temperature field: i numbers x-direction
      for j=2:N+1
         Tnew(i,j)=1/4*(Told(i-1,j)+Told(i+1,j)+Told(i,j-1)+Told(i,j+1));
      end
   end
   eps=0;                    %calculate convergence again
   for i=2:N+1
      for j=2:N+1
         eps=eps+abs(Tnew(i,j)-Told(i,j));
      end
   end
   Told=Tnew;                %replace old field with new one
end

x=linspace(0,1,N+2);         %make coordinates
y=x;

contourf(x,y,Tnew,19);       %show solution in graph form
```

Advanced methods for finding a solution exist for the above example. Also, the convergence criterion used is just one of the many possibilities. Numerical analysis provides the necessary tools for carrying out simulations of this kind in a responsible manner. Knowledge of these is obviously essential for numerical simulations of transport phenomena.

The example, see Figure 3.8 for the temperature field resulting from the given input data, illustrates that the balance method can be used to generate the necessary equations fairly simply. The big advantage of using the balance method is that with numerical solving the balances are always complied with at every (local) level. This easily vindicates the basics of transport phenomena. Many books that deal with numerical techniques in transport phenomena have been written. One of the standard works is Patankar's textbook[10] dating from as far back as 1980.

The example of the square can also be relatively easily analytically solved with the help of Fourier analysis. This is because there is still a fair amount of symmetry in

---

10 Patankar S.V., *Numerical Heat Transfer and Fluid Flow*, Hemisphere Publ. Corp., 1980.

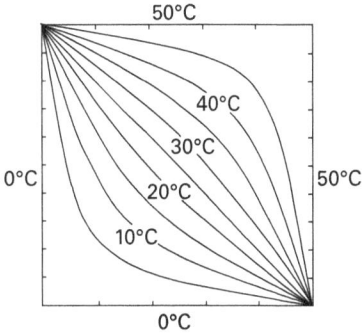

Figure 3.8: Temperature contours of the conducting square in a steady-state situation with two sides at 0 °C and two at 50 °C. Number of cells: 50 × 50.

the problem. Things look somewhat more tricky if production occurs in the square, but numerically this is not much more difficult. The balance equation will now contain an extra term as a result of this production. Consider the rectangle around P again. The production per unit of surface area is q''. The heat balance is now:

$$0 = -\lambda\frac{\partial T}{\partial x}\Big|_w \Delta y - \left(-\lambda\frac{\partial T}{\partial x}\Big|_e\right)\Delta y +$$

$$+ \left(-\lambda\frac{\partial T}{\partial y}\Big|_s\right)\Delta x - \left(-\lambda\frac{\partial T}{\partial y}\Big|_n\right)\Delta x + \iint\limits_{\Delta x, \Delta y} q''dxdy \tag{3.54}$$

For small Δx and Δy, q'' can be assumed to be constant over the control volume. The approximate algebraic equation is now (with $\Delta x = \Delta y$):

$$0 = \lambda(T_W + T_E + T_S + T_N - 4T_P) + q''_P \cdot \Delta x \Delta y \tag{3.55}$$

Here, too, the steady-state solution can be found by using a comparative iterative procedure.

By way of example, we assume that a heat source and a heat sink are present in the square. Both sources have the form of a 'delta peak' with strength q and $-q$, respectively. The source is at position $\{1/4 + 1/80, 3/4 - 1/80\}$, while the sink is at $\{3/4 - 1/80, 1/4 + 1/80\}$ (let the length of the sides of the square be equal to 1).

Once again, the square is divided up into smaller squares, two of which therefore contain an extra source term. None of the others do. Because of the delta peak, the production term in equation (3.54) is always zero, apart from the two smaller squares that contain the source and the sink. The production term for these is $+q$ and $-q$, respectively, regardless of the size of the smaller squares. For the sake of simplicity, the boundary conditions are now $T = 0$ °C for each side.

Equation (3.55) can be solved using the same iteration process as for equation (3.53). It is a good idea to include a source term for every cell: this is zero in every case apart from the two cells that contain the real source and sink. In terms of the algorithm, the central rule becomes

$$T_{new}(i,j) = 1/4 * \{T_{old}(i-1,j) + T_{old}(i+1,j) +$$
$$+ T_{old}(i,j-1) + T_{old}(i,j+1)\} + 1/4/\lambda * q(i,j)$$

In Figure 3.9, the numerical solution to this problem is shown for 40 × 40 elements.

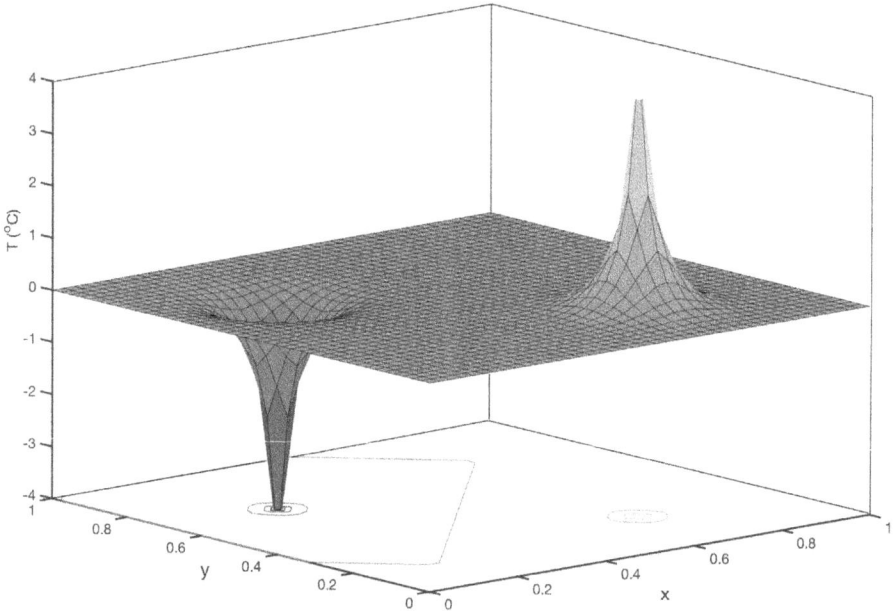

Figure 3.9

The examples discussed here use the simple geometry of a square. However, there is no problem in using the shape of a cooling fan, for example. Numerical techniques are often the only method if the geometry is complex.

Summary

In the case of steady-state conduction of heat, the link between the driving force ΔT and the heat transport is analogous to Ohm's law. The driving temperature difference causes a transport that is directly proportional to this temperature difference. The proportionality constant can be regarded as resistance to the transport of heat.

Both flat and non-flat geometries have been considered, and there is always the same link between driving force and heat transport. However, the heat flux is dependent on place in the case of curved geometries! This can be seen directly from heat balances over thin rings or shells.

We have also touched on numerical methods for solving thermal conductivity questions in multi-dimensional and complex geometries. Discretisation of the differential equation leads to an algebraic equation that links the values of the sought-after variables in nearby points to each another. An iterative solution strategy is needed in order to arrive at the solution.

3.2 Heat transfer coefficient and Nusselt number

3.2.1 Newton's law of cooling

The concept of a driving force that results in a flow that is proportional to the driving force has proved to be of great value in complex situations (unsteady and steady) as well. For engineers, the idea plays an important role in the complex situations they encounter in practice. Heat flow and driving force are generally linked to each other by **Newton's law of cooling** (which also described heating up). The law states that

$$\phi_q = hA\Delta T \tag{3.56}$$

Flow and driving force are therefore linked the other way around with regard to Ohm's law. Instead of working in terms of resistance, a measure for heat transport, or heat transfer, is used here. In equation (3.56), A is the surface area through which the heat flow passes, and coefficient h is called the **heat transfer coefficient** (the unit of which is W/m^2 K). It is therefore $1/h$ itself that can be interpreted as heat resistance. Surface area A is included explicitly in equation (3.56), because it is obvious that the heat flow increases (more or less) proportionally to the surface area across which the driving force is located.

The results of Section 3.1 can now be summarised in terms of an h-value for every situation (**check this!**):

Table 3.1

Heat transfer coefficient	Flat slab	Cylindrical wall (related to outer surface)	Sphere
h	$\dfrac{\lambda}{D}$	$\dfrac{2\lambda}{D_2 \ln(D_2/D_1)}$	$\dfrac{2\lambda}{D}$

3.2.2 The Nusselt number

As mentioned previously, it is often a good idea to use such links in a non-dimensional form. In the case of heat conduction, too, a non-dimensional form of h is defined. This number is known as the **Nusselt number**, or Nu for short. The Nusselt number is defined as

$$\text{Nu} = \frac{\text{heat resistance if one-dimensional steady-state conduction}}{\text{actual resistance (at prevailing conditions)}} =$$

$$= \frac{D/\lambda}{1/h} = \frac{hD}{\lambda} \tag{3.57}$$

This makes it possible, for example, to indicate quickly and easily the situation of steady-state heat conduction through a thin solid slice with thickness D, for example, as Nu = 1, or that from the surface of a sphere to 'infinity', see equation (3.45), as Nu = 2. Thanks to the definitions of the heat transfer coefficient, in (3.56), and the Nusselt number, in equation (3.57), it is possible to understand directly from Nu = 2, for example, that it concerns heat conduction from a spherical surface in steady-state conditions (and not time-dependent conditions or convective heat transport, for example).

3.2.3 Overall heat transfer coefficient

Many cases of heat flow involve two heat resistances in series. The conduction of heat through two thin slices has already been covered in Section 3.1, for example. The result of the analysis was

$$\phi_q'' = \left(\frac{D_1}{\lambda_1} + \frac{D_2}{\lambda_2}\right)^{-1} \Delta T = \left(\frac{1}{h_1} + \frac{1}{h_2}\right)^{-1} \Delta T \tag{3.58}$$

The items in brackets can also be replaced by one 'heat exchange coefficient'. This contains both heat transfer coefficients, h_1 and h_2. This coefficient is therefore also known as the **overall heat transfer coefficient** and is denoted as U. From equation (3.58) it therefore follows that

$$\frac{1}{U} = \frac{1}{h_1} + \frac{1}{h_2} \tag{3.59}$$

Equation (3.59) appears to have general validity and is easy to derive. To that end, look at the heat flow in Figure 3.10, which goes into medium 2 from medium 1. Denote the temperature of the interface by T_i. The following then applies to the heat flux from medium 1 to the interface:

$$\phi_q'' = h_1(T_1 - T_i) \tag{3.60}$$

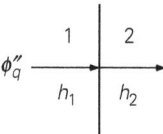

Figure 3.10

This flux also runs into medium 2 from the interface, as an interface cannot absorb or produce heat. The expression for the flux from the interface into medium 2 is

$$\phi_q'' = h_2(T_i - T_2) \tag{3.61}$$

Eliminating the temperature from the interface provides the sought-after link:

$$\phi_q'' = \left(\frac{1}{h_1} + \frac{1}{h_2}\right)^{-1} \Delta T \tag{3.62}$$

This result can be interpreted analogously to Ohm's law: the overall resistance ($1/U$) is the sum of the two secondary resistances, $1/h_1$ and $1/h_2$.

It is common practice among engineers to ignore the smallest of the two resistances to heat transfer when the ratio of the two resistances is beyond a certain value. Which critical value is used, determines of course how accurate the estimate may be upon ignoring the smallest resistance. A typical example of such a situation is heat transfer from a solid object to its ambient medium (or the other way around). The ratio of the object's internal resistance to heat transfer – in the case of a flat plat D/λ_i of course – to the external resistance $1/h_e$ is often represented by the **Biot number**, denoted as Bi:

$$\text{Bi} = \frac{D/\lambda_i}{1/h_e} = \frac{h_e D}{\lambda_i} \tag{3.63}$$

The Biot number should not be confused with the **Nusselt number**: Bi compares for an object the internal and external resistances to heat transfer, while Nu compares the resistance to heat transfer (external or internal) under the actual conditions with that in the case of one-dimensional steady-state heat transport across a flat slab (Nu = 1).

Example 3.3. Radioactive sphere II
In a sphere, 10 cm in diameter and consisting of uranium oxide, heat production q (in W/m^3) as a result of the radioactive decay takes place uniformly distributed over the volume. The sphere is now in a medium that in the bulk, that is, far away from the sphere, has a constant, lower temperature T_∞.
Derive an expression for the temperature at the centre of the sphere that comprises T_∞.

Solution
Inside the sphere there are no changes with respect to Example 3.2. Inside,

$$\frac{dT}{dr} = -\frac{q}{3\lambda_i} r \tag{3.64}$$

therefore still applies – see equation (3.49). Unlike in Example 3.2, however, now the temperature at $r = R$ is determined by the heat loss to the environment. In any steady state, the heat flux from the sphere's interior towards the sphere's surface equals the heat flux from the surface into the ambient medium:

$$-\lambda_i \frac{dT}{dr}\bigg|_{r=R} = \frac{qr}{3}\bigg|_{r=R} = \frac{qR}{3} = h_e(T_{r=R} - T_\infty) \tag{3.65}$$

Then it follows that

$$T_{r=R} = T_\infty + \frac{qR}{3h_e} \tag{3.66}$$

[Note that the latter result can also be obtained with the help of an overall heat balance over the sphere that expresses that under steady-state conditions the heat production inside the sphere equals the heat transfer to the ambient medium.]

Integrating equation (3.64) with boundary condition (3.66) at $r = R$ gives for the temperature T_0 at the heart of the sphere:

$$T_0 = T_\infty + \frac{qR}{3h_e} + \frac{qR^2}{6\lambda_i} = T_\infty + \frac{qR}{3}\left(\frac{1}{h_e} + \frac{R}{2\lambda_i}\right)$$

(3.67)

This equation illustrates that T_0 depends on both the internal and the external heat transport. When the internal resistance to heat transport may be ignored with respect to the external resistance, then equation (3.67) simplifies to equation (3.66):

$$T_0 \approx T_{r=R} = T_\infty + \frac{qR}{3h_e}$$

(3.68)

while for negligible external resistance an expression for T_0 is obtained that agrees with equation (3.50). Table 3.1 shows that if the external medium is stagnant and external heat transfer therefore is by conduction only, then $h_e = 2\lambda_e/D$ applies to equations (3.65)–(3.68). It is crucial to make a marked distinction between internal and external resistance, and between internal and external physical properties indeed.

Summary

Analogous to Ohm's law, the link between the temperature difference and the resulting heat flow is shown by Newton's law of cooling: transfer rate ϕ_q (in W) is linked to driving force ΔT via the heat transfer coefficient h:

$$\phi_q = hA\Delta T$$

The heat transfer coefficient (=reciprocal resistance) has different values for different situations. The non-dimensional form of h is the Nusselt number, Nu. For heat transfer through conduction round a sphere, Nu = 2 applies.

More complex situations (when an object exchanges heat with the ambient medium for instance) can be described more usefully with an overall heat transfer coefficient U, which is composed from the individual heat transfer coefficients according to $1/U = 1/h_1 + 1/h_2$. This relation expresses that the total resistance to heat transfer is the sum of two partial resistances in series, on the analogy of Ohm's law. Whenever one of these two resistances is much smaller than the other, the smallest is often ignored.

3.3 Unsteady-state heat conduction

3.3.1 Penetration theory: conceptual

In the above sections, we have looked at heat conduction in steady-state situations. It is now time to consider the more difficult case of unsteady-state (i.e. time-dependent, or transient) heat conduction: the temperature in the material or in the fluid-at-rest at a given location is now a function of time, while in general the temperature will vary from one place to another as well. We will first look in detail at the simplest geometry, the 'semi-infinite slab'.

To that end, consider a very large piece of material that takes up half of space, namely that part where $x \geq 0$. Initially, the temperature of the whole block is T_0. At time $t = 0$, the temperature of the left-hand side surface ($x = 0$) is suddenly raised to T_1, where it remains. In many practical situations, it is then important to know how the heat penetrates the material. Or put another way, what is the temperature profile in the material at any given time?

After all the previous examples, it will be obvious that in order to answer these questions, it will be necessary to draw up an unsteady-state heat balance for a random volume in the material. Given that the material in the directions of y and z stretches out into infinity, and that one uniform temperature is being imposed on the entire left-hand side, this is a one-dimensional problem; changes will therefore only be possible in direction x. A heat balance (=thermal energy balance) for a slice between x and $x + dx$ (see Figure 3.11) is therefore sufficient.

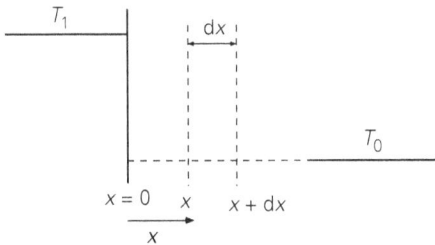

Figure 3.11

The volume for this area is $V = L \cdot W \cdot dx$ (L is the dimension in the direction of y, and W in the direction of z). The unsteady-state thermal energy balance is now:

$$\frac{\partial U}{\partial t} = \frac{\partial (\rho V c_p T)}{\partial t} = LWdx \; \rho c_p \frac{\partial T}{\partial t} = \phi_{q,\text{in}} - \phi_{q,\text{out}} \tag{3.69}$$

where the size of the control volume is constant, constant pressure and constant density have been assumed (see also Bird et al.[11]) and specific heat c_p (loosely replacing c_v) is taken constant as well.

The link between the heat flow and the temperature gradient is also shown by Fourier's law in the unsteady-state situation, namely:

$$\phi_{q,\text{in}} = -LW\lambda \frac{\partial T}{\partial x}\bigg|_x \tag{3.70}$$

A similar equation applies to the outgoing flow as well, of course. The thermal energy balance can therefore be written as follows:

---

11 Bird R.B., W.E. Stewart and E.N. Lightfoot, *Transport Phenomena*, Wiley, 2nd Ed., 2002, p. 337.

$$\frac{\partial T}{\partial t} = \frac{\lambda}{\rho c_p} \frac{\frac{\partial T}{\partial x}\big|_{x+dx} - \frac{\partial T}{\partial x}\big|_x}{dx} = \frac{\lambda}{\rho c_p} \frac{\partial^2 T}{\partial x^2} \tag{3.71}$$

As already shown in Chapter 2, the factor $\lambda/(\rho c_p)$ is usually abbreviated to the symbol a, the **thermal diffusivity**. This means equation (3.71) is

$$\frac{\partial T}{\partial t} = a \frac{\partial^2 T}{\partial x^2} \tag{3.72}$$

Differential equation (3.72) still requires appropriate boundary and initial conditions in order for it to describe the problem in full. These boundary conditions are

$$
\begin{aligned}
T(x=0) &= T_1 && \text{for } t \geq 0 \\
T(t=0) &= T_0 && \text{for } x \geq 0 \\
T(x \to \infty) &= T_0 && \text{for } x \to \infty \text{ for each } t
\end{aligned}
\tag{3.73}
$$

The first boundary condition means that exactly so much heat can be supplied from the domain $x < 0$ that at $x = 0$ the temperature can be kept at T_1; in other words, the resistance to heat transport lies entirely in domain $x > 0$ where the molecules account for the heat transport. The last boundary condition means that 'far away' temperature is not (yet) affected by the change to the temperature on the left. This will be dealt with in greater detail later.

3.3.2 Temperature profile and penetration depth

With the help of dimensional analysis, it is easy to gain an understanding of which combinations of variables ultimately determine the problem of differential equation (3.72). For this, the change in temperature from the initial temperature T_0 can be examined (note: like in many heat transport cases, the interest is in the temperature difference!). Owing to the above heat balance, the relevant parameters are already known: $T - T_0$ is dependent on $T_1 - T_0$, a, t and x; therefore:

$$T - T_0 = k(T_1 - T_0)^\alpha a^\beta t^\gamma x^\delta \tag{3.74}$$

Solving the resulting set of equations for α, β, γ, and δ produces $\alpha = 1$, $\beta = -\delta/2$, and $\gamma = -\delta/2$ for δ, which is yet to be determined (see Chapter 2). From the dimensional analysis, it therefore follows that this problem is determined by two non-dimensional groups, namely:

$$\frac{T - T_0}{T_1 - T_0} = F\left(\frac{x}{\sqrt{at}}\right) \tag{3.75}$$

As already discussed in Chapter 2, the form of function F cannot be determined through dimensional analysis. For that, the exact solution to the problem must be found. Apart from the exact solution, experience shows that in order to find a particular difference in temperature $T - T_0$ at a certain position x, more time has to elapse the greater x is. This experience-based fact can also be inferred from the result of the dimensional analysis. Imagine a particular value for $T - T_0$; the value of F is then established. Suppose that this is measured at time t_1 at place x_1, then it is logical that this value for $T - T_0$ at position x_2 ($>x_1$) will be measured later, specifically at time t_2 that complies with

$$\frac{x_2}{\sqrt{at_2}} = \frac{x_1}{\sqrt{at_1}} \rightarrow t_2 = \frac{x_2^2}{x_1^2} t_1 \tag{3.76}$$

It is only then that the same value for F is found again. Conversely, if the temperature is measured at fixed position x, it will appear that the rise in temperature will proceed increasingly slowly at that point.

As already mentioned, the exact temperature profile follows from the solution to differential equation (3.72) with boundary conditions (3.73). Written in two non-dimensional groups, the solution is

$$\frac{T - T_0}{T_1 - T_0} = 1 - \frac{2}{\sqrt{\pi}} \int_0^{\frac{x}{2\sqrt{at}}} \exp\left(-s^2\right) ds \tag{3.77}$$

The integral on the right-hand side cannot be solved analytically and is known as an error function. This **error function** is tabulated in various handbooks and is available in many computer programs such as Excel, Matlab, or Maple. The temperature profile according to equation (3.77) for a number of different points in time is shown in Figure 3.12.

With the help of temperature profile (3.77), it is now possible to calculate the heat flux through interface $x = 0$, thanks to Fourier's law. This, in other words, is the heat flux that penetrates the material. The result is

$$\phi_q''\Big|_{x=0} = -\lambda \frac{\partial T}{\partial x}\Big|_{x=0} = \lambda \frac{T_1 - T_0}{\sqrt{\pi a t}} \tag{3.78}$$

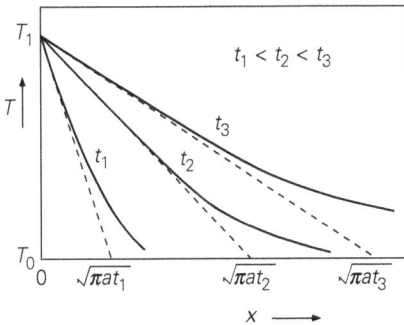

Figure 3.12

The flux decreases over time, as may be expected from the profiles in Figure 3.12 – this shows that the gradient in the temperature at $x = 0$ is becoming smaller and smaller over time so that the flux across the interface decreases. Equation (3.78) further expresses that the driving force for the heat transport is constant, namely $T_1 - T_0$ in agreement with the constant boundary conditions at $x = 0$ and at $x \to \infty$.

Equation (3.78) is very important and, luckily, easy to remember: the flux at $x = 0$ can be calculated by dividing the constant driving force $\Delta T = T_1 - T_0$ by the length scale $\sqrt{\pi a t}$ and then by multiplying by λ. It was already clear that $\sqrt{a t}$ had the dimension of a length scale in equations (3.75) and (3.76). Analogous to equation (3.17), the combination $\sqrt{\pi a t}/\lambda$ stands for the resistance to heat transport that increases over time because the thickness $\sqrt{\pi a t}$ to which ΔT relates, increases.

Length $\sqrt{\pi a t}$ is called the **penetration depth** because this is a typical length over which the temperature of the material noticeably changes within time interval t. This measure of length therefore expresses how far the heat has penetrated the material. There are two reasons for this choice of dimension: first, the tangent line to the temperature profile at $x = 0$ at point of time t cuts through the x-axis precisely at $\sqrt{\pi a t}$; second, the increase in temperature at that position is less than some 20% of the overall driving force.

The penetration depth is a useful quantity for quickly estimating how far the heat has penetrated a material, but it is clear from Figure 3.12 that at each moment the temperature is already rising even where x-values are greater than the penetration depth. The nature of the temperature profile does not allow an exact measure for the penetration of heat, however. Because of the aforementioned considerations, the heat transport model dealt with above is referred to as the **penetration theory**.

3.3.3 Applying penetration theory

An important question that remains is this – when can the penetration theory be applied in practical situations? Answer: in every case that involves unsteady-state heat conduction in a material, the initial temperature of which is uniform, and the temperature of one side of which is suddenly raised and kept at the new temperature. Also, the 'rear' of the material may not have undergone any notable change of temperature. After all, one of the boundary conditions is $T = T_0$ for $x \to \infty$ for every t. This latter condition for the penetration theory imposes a limitation on the duration of validity of the penetration theory. After some time, the temperature of the rear will also have noticeably changed. This can be roughly estimated with the help of the penetration depth.

Suppose that the dimension of the body (towards x) is L, then the penetration theory will be valid as long as the penetration depth is clearly less than L. To be sure that the temperature at the rear still is the original one (see Figure 3.12), the restriction $\sqrt{\pi a t} < 0.6\, L$ is used for security's sake, which translates into

$$\frac{at}{L^2} < 0.1 \tag{3.79}$$

In the case of unilateral heating of an object, it may be evident to select the thickness D of the object (towards x) as the relevant length scale and to define the non-dimensional Fourier number Fo with this D:

$$\text{Fo} = \frac{at}{D^2} \tag{3.80}$$

Fo is therefore the square of the ratio of two length scales: Fo compares the distance over which the temperature noticeably changes as a result of conduction with the actual dimension of the body in the direction of the heat transport. This number can also be interpreted in another way: Fo is also the ratio of two times, namely the actual process time t in relation to the time needed to noticeably change the temperature of the rear side of the material $(=D^2/a)$.

Note that the mathematics as discussed above relates to a flat geometry and does not apply directly to curved geometries such as a cylinder or a sphere. Nevertheless, this theory can then still be used as long as the penetration depth is so small that the curvature is not felt. For instance, heat penetrating into a sphere with radius R will first only be noticed in a very thin shell of thickness δ, providing that $\delta \ll R$ penetration theory may safely be applied. In this situation, the radius R is the relevant length scale to compare with the penetration theory. Yet, in such a case, Nu is also usually defined in terms of the diameter D.

Example 3.4. A copper slab.
A flat copper slab $(a = 1.17 \times 10^{-4}$ m^2/s) forms the top of a square channel through which water is flowing (see Figure 3.13). The thickness of the slab, D, is 3 mm. The water and the slice are both at a temperature of 20 °C. Suddenly, water at 40 °C flows through the channel. (Note that such an abrupt change is not easy to achieve in practice; also, it is assumed that the resistance to the heating up is located *in* the slab.)

Give an estimate of the time that elapses before the temperature of the top side of the slab noticeably changes.

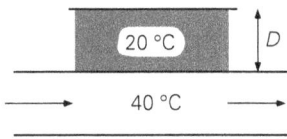

Figure 3.13

Solution
The solution to this transient problem can be found by using the error function. If we could apply a quantitative criterion for the qualitative 'noticeable' change of temperature, we could decide that the temperature had changed noticeably if such a change were 5% of the overall difference in temperature. In this case, it is 1 °C. Here, the relevant length scale is again the slab thickness D. With the help of equation (3.77), it therefore follows that

$$\text{erf}\left(\frac{D}{2\sqrt{at}}\right) = 1 - \frac{T - T_0}{T_1 - T_0} = 0.95 \tag{3.81}$$

Consulting an error function table produces

$$\frac{D}{2\sqrt{at}} = 1.386 \rightarrow t = 0.01s$$

If this time could be estimated by equating the penetration depth to the slab thickness, then the time could be found from $D = \sqrt{\pi at}$ and the result would be 0.025 s. The latter value – of course larger than the result found on the basis of the exact solution with the error function (see Figure 3.12) – illustrates once again the somewhat arbitrary nature of the definition of penetration depth. As a matter of fact, heat conduction always proceeds a bit faster than estimated by means of the concept of penetration depth.

Example 3.5. Bitumen on oak.
A very large flat sheet of oak ($\lambda = 0.19$ W/mK, $\rho = 800$ kg/m^3, $c_p = 2.4 \cdot 10^3$ J/kgK) has a thickness D of 1 cm and a uniform initial temperature $T_w = 20$ °C.
    At $t = 0$, a layer of bitumen ($\lambda = 0.74$ W/mK, $\rho = 1{,}300$ kg/m^3, $c_p = 920$ J/kgK) of the same thickness D is laid onto the wood (see Figure 3.14). This layer has a uniform initial temperature of $T_b = 50$ °C.
    What, after 5 s,
a)   is the penetration depth in both materials?
b)   is the temperature of the interface?
c)   is the heat flux through the interface?
d)   are the values of the temperature gradients on both sides of the interface between the wood and the bitumen?

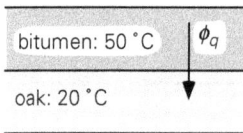

Figure 3.14

Solution
It can be assumed that no heat exchange occurs with the surroundings.
It is clear that this is an unsteady-state problem. In order to be able to answer the questions with the help of the penetration theory, it is first necessary to investigate whether the penetration depths are clearly smaller than the thickness of the wood and the bitumen. This effectively amounts to asking question a.

a) With subscript b for bitumen and w for oak (wood), it follows for both penetration depths that:

$$\sqrt{\pi a_w t} = 1.25 \text{ mm}; \quad \sqrt{\pi a_b t} = 3.1 \text{ mm} \tag{3.82}$$

Given that both penetration depths are clearly less than 0.6 times the thicknesses of the wood and the bitumen layer, the penetration theory may indeed be used for the rest of this problem.

b) To answer questions b and c, it is first a good idea to establish the terms for the heat flux through the interface. Call the temperature of the interface T_i. At a random point of time, the heat flux from the bitumen to the interface will be given by (providing that the penetration theory may be used)

$$\phi_q'' = \frac{\lambda_b}{\sqrt{\pi a_b t}} (T_b - T_i) \tag{3.83}$$

Similarly, the flux from the interface into the oak is shown by

$$\phi_q'' = \frac{\lambda_w}{\sqrt{\pi a_w t}} (T_i - T_w) \tag{3.84}$$

Deducting both flux equations (remember that both fluxes are equal) gives

$$T_i = \frac{\frac{\lambda_b}{\sqrt{a_b}} T_b + \frac{\lambda_w}{\sqrt{a_w}} T_w}{\frac{\lambda_b}{\sqrt{a_b}} + \frac{\lambda_w}{\sqrt{a_w}}} = 311\mathrm{K} = 38 \text{ °C} \tag{3.85}$$

This result illustrates that the interface temperature is not time-dependent and adjusts itself instantaneously at $t = 0$.

c) The fluxes on both sides of the interface are calculated using equations (3.83) and (3.85), and are therefore, where $t = 5$ s:

$$\phi_q'' = 2.7 \text{ kW/m}^2 \tag{3.86}$$

Both fluxes are therefore time-dependent, but always equal to each other.

d) The temperature gradients can be calculated from

$$\left. \frac{\partial T}{\partial x} \right|_{\text{interface}} = -\frac{1}{\lambda}\phi_q'' = -\frac{\Delta T}{\sqrt{\pi a t}} \tag{3.87}$$

After 5 s, equation (3.87) produces, for the bitumen

$$\left. \frac{\partial T}{\partial x} \right|_{\text{interface}} = -\frac{(T_b - T_i)}{\sqrt{\pi a_b t}} = -3.9 \times 10^3 \text{ K/m} \tag{3.88}$$

and for oak

$$\left. \frac{\partial T}{\partial x} \right|_{\text{interface}} = -\frac{(T_i - T_w)}{\sqrt{\pi a_w t}} = -14.4 \times 10^3 \text{ K/m} \tag{3.89}$$

This situation, which has now been calculated in full, is illustrated in Figure 3.15. The typical error function profiles (of Figure 3.12) that meet at the constant interface temperature T_i can clearly be seen on both sides of the interface. Of note also is that the penetration depths and the temperature gradients (both time-dependent) are different on the two sides, which is the result of the different properties of the two materials.

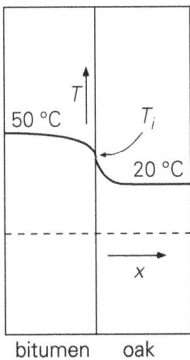

bitumen   oak   **Figure 3.15**

3.3.4 Heat transfer coefficient for short times

The result of the penetration theory – see equation (3.78) – can also be written in the form of Newton's law of cooling:

$$\phi_q'' = h\Delta T = \frac{\lambda}{\sqrt{\pi a t}}\Delta T = \sqrt{\frac{\lambda \rho c_p}{\pi t}}\Delta T \tag{3.90}$$

In other words (note that equation (3.90) includes the flux):

$$h = \sqrt{\frac{\lambda \rho c_p}{\pi t}} \tag{3.91}$$

In this case, therefore, h is **a function of time**! Note that again h is the reciprocal of the resistance to heat transfer already discussed in the context of equation (3.78). The related equation for the Nusselt number is now:

$$\mathrm{Nu} = \frac{hD}{\lambda} = \sqrt{\frac{\lambda \rho c_p}{\pi t}}\frac{D}{\lambda} = \frac{1}{\sqrt{\pi}}\sqrt{\frac{D^2}{at}} \tag{3.92}$$

or

$$\mathrm{Nu} = 0.564\,\mathrm{Fo}^{-1/2} \text{ providing that } \mathrm{Fo} < 0.1 \tag{3.93}$$

Many cases involve a penetration process that in total takes t_e seconds (starting from $t = 0$, the start of the penetration); it is then often a good idea to calculate the mean heat flow for this process. As the driving force ΔT is constant, this amounts to averaging $h(t)$ over interval $(0, t_e)$. In that case, the following applies to the mean transport rate:

$$\overline{\phi}_q = \frac{1}{t_e}\int_0^{t_e} h(t)\,A\Delta T\,dt = \frac{1}{t_e}\int_0^{t_e} h(t)dt\,A\Delta T = \overline{h}\,A\,\Delta T \tag{3.94}$$

With the help of equation (3.91), it then follows for the mean heat transfer coefficient $\overline{h}$ that

$$\overline{h} = \frac{1}{t_e}\int_0^{t_e}\sqrt{\frac{\lambda \rho c_p}{\pi t}}\,dt = 2\sqrt{\frac{\lambda \rho c_p}{\pi t_e}} = 2\,h(te) \tag{3.95}$$

As given in equation (3.95), the time-average h-value for the interval $(0, t_e)$ is equal to twice the h-value at point of time t_e.

Example 3.6. A copper slab II.

How much heat was supplied to the copper slab in Example 3.4 per unit of surface area during the first 4 ms and how large was the temperature rise after these 4 ms?

The physical properties of copper you may need: $\lambda = 403$ W/mK, $\rho = 8,960$ kg/m^3, $c_p = 386$ J/kg K.

Solution

The penetration depth at $t_e = 4$ ms is $\sqrt{\pi a t_e} = 1.21$ mm, so clearly less than $0.6D$. The penetration theory may therefore be used during the entire period of 4 ms. Now that equation (3.87) is available for $\bar{h}$, the question can be easily answered: the total amount of heat (per unit of surface area) Q'' that goes into the slab follows from integrating the heat flux during the 4 ms:

$$Q'' = \int_0^{t_e} \phi''_q \, dt = \bar{h} t_e \, \Delta T = 2h(t_e) \, t_e \Delta T = 5.33 \cdot 10^4 \text{ Jm}^{-2} \tag{3.96}$$

This may seem like a very large amount of heat, but the rise in mean temperature $\Delta(T) = Q''/(\rho c_p D)$ of the copper slab in those 4 ms is just 5.1 °C.

Summary

!

The partial differential equation for $T(x,t)$, which describes the unsteady-state penetration of heat through conduction in a flat slab, has been derived with the help of a micro-balance. Both the theory and a few applications of the penetration theory have been looked at, in relation to short periods of time.

The penetration theory, and its associated profiles and formulas, can be used effectively if:
a) only conduction is involved;
b) if the geometry is flat;
c) the temperature of the body is initially uniform, while at $t = 0$ one side suddenly 'acquires' a different fixed temperature;
d) if the penetration depth is $\sqrt{\pi a t} < 0.6D$ – or Fo $= \left(at/D^2\right) < 0.1$.

In curved geometries (with radius R), penetration theory is applicable as long as $\sqrt{\pi a t} \ll R$.

Under these conditions, the temperature profile is written in terms of the error function. You have to be able to sketch the typical temperature profiles that are associated with this and those that illustrate the course of the heat penetration. For the time-dependent heat flux that penetrates the material, the following applies:

$$\phi''_q = h\Delta T = \frac{\lambda}{\sqrt{\pi a t}}\Delta T = \sqrt{\frac{\lambda \rho c_p}{\pi t}}\Delta T$$

From this, it follows

$$h = h(t) = \frac{\lambda}{\sqrt{\pi a t}} = \sqrt{\frac{\lambda \rho c_p}{\pi t}} \text{ and } \bar{h} = 2h(t_e)$$

For short periods of time, therefore, $\Delta T =$ constant, but $h = h(t)$. The latter can also be expressed in terms of non-dimensional numbers: Nu $= 0.564$ Fo$^{-1/2}$ (providing that Fo < 0.1 for a unilaterally heated or cooled slab).

3.3.5 Long-term heating

When examining the penetration theory, it was stated expressly that the rear side of the slab should not initially be affected by the penetration of the heat. This requirement is reformulated as follows: at unilateral heating, the penetration depth must be clearly less than the thickness D of the slab; in other words, if Fo < 0.1. What course of action should now be taken if this is not (or no longer) the case? The problem can, of course(!), still be described and solved with the help of partial differential equation (3.72), for which three boundary conditions are still required. However, because the boundary condition at the rear side is now different, the mathematics become complex. Fortunately, though, it appears that in a certain category of problems, the situation can be simplified after some time. This is the case with bodies heated from both sides (double sided) or all round (in the case of a sphere or a cylinder). Here, too, the standard example is a flat geometry; see Figure 3.16.

Figure 3.16 shows a slab of finite thickness D. Before $t = 0$, the whole slab was at temperature T_0. At $t = 0$, the temperature of both surfaces is suddenly raised to, and kept at, temperature T_1. Now, though, heat is penetrating the material from both sides. The resistance to the heat transport is located entirely in the slab: heat exists in abundance outside it. Clearly (see Figure 3.16), the penetration theory is applicable for short periods of time, that is for $t < t_1$ with $\sqrt{\pi a t_1} < 0.6R$ as now $R = D/2$ is the relevant length scale, whence $\sqrt{\pi a t_1} < 0.3D$ or Fo < 0.03.

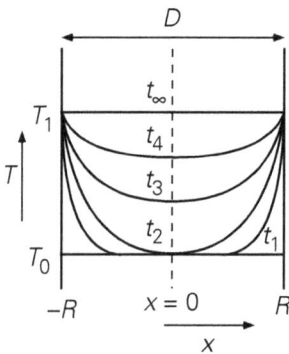

Figure 3.16

However, at a certain moment, $t = t_2$, the temperature will start to rise in the middle of the material as well and the penetration theory will no longer be valid: the temperature T_0 disappears, so that the third boundary condition of equation (3.73) no longer applies either. After some time, for example after $t = t_3$ (i.e. when Fo > 0.03), the situation simplifies itself again: it seems that the successive temperature profiles have become spatially similar. For $t > t_3$, separation of variables leads to a solution of the form

$$\frac{T - T_0}{T_1 - T_0} = 1 - A(t) \cdot f\left(\frac{x}{R}\right) \tag{3.97}$$

The associated boundary conditions are

$$T(t=0) = T_0 \quad \text{for } |x| \leq R$$
$$T(x=R) = T_1 \quad \text{for } t > 0$$
$$\frac{\partial T}{\partial x} = 0 \qquad \text{for } x = 0 \text{ and } t > 0$$

$$(3.98)$$

Substituting equation (3.97) into the partial differential equation (3.72), with regard to the boundary conditions (3.98), leads to an exact solution (first found by **Fourier**) that consists of a series of terms that all have the form of equation (3.97). Function A is always an exponential function of time with a negative argument in which k^2 occurs, while $\cos(k\pi x/D)$ is found for the f function with just the odd k-values. The first term in this series gives a sufficiently accurate result for the double-sided heated slab (provided Fo > 0.03) of Figure 3.16:

$$\frac{T - T_0}{T_1 - T_0} = 1 - \frac{4}{\pi} \exp\left(-\pi^2 \frac{at}{D^2}\right) \cdot \cos\left(\pi \frac{x}{D}\right)$$

$$(3.99)$$

Because the temperature in the slab is now changing **everywhere**, the term **long-term heating** is used.

Now that the temperature profile is known, it is possible to again look at the link between driving force and flux, according to the tested formula of Newton's law of cooling, from which an expression for the heat transfer coefficient h can be expected. Because the temperature is now no longer constant anywhere in the body, it is first necessary to select a temperature in the body on the basis of which the driving force ΔT can be established. In practice, there are two choices that are important here: the temperature (T_c) of the centre of the body and the mean temperature ($\langle T \rangle$) of the body. Both options are discussed here.

In terms of the **mean temperature** $\langle T \rangle$, Newton's law of cooling states that

$$\phi_q'' = h(T_1 - \langle T \rangle)$$

$$(3.100)$$

The mean temperature $\langle T \rangle$ follows by averaging equation (3.99) and is

$$\langle T \rangle = T_1 - \frac{8}{\pi^2}(T_1 - T_0)\exp\left(-\pi^2 \frac{at}{D^2}\right)$$

$$(3.101)$$

We then determine ϕ_q'' at position $x = -R$ with the help of temperature profile (3.99); remember that Fourier's law still applies to conduction, even in unsteady-state conditions:

$$\phi_q'' = -\lambda \frac{\partial T}{\partial x}\bigg|_{x=-R} = \frac{2\lambda}{R}(T_1 - T_0)\exp\left(-\pi^2 \frac{at}{D^2}\right)$$

$$(3.102)$$

Combining equations (3.100)–(3.102) produces (still on the basis of $\langle T \rangle$!):

$$h = \frac{2\lambda/R}{8/\pi^2} = \frac{\pi^2}{4}\frac{\lambda}{R} \tag{3.103}$$

or

$$Nu = \frac{h2R}{\lambda} = \frac{\pi^2}{2} = 4.93 \tag{3.104}$$

All in all, this is a very simple and compact result for the condition Fo > 0.03 for double-sided heating or cooling – due to the fact that the temperature profiles are similar in shape.

Similarly, in the case of long times (Fo > 0.03), the following apply:

$$\text{For a long cylinder: } Nu = 5.8$$
$$\text{For a sphere:} \qquad Nu = 6.6 \tag{3.105}$$

The second way of defining the driving force for the heating up is based on the **temperature of the centre** of the body. The analogy to equation (3.100) is then

$$\phi_q'' = h\,(T_1 - T_c) \tag{3.106}$$

Note: this concerns the same heat flux, but the driving force has been selected differently, which means there is also a different h. By eliminating $T_1 - T_c$ with the help of equation (3.99), and by comparing the resulting expression for ϕ_q'' with equation (3.102), it is now found that $Nu = \pi$. Here, too, h is a constant for the longer times, while the driving force for the heat transport, namely $\Delta T = T_1 - T_c$, is a function of time. However, the value of h when T_c is used differs from the h-value when $\langle T \rangle$ is used.

! **Summary**

When both sides of a slab (thickness D) are heated up (or cooled down), or when a cylindrical or spherical body is heated (or cooled down) all round, the temperature profiles become spatially similar at successive points in time, after some initial time has passed (Fo > 0.03 with still Fo = at/D^2). The solution to $T(x, t)$ for this situation has been presented. As this long-term heating continues, the heat flow through the surface can be described using a constant heat transfer coefficient, while the driving force is actually a function of time. The driving force that is used is either difference between the surface temperature and the mean temperature, or the difference in temperature between the surface and the centre:

The difference between penetration theory and the long-term heating concept for a double-sided or all round heated (or cooled) object can be summarised as follows:

Penetration theory	Fo < 0.03	$h = h(t)$	ΔT = constant
Long-term heating	Fo > 0.03	h = constant	$\Delta T = \Delta T(t)$

For long periods of time, and based on $\langle T \rangle$, the following constant Nusselt numbers should be used:

Sphere	6.6
Cylinder	5.8
Flat slab	4.93

The convention is to take the thickness or diameter of an object as the length scale in both Fo and Nu.

3.3.6 The overall heating up of an object

In the foregoing sections, we have looked at unsteady-state heating through conduction during short and long periods of time separately. For the intermediate range of Fo values – around 0.1 for unilateral, or around 0.03 for double-sided heating or cooling – the differential equation can also be solved, but this will not be dealt with here.

The exact solutions for the total heating process for a number of finite-size objects are shown in graphic form in Figures 3.17 and 3.18 for the driving force based on $\langle T \rangle$ and T_c, respectively. In both graphs, a non-dimensional difference in temperature is shown on the vertical axis, while the Fourier number Fo (the non-dimensional time) is on the horizontal axis. Notice that the gradients of corresponding lines are the same for longer periods of time **and** that all straight lines (for Fo > 0.03) do not go through point (0,1) on extrapolation.

Finally, it should be mentioned that Figures 3.17 and 3.18 are very practical for, say, Fo > 0.03, but that for smaller Fo values, reading the plot may lead to rather inaccurate results and it is better to use penetration theory anyhow.

Example 3.7. A glass sphere.
A solid glass sphere ($a_{glass} = 4.4 \cdot 10^{-7}$ m^2/s) with a diameter of 2 dm has a temperature of 20 °C. Suddenly, the temperature of the surface is increased by 10 °C.
   How long does it take until the mean temperature of the sphere has risen by 9 °C?

Solution
From the values it follows that

$$\frac{T_1 - \langle T \rangle}{T_1 - T_0} = 0.1$$

Taking this value in Figure 3.17 on the curve for the sphere produces Fo = 0.045. This means that it takes 4,090 s ≈ 1.1 h before $\langle T \rangle = T_0 + 9$ °C.

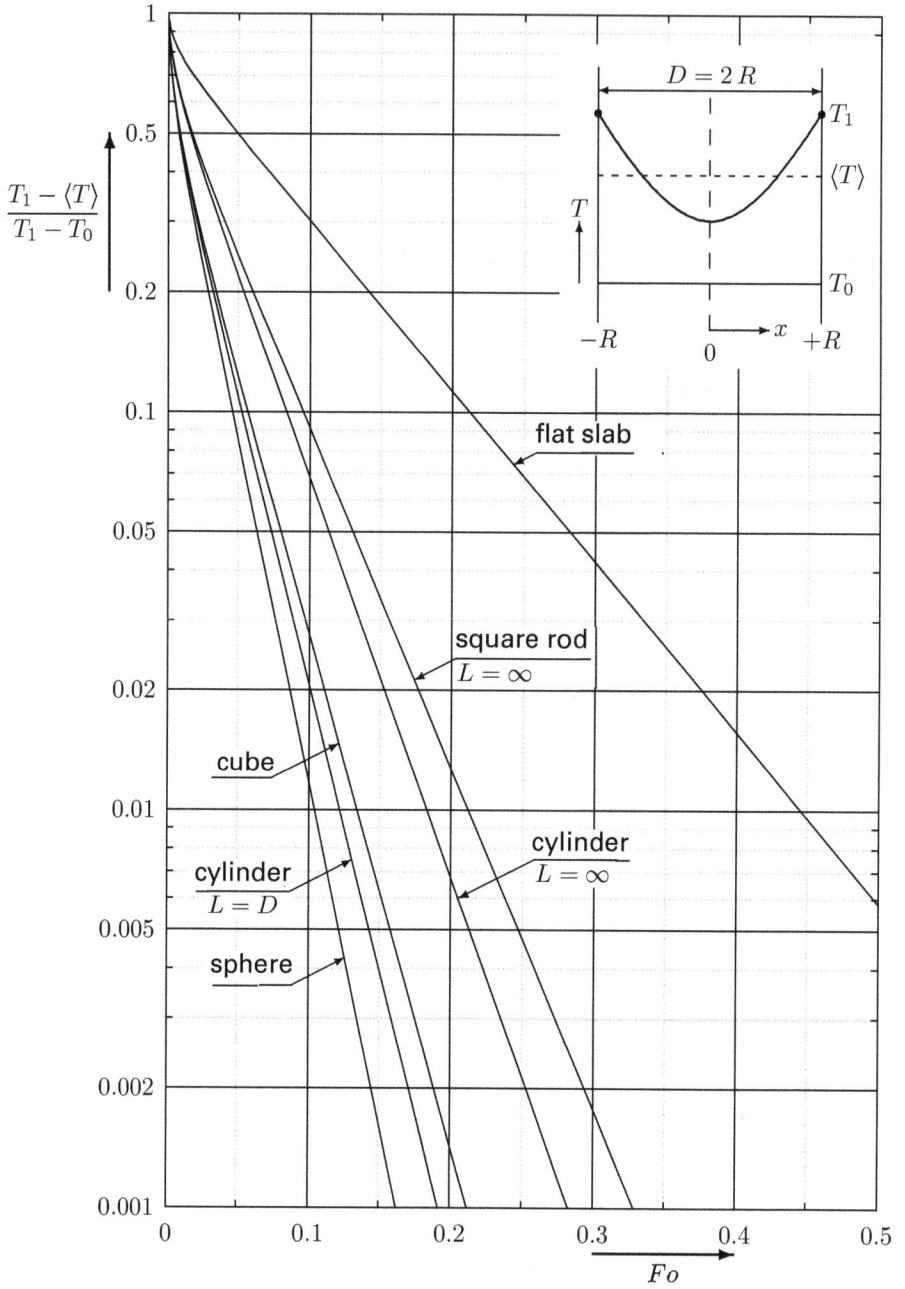

$$\frac{T_1 - \langle T \rangle}{T_1 - T_0}$$

flat slab

square rod
$L = \infty$

cube

cylinder
$L = \infty$

cylinder
$L = D$

sphere

Fo

$D = 2R$

T_1

$\langle T \rangle$

T

T_0

$-R$

0

x $+R$

Figure 3.17

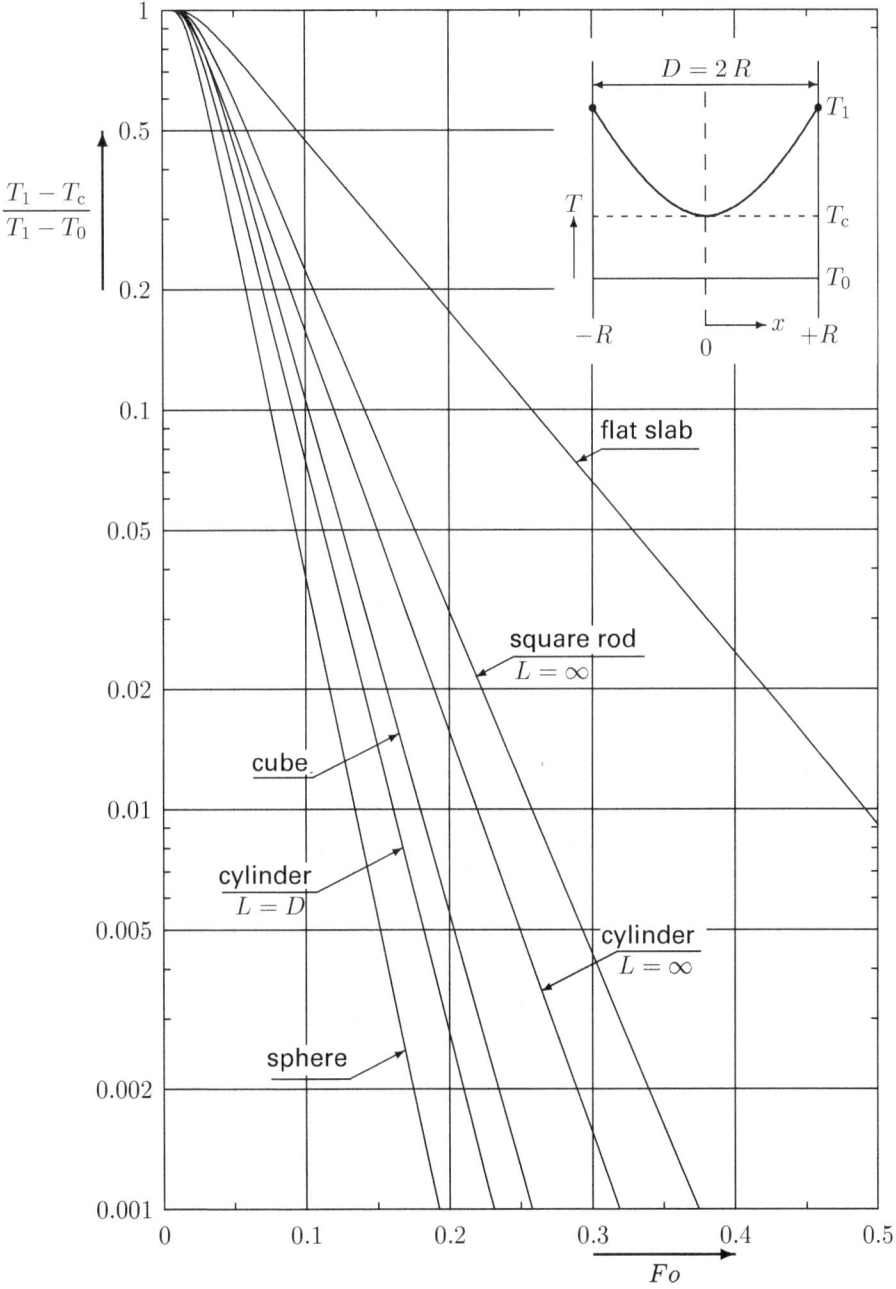

Figure 3.18

Example 3.8. The glass sphere II.
For the same glass sphere of Example 3.7, how long does it take until the temperature at the centre of the sphere has risen by 9 °C and calculate the mean temperature of the sphere at that moment in time.

Solution
First, when the temperature of the centre of the glass sphere has risen by 9 °C, the non-dimensional temperature difference

$$\frac{T_1 - T_c}{T_1 - T_0} = 0.1$$

Then it follows from Figure 3.18 that Fo = 0.075. It therefore takes 6,820 s ≈ 1.89 h before the centre of the sphere has risen by 9 °C. As expected, this is markedly longer than in the $\langle T \rangle$-example: the temperature of the centre is, after all, always the furthest removed from the temperature at the edge, T_1.

It is now easy to calculate what the mean temperature is if $T_c = 9$ °C $+ T_0$. Look now at Fo = 0.075 in Figure 3.17. This gives

$$\frac{T_1 - \langle T \rangle}{T_1 - T_0} = 0.03 \longrightarrow \langle T \rangle = T_1 - 0.03 \cdot (T_1 - T_0) = 29.7 \text{ °C}$$

Now that it has been established that for longer periods of time, the internal heat transfer coefficient h is constant, unsteady-state heating up and cooling down-related problems that only concern the mean temperature of a body can be solved relatively straightforwardly, based on a heat balance for the body in question. Because h is assumed to be constant (and this is incorrect at the start of the heating up or cooling down process), the solution and outcomes are only approximations.

This method will be illustrated for a sphere that is initially at a uniform temperature of T_0. At $t = 0$, the surface of the sphere is changed to temperature T_1. The heat balance for the sphere (based on $\langle T \rangle$) is

$$\frac{d(\rho V c_p \langle T \rangle)}{dt} = A h (T_1 - \langle T \rangle) \tag{3.107}$$

from which for constant physical properties:

$$\frac{d\langle T \rangle}{dt} = \frac{6h}{\rho c_p D} (T_1 - \langle T \rangle) \tag{3.108}$$

The general solution to this differential equation is, **if h is constant**:

$$\langle T \rangle - T_1 = K \exp\left(-\frac{6h}{\rho c_p D} t\right) \tag{3.109}$$

Here, K is an integration constant that has to be determined from the initial condition. This causes a problem, however. At $t = 0$, the temperature in the sphere was T_0, but T_0 cannot be used as a boundary condition, given that when determining solution (3.109) to the differential equation (3.108), it was assumed that h was a constant. This is only the case for the longer periods of time; for shorter periods, $h = h(t)$! This can also be clearly seen in Figures 3.17 and 3.18: for short times, that is for a range of low Fo, the lines are curved. As the vertical axis in the figure is logarithmic, this illustrates that the solution to the un-steady-state heat conduction problem for short periods of time does not indeed correspond to the form of equation (3.109). The line in the figures is straight and the temperature is an e-power only for the longer periods of time, as expressed by equation (3.109).

Nonetheless, boundary condition $t = 0 \rightarrow \langle T \rangle = T_0$ is used. In principle, this is incorrect. Whether the error that this leads to is acceptable is difficult to say, generally speaking – this has to be assessed in each individual case.

Example 3.9. The glass sphere III.
For the glass sphere in Examples 3.7 and 3.8, use differential equation (3.108) and the assumption that h is constant in order to estimate how long it takes before the mean temperature has risen by 9 °C.

Solution
From the exact solution obtained with the help of Figure 3.18, it is already known (see Example 3.7) that $\langle T \rangle = T_0 + 9$ °C after $t = 4{,}090$ s. Solving equation (3.108) with $h = $ constant for every t and with boundary condition $\langle T \rangle = T_0$ produces

$$\frac{T_1 - \langle T \rangle}{T_1 - T_0} = \exp\left(-\frac{6h}{\rho c_p D}t\right) \tag{3.110}$$

With the help of non-dimensional numbers, equation (3.110) is written as

$$\frac{T_1 - \langle T \rangle}{T_1 - T_0} = \exp\left(-\frac{6\,\mathrm{Nu}\,(\lambda/D)}{\rho c_p D}t\right) = \exp\left(-6\,\mathrm{Nu}\,\mathrm{Fo}\right) \tag{3.111}$$

In the case of long-term heating of a sphere, $\mathrm{Nu} = 6.6$. With the help of the values for a, D, and Nu, it now follows that $\langle T \rangle = T_0 + 9$ °C after $t = 5{,}290$ s. This approach results in a clearly longer time than found in Example 3.7 with the help of Figure 3.18.

Finally, we will demonstrate through an example how a long-term heating problem with one insulated wall can quickly be solved with the help of Figures 3.17 and 3.18.

Example 3.10. A drum cooler.
A drum cooler (diameter $D = 1$ m) has a surface temperature of 20 °C and rotates at a rate of 1 rev/min. A bituminous product at a temperature of 80 °C (bitumen just about flows at this temperature) is applied to the drum cooler, before being scraped off again after one quarter of a revolution. It is then required that the mean temperature of the cooled bitumen is no higher than 25 °C. Heat exchanged between the bitumen and the surrounding air is negligible. For the bitumen it is given that $c_p = 920$ J/kg K, $\rho = 10^3$ kg/m³, and $\lambda = 0.17$ W/mK.

The question to be answered is this: how thick may the layer of bitumen be in order for the requirement regarding the mean temperature at the point at which it is scraped off to be met?

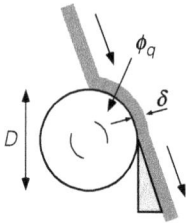

Figure 3.19

Solution
First of all, it should be realised that bitumen at the conditions of interest ('just about flowing') more or less behaves as a solid material moving in plug flow with the drum. When moving with a small volume element in the bitumen layer, one would notice how temperature decreases in time. This also implies that the heat to be removed via the drum can only be transported towards the drum by conduction.

It may further be assumed that the thickness of the layer of bitumen is much less than the radius of the drum cooler and this implies that the curvature of the bitumen layer may be ignored. The problem can therefore be regarded as the cooling off of a flat plate. Figure 3.19 shows the typical dimensions and flows.

Finally, it may safely be assumed that there is no heat exchange between the bitumen and the air – in reality it is not absolutely zero, but very small in comparison with the heat flow on the side of the drum cooler. But then, owing to Fourier's law, the temperature gradient on the surface of the bitumen is also zero (more or less):

$$\phi_q'' = -\lambda \frac{dT}{dx} = 0 \;\rightarrow\; \frac{dT}{dx} = 0 \tag{3.112}$$

If the layer of bitumen is now reflected in this interface and this mirror image is 'attached' to the actual layers, then the situation arises that was dealt with in the case of long-term heating of a thin layer (from both sides). Consequently, Figures 3.17 and 3.18 can be used **providing** that 2δ is used as the thickness of the layer. It was required that the cooled bitumen was no higher than 25 °C. Of course, the highest temperature is found on the airside of the bitumen. This is therefore the centre of the layer including its mirror image. Using Figure 3.18 then produces the following

$$\frac{T_1 - T_c}{T_1 - T_0} = \frac{20 - 25}{20 - 80} = 0.083 \;\rightarrow\; Fo = 0.275 \tag{3.113}$$

or $at/(2\delta)^2 = 0.275$. The time can be inferred from the fact that the bitumen spends one quarter of a revolution on the drum cooler, so $t = 15$ s. From this, it follows that the maximum layer thickness is $\delta_{max} = 1.6$ mm.

! **Summary**
When solving unsteady-state long-term heat conduction problems, it is possible to use two graphs in which the non-dimensional driving force has been plotted versus the Fourier number. These graphs represent the exact solution.

In addition, the technique of a heat balance with a constant heat transfer coefficient can be used for the heating up or cooling down of a body for an approximate solution.

3.3.7 Internal and external heat transport

It was assumed throughout the previous sections that the surface temperature of the object was changed to a fixed level at a given point in time ($t = 0$). In practice, however, it will *not* usually be the surface temperature that is changed to a constant level. In most cases, the object will suddenly enter an environment in which the temperature is different to that of the object. Heat has to be transported from the bulk of the environment to the surface of the object (or indeed vice versa) before the heat can flow into (or out of) the object. This will involve not just an internal heat resistance ($1/h_i$) in the object but also an external heat resistance ($1/h_e$) in the surrounding medium.

Analogous to the topic covered in Section 3.2.3, this produces an overall heat resistance ($1/U$), in accordance with

$$\frac{1}{U} = \frac{1}{h_i} + \frac{1}{h_e} = \frac{1}{Nu_i}\frac{D}{\lambda_i} + \frac{1}{Nu_e}\frac{D}{\lambda_e} \tag{3.114}$$

Equation (3.114) builds upon equation (3.59). An important difference is that now we have to do with a transient situation which should be reflected by the value of Nu_i. With a view to situations (to be dealt with later on) that the medium around the object is not stagnant, also the external resistance has been expressed in terms of an external Nusselt number Nu_e; in the case of a stagnant external medium around a spherical object, usually $Nu_e = 2$ (see Section 3.2.2) is used for the sake of simplicity. Here, too, a clear distinction has to be made between internal and external resistance to heat transport, between internal and external Nusselt numbers, and between the thermal conductivity coefficients of the internal and external media.

Again, in case one of the two partial resistances to heat transport is much smaller than the other, it is not unusual to ignore the smallest resistance of the two (either the internal or the external). Note that in the very beginning of the heating or cooling process the value of the internal heat transfer coefficient is very high for a short while: see equation (3.91). As a matter of fact, the external resistance to heat transport has been ignored in the earlier treatment of transient heat conduction. In Example 3.11, both resistances will be taken into account – resulting in a slower heating of the object than in the earlier examples.

Example 3.11. The glass sphere IV.
The initial temperature of the sphere we 'used' before was 20 °C. Now, the ambient temperature suddenly rises by 10 °C. The heat transfer coefficient for heat transport from the bulk of the surroundings to the sphere is $h_e = 10$ W/m² K.
 How long does it now take for the mean temperature of the sphere to rise by 9 °C?

Solution
Figure 3.17 cannot be used for finding the solution, as there is also an external heat resistance. However, a heat balance can still be drawn up, entirely analogous to that of equation (3.108). Here, boundary condition $\langle T \rangle = T_0$ at $t = 0$ is applied, which means we again act as if the overall heat transfer coefficient U is constant throughout the heating up period. The same function will then be found for the mean temperature as that in (3.110):

$$\frac{T_1 - \langle T \rangle}{T_1 - T_0} = \exp\left(-\frac{6\,U}{\rho c_p D}t\right) \tag{3.115}$$

The only thing to have changed is that h has been replaced by U. Again, however, this is **not** an exact solution (because U is not constant from the beginning). The overall heat transfer coefficient U follows from:

$$\frac{1}{U} = \frac{1}{h_e} + \frac{1}{h_i} = \frac{1}{h_e} + \frac{1}{Nu_i}\frac{D}{\lambda_i} \tag{3.116}$$

Unlike in the previous examples (where just the value of a_{glass} was needed), we now require $\lambda_{glass} = 0.8$ W/mK. This produces $U = 7.3$ W/m^{2}K, which means the time needed is $t = 1.9 \times 10^4$ s ≈ 5.3 h. As expected, this is longer than in the case without external heat resistance.

! **Summary**

Also in unsteady-state heat conduction problems, the internal heat transport resistance should be clearly distinguished from the external one. The overall heat transfer coefficient U follows from

$$\frac{1}{U} = \frac{1}{h_i} + \frac{1}{h_e}$$

This relation expresses that the total resistance is the sum of two partial resistances, the internal and the external. As a result, there also two Nusselt numbers, Nu_i and Nu_e, each defined with the thermal conductivity coefficient of the medium involved. Whenever one of the two partial resistances is (much) smaller than the other one, the smallest may be ignored.

3.3.8 Numerical approach

The unsteady-state penetration theory can also be tackled numerically by using a so-called discretisation scheme. The equation

$$\rho c_p \frac{\partial T}{\partial t} = \lambda \frac{\partial^2 T}{\partial x^2} \tag{3.117}$$

$$\text{with } T(t,0) = T_1, T(0,x) = T_0, T(t, x \rightarrow \infty) = T_0$$

is different in terms of structure to that for steady-state heat conduction. By again looking at the derivation of the equation in Section 3.3.1 with the thin slice dx in Figure 3.11, it is easy to derive the discretized form of the above differential equation here. The semi-infinite slab is divided into strips of finite thickness Δx. Consider a random strip, P, and name its neighbours West (W) and East (E). The heat balance drawn up for strip P for a time step Δt is

$$\rho c_p \frac{T_P(t + \Delta t, x)\,A\,\Delta x - T_P(t,x)\,A\,\Delta x}{\Delta t} =$$

$$= -\lambda \frac{T_P - T_W}{\Delta x} A - \left(-\lambda \frac{T_E - T_P}{\Delta x} A\right) \tag{3.118}$$

The time is now also discretised: the temperature distribution is only calculated on multiples of Δt ($t_i = i \cdot \Delta t$, where $i = 1, 2, 3, \ldots$). It is therefore sufficient to indicate the time using the superscript i. This means the above equation can be written as follows:

$$T_P^{i+1} = T_P^i + \frac{a\,\Delta t}{\Delta x^2}\left(T_W^i + T_E^i - 2T_P^i\right) \tag{3.119}$$

where $a = \dfrac{\lambda}{\rho c_p}$ as before.

In equation (3.119), we can put the terms together, using T_P^i:

$$T_P^{i+1} = \left(1 - \frac{2a\Delta t}{\Delta x^2}\right)T_P^i + \frac{a\,\Delta t}{\Delta x^2}\left(T_W^i + T_E^i\right) \tag{3.120}$$

The superscript refers to the point in time, and the subscript to the position. Solving the above equation is easy: for each new time step, the only thing that is needed is information from the previous time step. However, choices still have to be made for Δx and Δt, and they cannot be made entirely independently of each other. It is easy to understand that for $\frac{a\Delta t}{\Delta x^2} > \frac{1}{2}$, a negative (i.e. unstable) link arises between T_P^{i+1} and T_P^i.

In Figure 3.20, the numerical solution to the penetration theory problem is illustrated for a slab of copper ($a = 1.17 \times 10^{-4}$ m²/s), with thickness $D = 1$ m. At $t = 0$, the temperature on one side is raised by 10 °C. For the numerical simulations, the domain of the calculation must be finite, of course: the other side is subject to the boundary condition that the copper there is insulated. The penetration theory can be used if $\sqrt{\pi a t} < 0.6D$. This corresponds to $t < 980$ s.

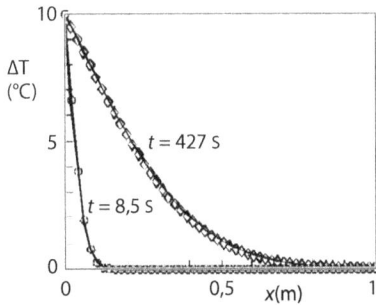

Figure 3.20

Figure 3.20 shows numerical solutions for two moments in time: $t = 8.5$ s and 427 s, respectively (where $\Delta x = 0.02$ m and $\Delta t = 0.85$ s). It also shows the analytical solutions as solid lines for these two moments in time. It is clear that the numerical solutions closely correspond to the analytical ones.

The numerical methods can of course also be used effectively in the case of long-term heating. The temperature distribution of the copper strip for $t = 2,654$ s is shown in Figure 3.21. The numerical solution is denoted by the symbols, while the solid line

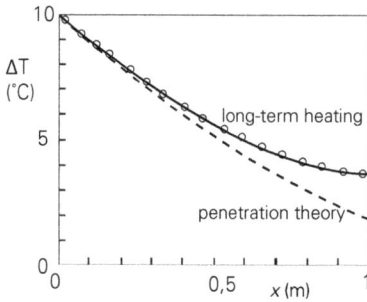

Figure 3.21

is the analytical long-term solution to equation (3.99). There is a good agreement between numerical and analytical solutions. The broken line is the solution according to penetration theory which, it is clear, is no longer valid here.

! **Summary**
The coverage of numerical methods has been extended to time-dependent problems. The discretisation according to time has been explained and illustrated. The choice of the time step is linked to the size of Δx.

3.4 The general micro-balance for heat transport

In general, pure heat conduction will not be the only factor at play. A liquid or a gas can of course also transport heat convectively. Moreover, heat does not always have to be transported in one direction. This section will therefore look at how situations are described in which both convection and conduction play a role. Just as with simple one-dimensional heat conduction cases, the starting point is a micro-balance for an elementary volume somewhere in the domain under consideration. In order for the micro-balance to be valid as generally as possible, as few assumptions as possible will be made. For the sake of simplicity, however, a Cartesian coordinate system will be used.

Consider in the three-dimensional domain the flow field, varying spatially from one place to another and in time, of a fluid the temperature of which also varies spatially and in time. Take any block from somewhere in the flow between x and $x + dx$, y and $y + dy$, and z and $z + dz$ in the fluid. This is the micro cubic control volume element of size $dxdydz$ for which we are going to draw up a heat balance. The situation is illustrated in Figure 3.22.

The various terms of a heat balance of this kind will now be discussed: the heat accumulation term that is always on the left-hand side of a balance, the heat transport terms 'in' and 'out' on the six surfaces of the control volume, and the heat production. A distinction must be made between convection and conduction when filling in for the heat transport terms. Because, in principle, each variable depends on time and position, all derivatives may be partial derivatives, denoted by the partial derivative (curled) symbol ∂.

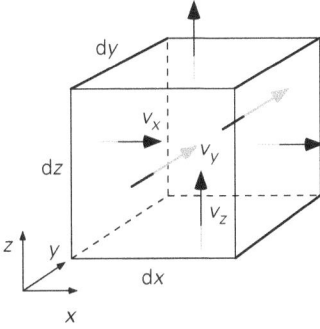

Figure 3.22

Accumulation

The unsteady-state term is

$$\frac{\partial U}{\partial t} = \frac{\partial}{\partial t}(\rho u\, dx\, dy\, dz) = \rho c_p \frac{\partial T}{\partial t} dxdydz \tag{3.121}$$

where the size of the control volume ($=dxdydz$) is constant, constant pressure and constant density have been assumed (see also Bird et al.[12]), and specific heat c_p (loosely replacing c_v) is taken constant as well, just like in equation (3.69).

Convective transport

Flow rate $v_x(x,y,z)\, dydz$ with an internal energy concentration of $\rho c_p T\,(x,y,z)$ in J/m^3 enters a volume through the surface on the left-hand side. Therefore, entering this side convectively is

$$\left[v_x \rho c_p T\right]_{x,y,\,z} dydz \tag{3.122}$$

However, thermal energy flows **out** convectively through the right-hand side of the volume. The magnitude of this flow is, of course:

$$\left[v_x \rho c_p T\right]_{x+dx,y,z} dydz \tag{3.123}$$

Together, these two flows make a net contribution to convective transport in the direction of x (the y and z subscripts can be left out: they have them in common anyway):

$$\left[v_x \rho c_p T\right]_x dydz - \left[v_x \rho c_p T\right]_{x+dx} dydz =$$

$$= -\frac{\partial}{\partial x}\left(v_x \rho c_p T\right) dxdydz \tag{3.124}$$

---

12 Bird R.B., W.E. Stewart and E.N. Lightfoot, *Transport Phenomena*, Wiley, 2nd *Ed.*, 2002, p. 337.

Entirely analogously, the convective flows through the front and rear sides produce, net:

$$\left[v_y \rho c_p T\right]_y dxdz - \left[v_y \rho c_p T\right]_{y+dy} dxdz =$$

$$= -\frac{\partial}{\partial y}\left(v_y \rho c_p T\right) dxdydz \tag{3.125}$$

and through the bottom and top sides:

$$\left[v_z \rho c_p T\right]_z dxdy - \left[v_z \rho c_p T\right]_{z+dz} dxdy =$$

$$= -\frac{\partial}{\partial z}\left(v_z \rho c_p T\right) dxdydz \tag{3.126}$$

Molecular transport

The molecular heat flow (conduction) can be described using Fourier's law. The following comes in through the left-hand side of the control volume, via conduction:

$$-\left[\lambda \frac{\partial T}{\partial x}\right]_{x,y,z} dydz \tag{3.127}$$

The following leaves through the right-hand side, again via conduction

$$-\left[\lambda \frac{\partial T}{\partial x}\right]_{x+dx,y,z} dydz \tag{3.128}$$

Merging these two flows in the direction of x produces

$$-\left[\lambda \frac{\partial T}{\partial x}\right]_{x,y,z} dydz - \left\{-\left[\lambda \frac{\partial T}{\partial x}\right]_{x+dx,z}\right\} dydz =$$

$$= \frac{\partial}{\partial x}\left[\lambda \frac{\partial T}{\partial x}\right] dxdydz \tag{3.129}$$

The front and rear sides contribute to the net flow in the direction of y in the same way:

$$-\left[\lambda \frac{\partial T}{\partial y}\right]_{x,y,z} dxdz - \left\{-\left[\lambda \frac{\partial T}{\partial y}\right]_{x,y+dy,z}\right\} dxdz =$$

$$= \frac{\partial}{\partial y}\left[\lambda \frac{\partial T}{\partial y}\right] dxdydz \tag{3.130}$$

while in the direction of z, the net contribution is equal to

$$-\left[\lambda\frac{\partial T}{\partial z}\right]_{x,y,z}dxdy - \left\{-\left[\lambda\frac{\partial T}{\partial z}\right]_{x,y,z+dz}\right\}dxdy =$$

$$= \frac{\partial}{\partial z}\left[\lambda\frac{\partial T}{\partial z}\right]dxdydz \tag{3.131}$$

Production

Finally, it is still possible for heat to be produced in the control volume. It is usual to indicate the production per unit of volume using the letter q (just as the transformation per unit of volume is usually used with chemical reactions). In the control volume, therefore,

$$q\,dxdydz \tag{3.132}$$

is now produced. In general, heat production is a function of position and time, therefore, of $\{t,x,y,z\}$.

The general micro-balance for heat transport

Combining the accumulation term, all the net flows (six in all), and the production term now gives us the heat balance for the micro control volume

$$\frac{\partial(\rho c_p T)}{\partial t}dxdydz =$$

$$-\frac{\partial}{\partial x}(v_x\rho c_p T)dxdydz - \frac{\partial}{\partial y}(v_y\rho c_p T)dxdydz - \frac{\partial}{\partial z}(v_z\rho c_p T)dxdydz +$$

$$+\frac{\partial}{\partial x}\left[\lambda\frac{\partial T}{\partial x}\right]dxdydz + \frac{\partial}{\partial y}\left[\lambda\frac{\partial T}{\partial y}\right]dxdydz + \frac{\partial}{\partial z}\left[\lambda\frac{\partial T}{\partial z}\right]dxdydz +$$

$$+q\,dxdydz \tag{3.133}$$

The method used here for drawing up the general micro-balance is known as the **cubic volume element method**.

Equation (3.133) can be simplified by dividing by $dxdydz$ and is usually written in a slightly different form, namely by transferring the convective terms to the left-hand side:

$$\frac{\partial(\rho c_p T)}{\partial t} + \frac{\partial}{\partial x}(v_x\rho c_p T) + \frac{\partial}{\partial y}(v_y\rho c_p T) + \frac{\partial}{\partial z}(v_z\rho c_p T) =$$

$$= \frac{\partial}{\partial x}\left[\lambda\frac{\partial T}{\partial x}\right] + \frac{\partial}{\partial y}\left[\lambda\frac{\partial T}{\partial y}\right] + \frac{\partial}{\partial z}\left[\lambda\frac{\partial T}{\partial z}\right] + q \tag{3.134}$$

If the differences in temperature are now not so great, it is possible to work with a constant λ and to allow equation (3.134) to be simplified to

$$\frac{\partial(\rho c_p T)}{\partial t} + \frac{\partial}{\partial x}\left(v_x \rho c_p T\right) + \frac{\partial}{\partial y}\left(v_y \rho c_p T\right) + \frac{\partial}{\partial z}\left(v_z \rho c_p T\right) =$$

$$= \lambda\left[\frac{\partial^2 T}{\partial x^2} + \frac{\partial^2 T}{\partial y^2} + \frac{\partial^2 T}{\partial z^2}\right] + q \qquad (3.135)$$

Equation (3.134) is the general form of heat transport in a fluid if the fluid conforms to Fourier's law. Remember that it is not just the temperature T that is unknown in this equation; the same applies to the entire velocity field $\{v_x, v_y, v_z\}$, and it **too** therefore has to be solved! In Chapter 5, we will derive what the general balance for the velocity (=momentum concentration) field looks like. It will also deal with how the balance equation for the density of the liquid generally looks. All these balances have to be solved at the same time in order that the three-dimensional temperature profile in the fluid can be determined. Numerical solving techniques are available for this purpose. Remember that for solving differential equations, initial and boundary conditions are always required.

In practical situations that many engineers have to deal with, the problems are mostly too complex to be calculated with these general micro-balances. In addition, it is by no means always necessary for the whole temperature profile or the whole velocity field to be solved exactly or numerically: reasonably accurate models and calculations will suffice in many cases. For that reason, a more phenomenological approach will be adopted in the sections that follow, firmly aimed at thinking in terms of driving force (cause) and its associated transport (consequence). Transfer coefficients and non-dimensional numbers play an important role here.

In everyday practice, the use of non-dimensional numbers and correlations between them is still the most suitable method. This does not mean, however, that such correlations are not theoretically substantiated, and it does not mean either that the use of numerical solutions to the general micro-balances will not become increasingly important in industrial practice in the future!

! **Summary**

The generally valid micro-balance for heat transport is derived with the help of the cubic volume element method. In this analysis, account has been taken of the possibility of transporting heat by both convection and conduction.

This micro-balance contains not only the temperature field but also the velocity field and the density field as unknown functions, and therefore cannot be solved without additional equations. However, the balance does contain every case of heat conduction in a fluid that complies with Fourier's law.

3.5 Heat transfer coefficient with convection

3.5.1 General observations

After the introduction of the generally valid micro-balance for heat transport in the previous section, the rest of this chapter will look at more specific heat transfer situations. First, the focus will be directed at the phenomenon of heat transport from a solid wall to a flowing medium (or in the reverse direction). This is a very common situation in the real world.

An example that comes to mind is that of an exothermic reaction in an ideally stirred tank. The heat that is released will, in many cases, have to be removed. This can be done by hanging a cooling spiral in the container, for example, through which cold water flows. In order to properly dimension and operate equipment of this nature, it is necessary to know how much heat the cooling water will remove under the different range of conditions that could occur. The transfer of heat from the contents of the tank to the wall of the spiral, for example, will certainly depend on the flow conditions in the tank. Similarly, the transfer of heat from the wall of the spiral to the cooling water will depend on the flow conditions in the cooling spiral.

It is therefore important to analyse in greater detail what heat transport from a solid wall to a medium flowing along it depends on. To that end, let us reduce this (and similar) situation to the case of a warm wall along which a cold liquid flows by. Two mechanisms are involved, which occur simultaneously as illustrated in Figure 3.23.

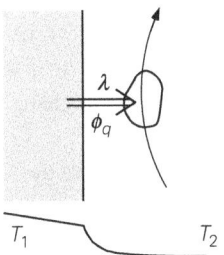

Figure 3.23

Firstly, due to viscosity, a layer of liquid immediately adjacent to the wall sticks more or less motionless to the wall. This means that heat can only be transferred to, or penetrate into, this layer by conduction. Secondly, outside this more or less stagnant layer, or film, convection is capable of transporting heat away, supported by eddies under turbulent-flow conditions. The thickness of this layer, or film, is the result of the dynamic interplay between viscosity and (laminar or turbulent) convection, and therefore may depend on the Reynolds number. There may be another (dynamic) balance between heat penetration through conduction into this film and convective heat transport, resulting in a more or less constant penetration depth (which is not necessarily equally thick as the above hydrodynamic film: to be discussed further on, see Figure 3.24). This is also suggested in Figure 3.23 by the sketched temperature profile typical of penetration theory.

Newton's phenomenological law of cooling is used to describe this complex heat transfer process:

$$\phi_q = h A \, \Delta T \tag{3.136}$$

It is therefore important to find h and/or Nusselt values for the various situations and geometries. It is about an extension of the simpler cases treated in Sections 3.1.2, 3.2.3, and 3.3.7. Because account now has to be taken of the flow of the fluid, the situation becomes much more complex than if only conduction were to play a role. An exact solution using analysis is therefore only (easily) achievable in a small number of cases. However, it is possible to derive approximate solutions via the so-called boundary layer theory, which are also supported by experiments, but for this it is necessary to consider the energy balance and the momentum and mass balances for the flowing liquids. Here, only the phenomenological approach in terms of h and Nu will be pursued.

First, using dimensional analysis, an examination will be made of which nondimensional groups can play a part. It will be clear from the start that the Reynolds number will occur, as well as Nu, given that flow is involved. The analysis starts with an inventory of the variables on which the heat transfer coefficient h may depend. In general, h depends on:

λ	There is conduction from the wall into the fluid (see Figure 3.23).
$D, \rho, \langle v \rangle$	This can be used to determine the mass flow rate of the fluid that flows past a wall or object.
x	Distance x from the start of the tube or from the point at which heat transfer occurs: potentially of importance in the case of flow over a wall or in a pipe if the (flow) conditions in the vicinity of the (pipe) wall are still changing (see e.g. Figure 2.15).
c_p	Co-determines how much heat the fluid can absorb and carry off (or give up).
μ, μ_w	The viscosity of the bulk of the fluid and the viscosity on the wall, respectively: after all, the viscosity is a measure of the mobility of a fluid and co-determines the distance from the wall at which convection starts to predominate over conduction; in addition, the viscosity is needed to describe the flow, such as with Re, but with a (large) difference in temperature between the bulk of the liquid and the liquid at the wall, μ and μ_w can differ somewhat.
ΔT	The driving force behind the heat flow from the wall to the bulk can sometimes actually affect the value of h itself, namely by introducing extra flow through changes in density.
β	The thermal expansion coefficient $\beta = - (1/\rho) \, (d\rho/dT)$: in combination with a difference in temperature ΔT, a difference in density may occur in a fluid, thereby inducing flow, thanks to gravity g (known as free convection).
g	The gravitational acceleration, only relevant in the case of free convection.

In other words:

$$h = f(\lambda, D, \rho, \langle v \rangle, x, c_p, \mu, \mu_w, \Delta T, \beta, g) \tag{3.137}$$

All these variables should be included in a dimensional analysis: 12 variables with 4 basic units. According to the Buckingham Π theorem, therefore, eight non-dimensional groups play a role. Performing the dimensional analysis, perhaps after some rearranging, results in the following groups:

$$Nu = f(Re, Pr, Br, \beta\Delta T, Gz, Gr, Vi) \tag{3.138}$$

where Nu stands for the **Nusselt** number, Re for the **Reynolds** number, Pr for the **Prandtl** number, Br for the **Brinkman** number, Gz for the **Graetz** number, Gr for the **Grashof** number, and Vi for the **'viscosity group'**. The various numbers are given below, with their physical meanings:

$$Nu = \frac{hD}{\lambda} = \frac{\text{actual heat transfer}}{\text{heat transfer caused only by conduction}}$$

$$Re = \frac{\rho \langle v \rangle D}{\mu} = \frac{v \cdot \rho v}{\mu v / D} = \frac{\text{inertia}}{\text{friction}} \quad \text{(see Section 2.2.2)}$$

$$Pr = \frac{c_p \mu}{\lambda} = \frac{v}{a} = \frac{\text{rate of molecular momentum transfer}}{\text{rate of molecular heat transfer}} \quad \text{(see Section 2.2.1)}$$

$$Br = \frac{\mu \langle v \rangle^2}{\lambda \Delta T} = \frac{\mu \frac{\langle v \rangle^2}{D^2} D}{\frac{\lambda}{D} \Delta T} = \frac{\text{heat production through dissipation (friction)}}{\text{heat transport through conduction}}$$

$$Gz = \frac{ax}{\langle v \rangle D^2} = \frac{\text{heat transfer through conduction}}{\text{heat transfer through convection}} \quad \text{(also: a kind of Fourier number)}$$

$$\beta\Delta T = \frac{\Delta \rho}{\rho} = \text{relative density difference as a result of } \Delta T$$

$$Gr = \frac{\beta \Delta T g x^3}{v^2} = \frac{\Delta \rho g x}{\rho \langle v \rangle^2} \left(\frac{\rho \langle v \rangle D}{\mu} \right)^2 \left(\frac{x}{D} \right)^2 = \frac{\text{buoyancy}}{\text{friction}}$$

$$Vi = \frac{\mu}{\mu_w} = \text{measure for viscosity differences as a result of } \Delta T$$

As can be seen, all the numbers stand for ratios of driving forces, transport velocities, or transport quantities. It is also evident that some of the widely used numbers are actually composed of several others; in many cases, this is a question of preferences that have evolved over time. Not all these numbers play a role all the time. The Grashof number, for example (a kind of Reynolds number for free convection) plays no part in conditions that are fully dominated by forced convection. Forced and free convection are discussed separately below, for that reason. The Brinkman number will not feature in the remainder of this book: it is relevant primarily in the case of flows of highly viscous polymers in which a large amount of frictional heat develops.

Finally, one should realize that in the example of Figure 3.23 the focus was on the convective heat transport from the wall surface into the fluid flow. This ignored the heat transport within the wall towards the wall surface. As a matter of fact, the overall heat transfer rate depends on the internal heat resistance inside the wall as well as the external heat resistance on the fluid side. The discussion of Section 3.3.7 and Equation (3.114) fully apply here as well. Whenever one of the two partial heat resistances is (much) smaller than the other one, the smallest may still be ignored.

Summary
Two mechanisms play a role in the case of heat transfer from a solid wall to a flowing medium: conduction from heat at right angles to the wall through the initial layers of the liquid, which are located immediately next to the wall, and convective transport of heat parallel to the wall through the flowing liquid (or gas). Dimensional analysis can be used to determine which non-dimensional numbers can determine heat transfer. Problems of this kind are generally governed by eight numbers, but in practice it is rarely the case that all eight are needed at the same time.

3.5.2 Heat transfer in the case of forced convection

Heat transfer in the case of forced pipe flow
In the case of heat transfer from or to a fluid that is **forced** to flow through a tube, under the influence of a difference in pressure, for example, no more than five of the aforementioned eight numbers play a role, namely.

$$\mathrm{Nu} = f(\mathrm{Re}, \mathrm{Pr}, \mathrm{Gz}, \mathrm{Vi}) \tag{3.139}$$

One role of the Reynolds number is certainly clear here: it very much matters for the heat transfer whether the flow through the tube is laminar or turbulent. With turbulent flow, the eddies also transport the heat across the main direction of flow, while with laminar flow, only conduction is responsible for transport in that direction. These two types of flow therefore have to be treated separately.

Below, Nu relations will be presented for a variety of situations. The first remark to be made is that this selection is not exhaustive: just typical cases have been identified. After all, the bottom line is about how such relations are used in heat balances for solving heat transfer problems. Further, these correlations are often empirically determined – see also the discussion after equation (2.53) – although in a number of cases the exponents of the non-dimensional numbers are also found with the help of analytical derivations or models (which are beyond the scope of this introductory textbook).

Laminar pipe flow

As said, in laminar pipe flows, radial heat transport only takes place through conduction within and between the layers (**laminae**, plural of *lamina*) that are typical of laminar flow; the radial velocity profile then affects the speed at which these **laminae** slide over each other and in this way also affects the conductive radial heat transport. As a result, laminar flows exhibit radial temperature profiles. In many cases, however, one is not really interested in these radial temperature profiles; rather, information about the average temperature as a function of the downstream coordinate x may be sufficient. Conceptually, this average temperature $\langle T(x) \rangle$ could be obtained by collecting at any position x the flow of liquid in a cup and mixing the collected liquid, and may therefore be denoted as the **mixing cup temperature**.

If a fluid flows into a tube at a particular flow rate, then the velocity profile at the tube entrance depends strongly on the situation prior to this – that is, upstream from the entrance. The velocity profile then has to adapt to the circumstances of the tubular flow (as described by the Reynolds number, for example). The boundary condition for the pipe wall is especially important here. This entry section, in which the velocity profile is still adapting, will be disregarded below with the view of the heat transfer.

Starting at a position $x = 0$ downstream of which the velocity profile is no longer changing, follow a fluid element that is flowing close to the wall and which is exchanging heat with it. It is important to distinguish two situations here:

a) initially, the following applies:

$$\text{Nu} = 1.08 \ \text{Gz}^{-1/3} \qquad \text{as long as Gz} < 0.05 \qquad\qquad (3.140)$$

Nusselt is only a function of the Graetz number Gz (which contains x). On closer consideration this is not very surprising, as the Graetz number is actually a Fourier number. It should be remembered that $x/\langle v \rangle$ is an estimate of the average time that a liquid package needs to get from the aforementioned point $x = 0$ to position x in the tube. Replacing $x/\langle v \rangle$ by t in the Graetz number does indeed produce Fo. This means that $x/\langle v \rangle$ is a measure of the time during which such a package is exposed to these new conditions.

In fact, this is a case reminiscent of penetration theory: for short periods of time, the heat penetrates the fluid element from the tube wall and the fluid at the wall therefore increases or decreases in temperature, while the bulk of the liquid notices nothing of this. As a result, the radial temperature profile keeps changing all the time (in the way that it always does in the case of the penetration theory for short periods of time). Note that, as expressed by the negative exponent of Gz, the heat transfer and therefore also Nu decrease with x – quite comparable with equation (3.93).

Equation (3.140) can be defined via integration in relation to x according to the mean Nusselt number for distance L (measured from $x = 0$). The result is

$$\langle \mathrm{Nu} \rangle = 1.62 \left(\frac{aL}{\langle v \rangle D^2} \right)^{-1/3} \text{ providing that } \frac{aL}{\langle v \rangle D^2} < 0.05 \qquad (3.141)$$

This latter equation can also be written as

$$\langle \mathrm{Nu} \rangle = 1.62 \, \mathrm{Re}^{1/3} \, \mathrm{Pr}^{1/3} \left(\frac{L}{D} \right)^{-1/3} \text{ providing that } \frac{aL}{\langle v \rangle D^2} < 0.05 \qquad (3.142)$$

b) At a greater distance from $x = 0$, there will again be a kind of long-term heating where the form of the radial temperature profile no longer changes (as is the case with long-term heating of a solid body – see Figure 3.16), and where h and also Nu become constant as well (meaning they are no longer a function of time $x/\langle v \rangle$). This expectation is correct: the following applies

$$\mathrm{Nu} = 3.66 \ \text{ providing that } \ \mathrm{Gz} > 0.1 \qquad (3.143)$$

Obviously, the Graetz number has assumed the role of the Fourier number in determining the question of when Nu is constant. The domain $x < 0.05 \cdot (\langle v \rangle D^2/a)$ is called the thermal entry section; if $x > 0.1 \cdot (\langle v \rangle D^2/a)$, then it is customary to speak of thermally developed.

Turbulent pipe flow
As already mentioned, in the case of turbulent tubular flow, heat transfer to the bulk of the liquid is much more effective due to the presence of eddies. However, the eddies can never penetrate as far as just next to the wall of the tube: they succumb to the viscosity. This means that, next to the wall, there will always be a thin and largely laminar flowing interface layer without eddies, a 'film'. The thickness of the film depends very much on the turbulence intensity of the fluid. Given that the turbulent eddies outside this film transport momentum much more effectively than the molecules in the film, the velocity gradient is often thought to be concentrated in the film. For the same reason, the resistance to heat transport is often ascribed to such a film.

The temperature profile shown in Figure 3.23 corresponds to this picture: at the wall, the temperature gradient is very steep, and more or less zero in the bulk of the fluid. In fact, the profile looks similar to that of unsteady-state heat conduction (for short periods of time – see Figure 3.12); the convection prevents the penetration depth from being able to increase over time. There is competition between the supply and removal of heat, with the turbulent flow being responsible for the latter. This competition determines the thickness of the film.

Working with an interface or film in which the whole gradient is thought to be concentrated (and outside of which the transport is provided entirely by eddies) is a much-used method and is known as **film theory**. When using the film theory, it is assumed that the profile in the film is always a straight line (even if the whole situation is unsteady: this assumes that the film is always able to adapt to changes very quickly). This image of a film along the wall is used both for momentum management and for heat transport. Strictly speaking, the **hydraulic film thickness** δ_h for momentum transport and **thermal film thickness** δ_q for heat transport do not have to be equal: in practice, the two film thicknesses appear to be in a ratio of $\text{Pr}^{1/3}$, providing that $\text{Pr} \geq 1$. Both film thicknesses are illustrated in Figure 3.24. Of course, both $\delta_q \ll D$ and $\delta_h \ll D$ (different from what is sketched in the figure). The fact that Pr appears here should not be surprising: the thickness δ_h can be expected to depend on υ, while a will determine thickness δ_q to a significant degree.

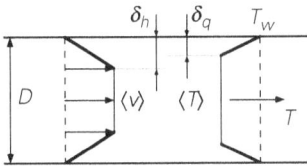

Figure 3.24

This way, and with a view to modelling heat transfer at a wall where a turbulent flow exists parallel to the wall, the entire temperature profile is schematised: the bulk of the liquid is at a uniform temperature (as a result of the activity of the eddies), while the progress of the temperature from the bulk value to the wall value is a straight line. The heat flux can then be modelled as

$$\phi_q'' = \lambda \frac{T_W - \langle T \rangle}{\delta_q} \tag{3.144}$$

The resistance to heat transport is therefore entirely in the film; this is expressed in the statement

$$h = \frac{\lambda}{\delta_q} \tag{3.145}$$

for the heat transfer coefficient. Incidentally, equation (3.144) is very similar to equation (3.78), and equation (3.145) ties in excellently with the h values in Table 3.1. In fact, the problem has now shifted to finding the correct expression for δ_q, which ensures that equation (3.144) does indeed represent the heat flux correctly. Remember that the above is a schematised version of reality: the temperature and velocity profiles are not really flat and neither δ_q nor δ_h is clearly defined; moreover, thicknesses δ_q and δ_h depend on the degree of turbulence in the bulk of the flow (in other words, on the Reynolds number).

   There are several ways of expressing the Nusselt number for heat transfer calculations involving turbulent flow through pipes; one of the most commonly used is

$$\mathrm{Nu} = 0.027\, \mathrm{Re}^{0.8}\, \mathrm{Pr}^{1/3}$$

$$\text{providing that } \mathrm{Re} > 10^4 \text{ and } \mathrm{Pr} \geq 0.7 \tag{3.146}$$

in which the factor $\mathrm{Pr}^{1/3}$ stands for the film thickness ratio δ_h/δ_q. Turbulent tubular flows also involve an entry section L_t, which is shorter than in the case of a laminar flow. As a result, the heat transfer coefficient in this entry section is higher: the value averaged over L_t is found by multiplying with the factor $\{1 + (D/L_t)^{0.7}\}$.

Example 3.12. Heating up water flowing through a tube.
A flow of water of 1 kg/s flows through a tube (length $L = 5$ m, diameter $D = 2$ cm). The liquid enters the tube at a temperature of 20 °C, while the temperature of the tube wall is 40 °C. The situation is steady.
   What is the mean temperature of the liquid as it flows out?

Solution
It is first of all necessary to determine whether the flow is laminar or turbulent, for which the velocity must first be calculated. This follows from $\phi_m = \frac{\pi}{4} D^2 \rho \langle v \rangle$, so that $\langle v \rangle = 3.2$ m/s. This means the Reynolds number is $\mathrm{Re} = 6.4 \times 10^4$, so the flow is clearly turbulent. The question about the average water temperature (as a function of x) is a quite natural one after the above explanation about the film model and the bulk mixing for the case of turbulent flow. The Prandtl number for water (e.g. at 30 °C) is $\mathrm{Pr} = 5.45$. This means that Nu relation (3.146) can be used.

   In order to now determine the temperature at the exit, it is necessary to answer the question of how much heat is absorbed by the water as it flows through the tube. Given that the water heats up, the driving force ΔT is position-dependent, so the temperature profile varying with x has to be calculated. This can be done on the basis of a heat balance for a small slice of the tube between x and $x + dx$ (see Figure 3.25).

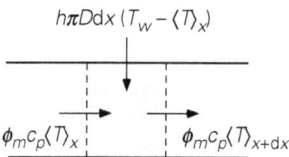

Figure 3.25

The (steady-state) heat balance for this slice is

$$0 = \frac{\pi}{4} D^2 \rho \langle v \rangle c_p \langle T \rangle \Big|_x - \frac{\pi}{4} D^2 \rho \langle v \rangle c_p \langle T \rangle \Big|_{x+dx} + h\,\pi D\, dx (T_w - \langle T \rangle) \tag{3.147}$$

where $\langle v \rangle$ and $\langle T \rangle$ denote the local velocity and the local temperature, respectively, averaged over the cross-section of the tube (so-called mixing cup variables). The associated boundary condition runs as $x = 0 \rightarrow \langle T \rangle = T_0 = 20$ °C. For the temperature profile, this gives

$$\ln \frac{T_w - \langle T \rangle}{T_w - T_0} = -\frac{4h}{\rho \langle v \rangle c_p D} x \tag{3.148}$$

The heat transfer coefficient h can be calculated from equation (3.146):

$$h = \mathrm{Nu} \frac{\lambda}{D} = 0.027 \frac{\lambda}{D} \mathrm{Re}^{0.8} \mathrm{Pr}^{1/3} = 1.0 \times 10^4 \text{ W/m}^2\text{K} \tag{3.149}$$

This means that the mean temperature at the end of the tube is 29.5 °C. In making this calculation, it should be noted that, strictly speaking, h is not constant: because T changes along the tube, Pr is not constant, so h is not either, and in fact this dependence on x should be taken into account in the integration as well. However, this is very much a secondary effect that can generally be ignored (it should be verified numerically). The same remark applies to ρ and c_p which generally are (weak) functions of temperature.

Heat transfer in the case of forced flow around objects

Flat plate
As discussed previously during the introduction of the concept of **friction drag** (see Figure 2.15 in Section 2.3.1), a laminar hydrodynamic boundary layer of increasing thickness $\delta_h(x)$ builds up along a flat plate located in a liquid parallel to the flow in the direction of x. In this interface, the velocity decreases from the undisturbed velocity v just at the edge $y = \delta_h(x)$ of the interface to zero (motionless liquid) just by the wall ($y = 0$). Analogous to the description of turbulent tubular flows, in the case of heat transfer to/from the wall, there is here too a thermal boundary layer of thickness $\delta_q(x)$, which again is not necessarily as thick as the hydrodynamic boundary layer. As a result, the local heat transfer resistance will depend very much on the local thickness $\delta_q(x)$. Calculations using the laminar boundary layer theory (substantiated by experiments) produce the following for the local value of the Nusselt number:

$$\text{Nu}_x = \frac{hx}{\lambda} = 0.332 \left(\frac{\rho v x}{\mu}\right)^{1/2} \left(\frac{\nu}{a}\right)^{1/3} = 0.332\,\text{Re}_x^{1/2}\text{Pr}^{1/3}$$

$$\text{providing that } \text{Re}_x < 3 \times 10^5 \tag{3.150}$$

Notice that both Nu and Re are defined here with the help of distance x from the point under examination, to the start of the slice. The background to this is that there is no other length scale available and that the thickness of the interface δ_q (which can therefore be regarded as a direct measure of the heat transfer resistance) happens to be a function of x. The restriction $\text{Re} < 3 \times 10^5$ is necessary because with greater Re values, the interface will become turbulent.

Sphere
The interface that builds up in the case of a flow around a sphere and the wake that forms behind the sphere were discussed when determining the resistance coefficient for such circumstances. The so-called dead 'water' that is located here makes only a limited contribution to the heat transfer. It is 'refreshed' far less frequently than the fluid that flows alongside the sphere at the front. A commonly used Nu relation for convective heat transfer to spheres is

$$Nu = 2.0 + 0.66 \, Re^{1/2} Pr^{1/3}$$

$$\text{providing that } 10 < Re < 10^4 \text{ and } Pr \geq 0.7 \tag{3.151}$$

In the first term on the right-hand side, it is possible to see the contribution of pure conduction to the heat transfer (see Section 3.2.2). The second term on the right is the increase in the heat transfer as a result of the flow around the sphere (convection).

Cylinder: long and across the direction of flow

A similar thing applies to a long cylinder across the main direction of flow:

$$Nu = 0.57 \, Re^{1/2} Pr^{1/3}$$

$$\text{providing that } 10 < 10^4 \text{ and } Pr \geq 0.7 \tag{3.152}$$

Example 3.13. Cooling of falling grains.
Round copper grains of diameter $D = 3$ mm, $\rho = 9 \times 10^3$ kg/m³, and $c_p = 386$ J/kgK have to be cooled from 80 °C to a mean temperature of 35 °C. This is done by dropping the grains into water, and the temperature of which is 30 °C. The falling velocity v is 0.88 m/s. The resistance to heat transport is entirely in the water. It can be assumed that the water does not heat up to any notable degree. How high must the column be through which the grains fall?

Solution
In order to be able to calculate the rate of cooling, it is necessary to draw up a heat balance for a whole grain. As it is accepted that the heat resistance is entirely in the water, it can be assumed that a grain is practically at uniform temperature T (metal is an excellent conductor of heat). The balance is (T_w is the temperature of the water)

$$\frac{d\left(\frac{\pi}{6} D^3 \rho c_p T\right)}{dt} = -h \, \pi D^2 (T - T_w) \tag{3.153}$$

The boundary condition associated with this is: $t = 0 \rightarrow T = T_0 = 80$ °C. This means that the solution is (providing that h is constant over time)

$$\ln \frac{T - T_w}{T_0 - T_w} = -\frac{6h}{\rho c_p D} t = -\frac{6hx}{\rho c_p D v} \tag{3.154}$$

Here, with the second equality of equation (3.154), the time has been replaced by the route covered divided by the velocity of a grain.

Heat transfer coefficient h now follows from an Nu relation for the flow around an immersed sphere. Based on the substance constants of water, it follows that $Re = 2{,}650$ and $Pr = 5.45$. This means that the use of Nu relation (3.151) is permitted, which produces $Nu = 61.8$ and therefore $h = Nu \, \lambda_w/D = 1.24 \times 10^4$ W/m²K (where $\lambda_w = 0.60$ W/mK). Entering the values into equation (3.154) tells us that the column must be at least 28 cm in length.

Heat transfer with differences in viscosity

In the general discussion on the heat transfer and on the Nu dependency of the other non-dimensional groups in equation (3.138), the **viscosity group** Vi was also mentioned. With this number, it is possible to account for the creation of a temperature gradient perpendicular to the wall as a result of the heat transfer, which in turn creates a viscosity gradient. In the case of turbulent tubular flow and flow around bodies, we therefore include $Vi^{0.14}$ on the right-hand side of the Nu statements. This means that equation (3.146) now becomes

$$Nu = 0.027\, Re^{0.8}\, Pr^{1/3}\, Vi^{0.14} \tag{3.155}$$

This correction by Vi is known as the **Sieder and Tate correction**.

3.5.3 Heat transfer in the case of free convection

In Section 3.5.1, it was derived through dimensional analysis that Nu generally depends on seven other non-dimensional groups, one of which was the Grashof number Gr, which is not needed when describing forced convection. After all, the Grashof number takes into account the fact that a difference in density in the fluid can arise as a result of local heating or cooling at a wall. Under the influence of gravity, this difference in density can lead to flow, known as **free convection**. This free convection is only relevant if a considerable forced flow was not already present as a result of differences in pressure.

One of the best-known examples of free convection is that of rising air along and above heating radiators. The warm radiator gives off its heat to the layer of air adjacent to it (via conduction). This heated air expands slightly, of course, which means it has a lower density than the air in the rest of the room. The warm air will rise and the space and the heat transferring surface area it 'vacates' will be taken over by cooler air, which itself will in turn be heated. The heat transport is of course much more effective through this free convection than when the same amount of heat has to be transported through conduction alone. Notice that Gr can also be defined with the help of the difference in density instead of the difference in temperature; after all,

$$\beta \Delta T = \beta (T_w - T_\infty) = \frac{\rho_\infty - \rho_w}{\langle \rho \rangle} \tag{3.156}$$

In the above statement, $\langle \rho \rangle$ is calculated at $\langle T \rangle = \frac{1}{2}(T_w + T_\infty)$, that is, under average conditions. In that case, Gr is

$$Gr = \frac{\beta \Delta T g x^3}{v^2} = \frac{x^3 g}{v^2}\frac{\Delta \rho}{\langle \rho \rangle} \tag{3.157}$$

The length scale to be used in this Grashof number is here denoted as x, because the flow along or over the heat transferring surface may develop as a result of which the heat transfer coefficient may vary; in addition, the length scale may depend on the orientation of the heat transferring surface and on whether the flow is laminar or turbulent.

There are three situations at which we therefore will look in more detail: free convection along a vertical plate, free convection above a horizontal plate, and free convection between two large horizontal plates.

Vertical plate

Consider a vertical plate with a (fixed, uniform) temperature T_w which is higher than temperature T_∞ of the air far away from the plate. The air a long distance away from the plate (in the bulk) is motionless. The situation is illustrated in Figure 3.26.

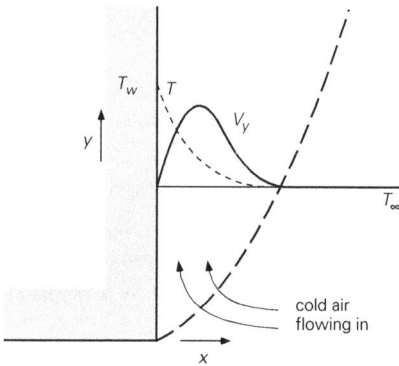

Figure 3.26

At first, the layer of air directly adjacent to the plate will be heated up and will therefore expand. This air will therefore rise and be replaced by colder air. This means that a velocity profile will develop along the plate, in the same way that it does in the case of forced flow along a plate, for example (see Figure 2.15). At the same time, a temperature profile will also be created. Both are illustrated in Figure 3.26. Here, too, the velocity boundary layer and the temperature boundary layer do not have to be of the same thickness (see the discussion in Section 3.5.1 about Figure 3.23 and Section 3.5.2 about Figure 3.24).

Because the velocity profile and the temperature profile develop alongside the slice, the heat transfer coefficient h is a function of y and therefore varies from one place to another. The greater the y coordinate, the thicker the boundary layer becomes and the smaller the local h is there.

For a vertical plate of height L, it appears that the mean (i.e. averaged over height L) heat transport coefficient $\langle h \rangle$ can be written as

$$\frac{\langle h \rangle L}{\lambda} = f\left(\frac{L^3 g}{a\upsilon} \beta |T_w - T_\infty|\right) \tag{3.158}$$

With the help of the non-dimensional groups covered earlier, this can be written as

$$\langle Nu \rangle = f(Gr \cdot Pr) \tag{3.159}$$

In order to be able to specify the form of function f, it is again necessary to make a distinction between **laminar** and **turbulent** conditions. Unlike the situation involving forced convection, the flow in the case of free convection cannot be characterised using a Reynolds number. The reason for this is that with free convection, no characteristic velocity is known in advance. Here, it is the case that the difference between laminar and turbulent can be made on the basis of the product of $Gr \cdot Pr$.

Recent research has produced the following statements for $\langle Nu \rangle$:

Laminar

$$\langle Nu \rangle = 0.52 \, (Gr \cdot Pr)^{1/4} \quad \text{providing that } 10^4 < Gr \cdot Pr < 10^8 \tag{3.160}$$

Turbulent

$$\langle Nu \rangle = 0.12 \, (Gr \cdot Pr)^{1/3} \quad \text{providing that } Gr \cdot Pr > 10^8 \tag{3.161}$$

It should be noted here that under laminar flow conditions, $\langle h \rangle$ is a function of height L as a result of the development of the laminar boundary layer along the plate, while in turbulent flows $\langle h \rangle$ does not depend on height L as L cancels out in equation (3.161). In the latter case, the heat transfer is governed by the turbulent eddies, the magnitude of which has nothing to do with L.

A horizontal tube
Laminar free convection from/around a horizontal cylindrical tube may be described in a good approximation with the help of equation (3.160) in which the tube diameter is now selected as the height L.

One horizontal plate
Free convection can also occur either below or above a horizontal plate as a result of heat transfer (above an electric hob, for example, or a fire). This free convection cannot develop here into a developing boundary layer, but will instead lead to 'bursts' that free themselves from the plate and rise or fall, as the case may be. The resulting turbulent eddies assist considerably in the removal of the heat. It cannot be assumed from this scenario that $\langle h \rangle$ depends on the dimension L of the plate. Consequently, it follows that

$$\langle Nu \rangle = 0.17 \, (Gr \cdot Pr)^{1/3} \quad \text{providing that } Gr \cdot Pr > 10^8 \tag{3.162}$$

Two horizontal plates

Consider now the heat transport through a fluid that is located between two very large horizontal plates, one above the other. The distance between the plates is D. If the temperature of the upper plate is greater than that of the lower one, a stable situation will generally be created (water at approximately $T = 4\,°C$ is an exception to this). No free convection will occur and heat transport will take place as a result of conduction. However, if the upper plate is the colder of the two and the lower plate the warmer, then free convection may occur between them. Here, too, there is a distinction between the various flow regimes based on $Gr \cdot Pr$. The length scale that has to be used in Gr is plate spacing D.

Hardly any free convection

If $Gr \cdot Pr < 1800$, the flow will be so slight that the heat transport is caused almost entirely through conduction:

$$\langle Nu \rangle = 1 \qquad \text{providing that } Gr \cdot Pr < 1{,}800 \qquad (3.163)$$

Laminar

$$\langle Nu \rangle = 0.15\,(Gr \cdot Pr)^{1/4} \quad \text{providing that } 10^4 < Gr \cdot Pr < 10^7 \qquad (3.164)$$

In this case, a remarkable phenomenon occurs: the flow takes place in a pattern of small cells, which together fill up the whole area between the plates. The height of the cells is equal to the distance between the plates, while the horizontal dimensions are approximately the same as this. This pattern is known by the name of **Bénard cells** and is illustrated in Figure 3.27.

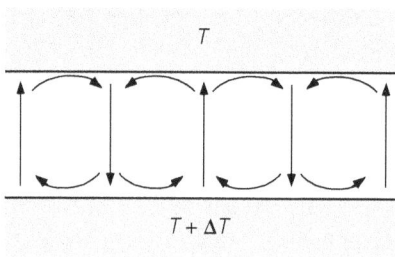

Figure 3.27

Turbulent

$$\langle Nu \rangle = 0.17\,(Gr \cdot Pr)^{1/3} \quad \text{providing that } Gr \cdot Pr < 10^7 \qquad (3.165)$$

The neat, steady-state structure of the Bénard cells has now disappeared of course: the turbulent eddies thoroughly even out the differences in velocity and temperature, but turn the flow structures transient and complicated.

Example 3.14. Heat transport between two horizontal plates.
Consider two large, horizontal plates, one above the other, and separated by distance D, of 4 cm. The space between them is filled with air (pressure is 1 bar). Consider the two following situations:
a)   the upper plate has a temperature of 40 °C; that of the lower is 20 °C;
b)   the lower plate has a temperature of 40 °C; that of the upper is 20 °C.
The question now is to determine the heat flux for both steady-state situations.

Solution to situation a
No free convection is involved here: the warmest layer with the least density is at the top. Therefore, the heat transport is determined by conduction in the air, and Nu = 1. The heat flux now follows from:

$$\phi_q'' = h\Delta T = \frac{Nu \cdot \lambda}{D}\Delta T = \frac{1\times 0.026}{0.04} 20 = 13 \text{ W/m}^2 \tag{3.166}$$

Solution to situation b
Free convection may well be involved here. In order to assess the conditions, it is first necessary to calculate Gr · Pr. The expansion coefficient β can be estimated with the help of the ideal gas law:

$$\beta = \frac{1}{V}\left(\frac{\partial V}{\partial T}\right)_p = \frac{1}{V}\frac{nR}{p} = \frac{1}{T} \tag{3.167}$$

This means that, for Gr · Pr

$$Gr \cdot Pr = \frac{D^3 g \, \Delta T \, \upsilon}{\upsilon^2 \, \langle T\rangle \, a} = 1.2\times 10^5 \tag{3.168}$$

and therefore laminar free convection is involved. For Nu, it now follows that

$$\langle Nu\rangle = 0.15 \, (Gr\cdot Pr)^{1/4} = 2.79 \tag{3.169}$$

which means the heat flux is greater than is the case with *a*, namely:

$$\phi_q'' = h\Delta T = \frac{Nu\cdot \lambda}{D}\Delta T \cdot = 36 \text{ W/m}^2 \tag{3.170}$$

Summary
If flow occurs as a result of differences in density caused by differences in temperature in a fluid, this is known as free convection. This type of flow results in a greater heat transport than in the case of conduction alone. There is a distinction between laminar and turbulent free convection. In the case of free convection, the regimes are classified with the help of Gr · Pr instead of with a Reynolds number. In every regime, the Nusselt number is a simple function of Gr · Pr. This applies to a number of elementary geometries.

3.5.4 The numerical approach of convective heat transport

A numerical approach to heat transport to and from the wall of a tube through which a laminar flow of liquid is passing is more complicated than the earlier examples in Sections 3.1.5 and 3.3.8 that involved only conduction as a transport mechanism. Now, we must also consider convective transport. The heat transport is still described by the heat balance. Because the medium is now flowing, we also need to know about the field of flow. For laminar flow (of a Newtonian liquid) in a straight, cylindrical-shaped tube, we know the exact solution to the velocity field (see Chapter 5). The velocity of a liquid element in the tube is parallel to the tube's axis and depends only on its radial position in the tube:

$$\frac{v_z(r)}{\langle v \rangle} = 2\left(1 - \left[\frac{r}{R}\right]^2\right) \tag{3.171}$$

where $\langle v \rangle$ denotes the mean velocity of the liquid, R is the radius of the tube, r is the radial coordinate, and z is the axial coordinate.

In the numerical analysis of heat transport, we use the symmetry in the problem. The elementary volumes into which we divide up the domain are now ring-shaped, with radii between $\{r, r + \Delta r\}$ and a position between $\{z, z + \Delta z\}$. For a steady-state situation, we can again consider a random annular element and its immediate neighbours inside the flow through the tube; see Figure 3.28.

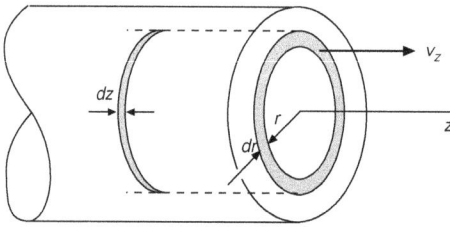

Figure 3.28

Here, the direction of flow of the liquid is 'west-east', while 'south-north' is the direction from the pipe centre-line to the pipe wall. We should now consider that both the west and east faces of cell P have a cell face area of $A_w = A_e = 2\pi\, r_p\, \Delta r$, while the north (outer) cell face area has a magnitude of $A_n = 2\pi\, (r_p + \Delta r/2)\, \Delta z$, and the south (inner) cell face area is $A_s = 2\pi\, (r_p - \Delta r/2)\, \Delta z$.

A steady-state heat balance for 'ring' P is now:

$$0 = A_w \left(v_w T_w - a \frac{dT}{dz}\Big|_w \right) - A_e \left(v_e T_e - a \frac{dT}{dz}\Big|_e \right) +$$

$$+ A_s \left(-a \frac{dT}{dr}\Big|_s \right) - A_n \left(-a \frac{dT}{dr}\Big|_n \right) \tag{3.172}$$

Note that the indices w, e, s, n show that the quantities at the interfaces of 'cell' P must be calculated (in other words, it is not the velocity in cell W but **at** the boundary of cells W and P that matters). This equation can be solved in the same way as with the steady-state heat conduction. Because both convection and conduction are now at play, the accuracy of the 'discretisation scheme' that we use is much more important.

An initial, naïve attempt could perhaps involve discretising the temperature gradient in a normal way; for example

$$\left.\frac{dT}{dz}\right|_w \approx \frac{T_P - T_W}{\Delta z} \tag{3.173}$$

For the convective term, we take the upstream values. That, after all, is what is flowing into the cell, therefore (note the difference in the subscripts between uppercase and lowercase letters):

$$v_w T_w \approx v_W T_W \quad \text{and} \quad v_e T_e \approx v_P T_P \tag{3.174}$$

This method of discretisation is known as 'upwind'. It does indeed seem to be a logical choice, although it is not without its problems. The accuracy of this method of discretisation is not very great: the scheme introduces a degree of numerical diffusion . This means that the diffusion coefficient becomes artificially greater. In this case, this is rather annoying because the problem depends on the relationship between convective and diffusive transport. By using the upwind scheme, we are causing huge disruption to this relationship.

Fortunately, there are discretisation schemes that are far less affected by this problem. To describe these in detail here would be to go beyond the remit of this book. Figure 3.29 shows the progress of the local Nusselt number as a function of the (equally local) Graetz number, as follows from a numerical simulation. It demonstrates that the numerical solution first follows equation (3.140) and then switches smoothly to the constant Nu value of 3.66 as given by equation (3.143).

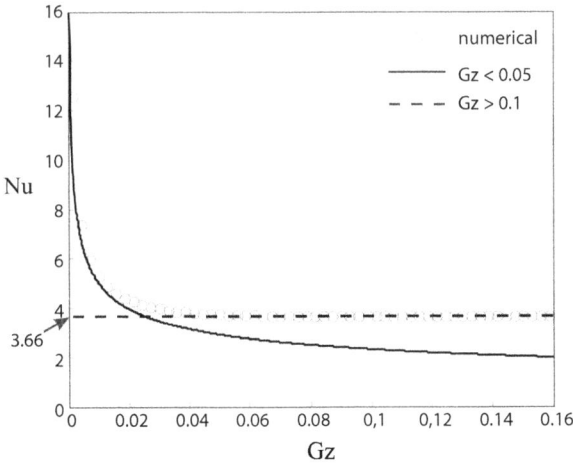

Figure 3.29

! **Summary**

For forced tubular flow, a distinction has to be made between laminar and turbulent flow. In the case of laminar flow, Nu is a function of the Graetz number and therefore of the place in the tube, while in the case of the turbulent flow, Nu is function of the Reynolds number and is constant along the tube. For convective heat transfer to and from bodies, Nu statements have been discussed that are analogous to those for forced turbulent tubular flows. Finally, reference has been made to the Sieder–Tate correction for important gradients in the viscosity.

Assuming that the field of flow is known, the numerical treatment of heat transport is extended to include the convective terms in the heat balance for an elementary volume. At the edges, a choice has to be made during discretisation of the values of the local velocity and the temperature. The upwind scheme has been introduced for that purpose; however, this can cause numerical diffusion.

3.6 Heat exchangers

3.6.1 Introduction

This chapter does not deal with any new correlations for heat transfer coefficients or Nusselt numbers, but it does examine devices that have been expressly designed for effectively realising heat transfer, that is, heat exchangers. Heat exchangers are found in the processing industry in very many shapes and sizes. An exhaustive examination of how they work and how they are constructed goes beyond the remit of a book on transport phenomena such as this one. Instead, we will restrict ourselves to some general comments and a few illustrations regarding how heat exchangers work.

Previous coverage of heat transfer has shown every time that heat flow ϕ_q (the purpose of a heat exchanger) is directly proportional to surface area A that is exchanging heat. For this reason, the heat exchanging surface in most heat exchangers is made as large as possible. It should be borne in mind that this surface becomes smaller in relative terms, the larger the device becomes, because the surface area per unit of volume is proportional to L^2/L^3, that is, to L^{-1}.

It is also important with heat exchangers to let the heat transfer coefficients be as great as possible. High velocities are advantageous for convective heat transfer (but also cost energy!), which is why devices with stirrers and circulation systems with pumps are used extensively. In general, free convection cannot be expected to be very useful. It is also important, generally speaking, to use favourable construction materials with a high heat conduction coefficient for the heat exchanging surface. Contamination of the surface by the liquids involved in the process must be prevented as much as possible, or periodically removed. Sometimes (in the case of non-mixable phases) it is possible to completely avoid having a partition between the process flows.

Any list of examples of heat exchangers should at least include the following (list is not exhaustive):

- 'Shell and tube heat exchangers': two process flows exchange heat, where one flow usually flows through a large number of parallel pipes, while the other passes over the outside of the pipes when going through the heat exchanger. This can be either concurrently or countercurrently. Baffle plates are often used to make sure the fluid on the outside of the pipes flows more or less in a zigzag pattern through the heat exchanger (dead corners should be avoided). The other flow that runs through the pipes is often passed through the heat exchanger in several tube passes.
- Stirred containers in which (reaction) heat is developed that has to be removed are fitted for that purpose either with a water jacket or a spiral through which a coolant is circulated by means of a pump.
- Scraped heat exchangers: in order to frequently remove the layers of crystals that grow on the heat exchanging surface, crystallisers are often fitted with fast-moving rotors with one or more scrapers. This scraping is not only needed for preventing the device from being covered with crystals, but in many cases, crystallisation occurs on an externally cooled wall (because of the removal of crystallisation heat). This wall has to be scraped in order not to undermine the cooling process.
- Stoves: process liquids are often heated up by conducting them through tube bundles (spirals, hairpins) placed in stoves (above the hearth and/or in the chimney). Steam is generally produced in this way as well.
- Chemical reactors are sometimes cooled by allowing one of the components of the reaction mix (often the solvent) to evaporate and then to condense again externally: an effective method of using a phase transition for removing heat – after all, the heat of evaporation is often large (in comparison with an achievable $c_p\Delta T$ effect). Variants of this are the use of boiling liquid in a water jacket and the cooling of a nuclear reactor with water/steam. The temperature in a reactor is always more or less constant in such cases: greater production of heat can be dealt with through greater evaporation and more intensive recirculation.

3.6.2 Heat transfer without phase transition

The example we will take of the use of the calculation of a heat flow from one flowing medium to another is that of an elementary heat exchanger. For the sake of simplicity, this consists of a straight tube through which liquid A flows, which has to be cooled down. This cooling is achieved by fitting a second tube concentrically around the first one, through which a colder liquid, B, flows. As a result of the difference in temperature between the two liquids, a heat flow from liquid A to liquid B will be created.

In principle, there are two versions: the A and B flows run parallel (concurrent flow), or they run in opposite directions (countercurrent flow). We will examine only the **concurrent flow** here. This situation is illustrated in Figure 3.30, and is steady.

An important practical question is this: how large is the heat transport rate ϕ_q through the tube wall from flow A to flow B?

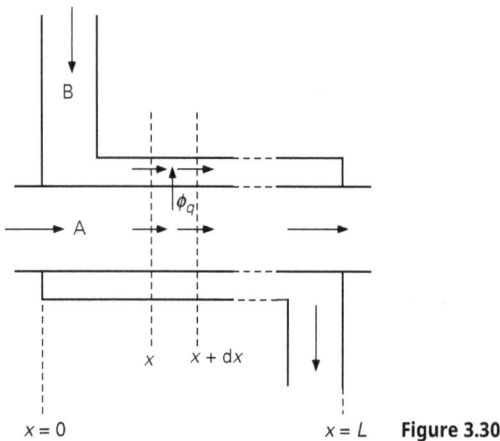

Figure 3.30

The heat transfer between the two liquids depends on the exchanging area A between the two liquids streams and the overall heat transfer coefficient U. The reciprocal of the latter quantity stands for the total resistance to heat transfer which in this case comprises three partial resistances: one in liquid A (denoted by $1/h_i$), one in liquid B (denoted by $1/h_e$), and one of the wall keeping the two liquids apart (denoted by d/λ_w where d stands for the thickness of the tube wall and λ_w for the thermal conductivity coefficient of the wall material):

$$\frac{1}{U} = \frac{1}{h_i} + \frac{d}{\lambda_w} + \frac{1}{h_e} \tag{3.175}$$

In comparison with equations (3.59) and (3.114), $1/U$ now also comprises a term that stands for the heat transport resistance of the tube wall. The two heat transfer coefficients h_i and h_e depend on the type of flow (laminar and turbulent) and should each be calculated from an appropriate Nu–Re correlation. For the sake of simplicity, it is assumed that U is constant (so independent of position x) and also that the tube through which A flows is constant in diameter.

The transfer of heat depends on the difference in temperature of liquids A and B. But because heat is transferred all the time as the liquids flow through the tube (see also Example 3.12), this ΔT is a function of position along the heat exchanger. It is therefore necessary to draw up a micro-balance for a slice of the heat exchanger between x and $x + dx$. This can be done twice – once for the liquid A inside the innermost tube and once for liquid B in the outer tube:

Liquid A:

$$0 = \phi_m^A c_p^A \left(T^A \big|_x - T^A \big|_{x+dx} \right) - U \cdot \pi D \ dx \cdot \left(T^A \big|_x - T^B \big|_x \right) \tag{3.176}$$

Liquid B:

$$0 = \phi_m^B c_p^B \left(T^B \big|_x - T^B \big|_{x+dx} \right) + U \cdot \pi D \ dx \cdot \left(T^A \big|_x - T^B \big|_x \right) \tag{3.177}$$

Both balances can be simplified to

$$\frac{dT^A}{dx} = -\frac{U \pi D}{\phi_m^A c_p^A} \left(T^A - T^B \right) \tag{3.178}$$

$$\frac{dT^B}{dx} = +\frac{U \pi D}{\phi_m^B c_p^B} \left(T^A - T^B \right) \tag{3.179}$$

Equations (3.178) and (3.179) form a coupled system. Deducting equation (3.179) from equation (3.178) produces an equation in $(T^A - T^B)$ which is easy to solve:

$$\frac{d}{dx} \left(T^A - T^B \right) = - U \pi D \left\{ \frac{1}{\phi_m^A c_p^A} + \frac{1}{\phi_m^B c_p^B} \right\} \left(T^A - T^B \right) \tag{3.180}$$

This differential equation, with the addition of the boundary condition,

$$x = 0 \rightarrow \left(T^A - T^B \right) = T_0^A - T_0^B \tag{3.181}$$

has the following solution for the difference in temperature as a function of x:

$$\ln \frac{T^A - T^B}{T_0^A - T_0^B} = - U \pi D \left\{ \frac{1}{\phi_m^A c_p^A} + \frac{1}{\phi_m^B c_p^B} \right\} x \tag{3.182}$$

The solution to $x = L$ is therefore now known as well. However, this does not include heat flow ϕ_q! This can be brought in by eliminating the term between brackets with the help of the two macro-balances. This takes place as follows:

The balance for the entire innermost tube:

$$0 = \phi_m^A c_p^A T_0^A - \phi_m^A c_p^A T_L^A - \phi_q \tag{3.183}$$

The balance for the entire outer tube:

$$0 = \phi_m^B c_p^B T_0^B \,{}' - \phi_m^B c_p^B T_L^B + \phi_q \tag{3.184}$$

Combining the last two equations produces

$$\frac{1}{\phi_m^A c_p^A} + \frac{1}{\phi_m^B c_p^B} = \frac{1}{\phi_q}\{(T_0^A - T_0^B) - (T_L^A - T_L^B)\} \tag{3.185}$$

This means that the heat flow can be expressed in the differences in temperature between the two liquids where the heat exchanger starts and ends:

$$\phi_q = UA\frac{(T_0^A - T_0^B) - (T_L^A - T_L^B)}{\ln(T_0^A - T_0^B) - \ln(T_L^A - T_L^B)} \tag{3.186}$$

The two mass flow rates and the specific heats no longer feature in this statement. Equation (3.186) is usually written in a much more compact form. Two abbreviations are introduced to that end, namely:

$$\Delta T = T^A - T^B \tag{3.187}$$

and

$$(\Delta T)_{\ln} = \frac{\Delta T_L - \Delta T_0}{\ln\frac{\Delta T_L}{\Delta T_0}} \tag{3.188}$$

Definition (3.187) is known as the **logarithmic mean temperature difference**. This means that result (3.185) of the whole calculation can be written in the form of Newton's law of cooling:

$$\phi_q = UA\,(\Delta T)_{\ln} \tag{3.189}$$

If the heat exchanger were to be operated countercurrently, a similar analysis would obtain exactly the same result. However, it should not be concluded from this that it does not matter whether the heat exchanger is operated concurrently or countercurrently. The values of the exit temperatures in the case of countercurrent flows will actually differ from those in the case of concurrent flows. With the former, the change in temperature along the heat exchanger also depends on what is known as the **heat extraction coefficient** $E_q = \phi_m^A c_p^A / \phi_m^B c_p^B$. This is all illustrated in Figure 3.31.

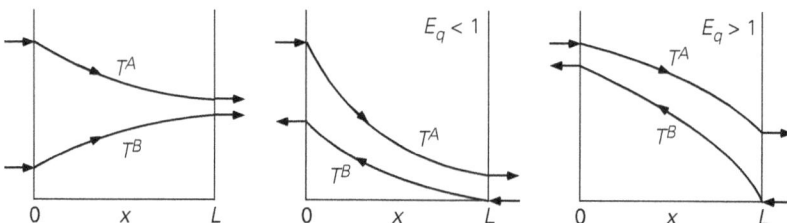

Figure 3.31

Summary

In this section, we have looked at a highly simplified heat exchanger. An important parameter in such a heat exchanger is the overall heat transfer coefficient that comprises three contributions:

$$\frac{1}{U} = \frac{1}{h_i} + \frac{d}{\lambda_w} + \frac{1}{h_e}$$

Apart from the elementary geometry, an important simplification is that this U does not depend on its position in the heat exchanger. This allows the micro-balances of the heat exchanger to be solved. However, the ΔT between the two liquids is a function of place.

With the help of two macro-balances, the heat flow has been brought into the expression for the temperature difference. The result is a relation for the heat flow that is given solely in terms of the liquid temperatures at both ends of the heat exchanger. The two mass flow rates and the specific heats no longer feature in this relation. Using the term 'logarithmic mean temperature difference' enables us to express the heat flow in the form of Newton's law of cooling.

Concurrent and countercurrent are two modes of operation of a heat exchanger. The temperature profiles along a heat exchanger depend on the mode of operation and on the heat extraction coefficient.

3.6.3 Heat transfer with phase transition

It was explained in the introduction to Section 3.6 how the heat effect of a phase transition can be used effectively for removing heat and of course for supplying it. Another factor is that the phase formed during the phase transition is generally of a different density to the original phase. This difference in density can, depending on the geometrical situation, lead to a difference in speed and therefore to more rapid removal of heat. This means that vapour that has formed from liquid in a vertical evaporation pipe is able to rise more quickly in the form of vapour bubbles, and therefore bring about extra mixing and extra heat transfer in the liquid phase. Conversely, in the case of a condenser pipe, condensed liquid can flow away more quickly under the influence of gravity. These are in fact examples of free convection as a result of a difference in density.

Below are two examples: one demonstrates how condensation heat features in a concrete heat transfer problem, and the other deals with how condensation influences the heat transfer coefficient on the steam side.

Example 3.15. A steam condenser.

In a condenser, steam condenses on the outside wall of a bundle of parallel pipes. The total surface area of the outside of the pipes, A, is 5 m². In continuous, steady-state conditions, 1,200 kg of steam condenses every hour (denoted by $\phi_{m,st}$). The pressure in the condenser is 1 bar. A flow rate of $\phi_{m,w}$ of 14,000 kg of cooling water per hour passes through the pipes. The temperature T_i of the cooling water that is added to the condenser is 15 °C. The condensation heat Δh_v of steam is 2,685 kJ/kg. The condenser is well insulated: there is no heat exchanged with the local environment.

What is the overall heat transfer coefficient U between the cooling water and the steam?

Solution
Because of the condensation heat that is released, the temperature of the cooling water moving down-stream rises. As a result, the driving force for heat transport along the pipes is not constant. It is therefore obvious that the logarithmic mean temperature difference $(\Delta T)_{\text{ln}}$ should be used so that the overall heat transfer coefficient U can be calculated from

$$U = \frac{\phi_q}{A(\Delta T)_{\text{ln}}} \tag{3.190}$$

In order to be able to work with this statement, both the temperature T_e of the cooling water at the exit of the pipes and the overall level of transferred heat flow ϕ_q must first be calculated. Both these quantities follow from a heat balance for the whole condenser:

$$\phi_{m,w} c_{p,w} (T_e - T_i) = \phi_q = \phi_{m,st} \Delta h_v \tag{3.191}$$

From this, it first follows that $\phi_q = 3,222$ MJ/h and that $T_e = 69.8$ °C. This makes it possible to calculate that $(\Delta T)_{\text{ln}} = 53.0$ °C and finally that $U = 3,384$ W/m²K.

Example 3.16. Film condensation.
Consider condensation of a pure vapour on a vertical plane of height L and width b (see Figure 3.32). The plane is entirely covered by a film which, under the influence of gravity, neatly flows laminarly downward. The liquid has density ρ, heat of evaporation Δh_v (in J/kg), and heat conduction coefficient λ. For the sake of simplicity, it can be assumed that the wall temperature T_w and temperature T of the condensing vapour are both constant.
  Derive how the mean heat transfer coefficient depends on height L.

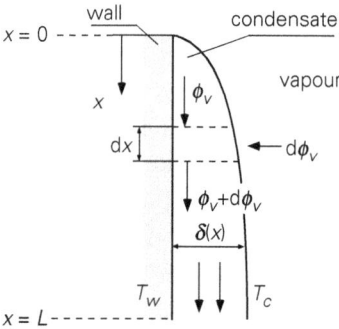

Figure 3.32

Solution
Because condensation occurs over the entire height L, film thickness δ, and liquid flow rate ϕ_v increase in the direction of x. Condensation takes place on the vapour side of the film leading to a film growing in thickness in downstream direction; the removal of the heat of evaporation through the film to the wall depends on the local thickness $\delta(x)$. This heat removal is shown locally by

$$\frac{\lambda}{\delta}(T - T_w)\, b\, dx \tag{3.192}$$

where $\delta = \delta(x)$. For the local condensation velocity (i.e. the local increase in liquid flow rate), the following local heat balance applies:

$$\Delta h_V \, \rho \, d\phi_V = \lambda \, (T - T_w) \frac{b}{\delta} \, dx \tag{3.193}$$

with $\delta = \delta(x)$, still. Further elaboration of equation (3.193) requires familiarity with the link between ϕ_V and δ (which will be dealt with in Chapter 5). Ultimately, the mean heat transfer coefficient $\langle h \rangle$ for height L is found thus:

$$\langle h \rangle = 0.94 \left[\frac{\Delta h_V \, \rho^2 \lambda^3 g}{L \, \mu \, (T - T_w)} \right]^{1/4} \tag{3.194}$$

It can be seen that $\langle h \rangle$ is inversely proportional to $L^{1/4}$. For this reason, condensers are often oriented horizontally; increasing the capacity is easily achieved by enlarging width b.

Summary
Using changes in enthalpy during phase transitions is a widely used and effective means of heat transport. It is sometimes possible to calculate the relevant heat transfer coefficients through modelling.

3.7 Heat transport through radiation

We have previously looked in detail at heat transport maintained by molecules, or by collective behaviour (convection), or by individual behaviour (conduction). However, there is another important form of heat transport: that caused by (electromagnetic) radiation – by photons, in other words. In fact, it is more accurate to refer to energy transport, although in many practical applications the radiation will only manifest itself when it is absorbed and converted into heat.

It should also be mentioned that this type of energy transport occurs independently of and parallel to transport caused by conduction and convection. It can therefore be described as a parallel circuit of resistances to heat transport. Analogous to the electric analogue, this results in the following total resistance:

$$\frac{1}{R_{\text{tot}}} = \frac{1}{R_1} + \frac{1}{R_2} \tag{3.195}$$

In the event that, in addition to thermal radiation, convection plays a role but conduction can be ignored, the following applies:

$$U = h_{\text{radiation}} + h_{\text{convection}} \tag{3.196}$$

The radiant heat transfer coefficient $h_{radiation}$, or simply h_r, i.e. the contribution of radiation to the heat balance of a particular control volume, is the subject of Section 3.7. Describing heat transfer by radiation in precise terms is a complex matter, one reason being the complex way in which the heat radiation that is emitted and absorbed by objects depends on the material properties of those objects, the angle at which the radiation is emitted, and the wavelength of the radiation. Below is an initial simple introduction to heat transfer by radiation, which can be used for approximate technical calculations.

3.7.1 Emission

All objects emit electromagnetic radiation continuously. For a 'completely black' body, or object, it can be derived in theory that the radiation emitted is proportional to the absolute temperature (in K) of (the surface of) the object to the power of4:

$$\phi_{black}'' = \sigma T^4 \qquad (3.197)$$

This is the **Stefan–Boltzmann Law**. The proportionality constant σ is called the **Stefan–Boltzmann constant** and has a value of 5.67×10^{-8} W/m² K⁴. Nusselt's notation[13] is very useful for making calculations:

$$\phi_{black}'' = \sigma' \left(\frac{T}{100} \right)^4 \qquad (3.198)$$

with T still in K and $\sigma' = 5.67$ W/m² K⁴.

[The derivation of this law belongs in the field of statistic mechanics and tells us that σ is exactly equal to $\pi^2 k_B^4 / 60 \hbar^3 c^2$ where k_B is the Boltzmann constant, $\hbar$ the Planck constant, and c the velocity of light. The Stefan–Boltzmann law follows from **Planck's law**, which gives the energy flux that is emitted in the field of wavelength between λ and $\lambda + d\lambda$ by an object of surface temperature T.]

However, there is hardly any object that is completely black. This problem is resolved through the addition of the **emission coefficient** e, which has a value between 0 and 1. Generally speaking, an object emits radiation as follows:

$$\phi_q'' = e \sigma T^4 \qquad (3.199)$$

If $e = 1$, then the object is a completely **black radiator**. In the case of the other extreme, $e = 0$, the object in question emits no radiation. If $0 < e < 1$ and with the emission coefficient being independent of the wavelength of the emitted radiation, then this is referred

---

[13] Nusselt W., *Technische Thermodynamik II*, Ed. Walter de Gruyter, Berlin, 1951.

to as a **grey body** (or object). An example of this is given in Figure 3.33, which shows the spectrum of a completely black object ($e = 1$) and a grey object ($e = 1/3$ in this example).

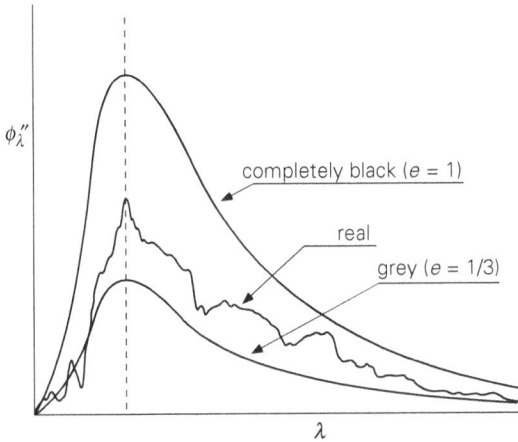

Figure 3.33

For a black or grey object, wavelength λ_{max}, where the maximum occurs in the spectrum, depends solely on the surface temperature of the object:

$$\lambda_{max}T = 2.898 \times 10^{-3}\text{m} \cdot \text{K} \qquad\qquad (3.200)$$

This is called **Wien's displacement law**. The Sun, with a surface temperature T_s of some 5,800 K, emits most of its energy in the visible light spectrum at wavelengths of 400–800 nm, while most of the energy emitted by warm objects on the Earth is in the range of infrared, at wavelengths of 2–10 μm.

The actual spectrum emitted or absorbed by a body is much more complex than Planck's law might lead us to expect: see Figure 3.33 again. In fact, it is necessary to work with a wavelength-dependent emission coefficient e_λ. In the real technological world, however, it is possible to obtain a good first-order approximation of heat radiation by using the grey-body approach with equation (3.199), where $0 < e < 1$. A few typical values of the emission coefficients of some materials for infrared thermal radiation are given in Table 3.2.

3.7.2 Reflection, transmission, absorption

When radiation falls on an object, it may be absorbed, reflected, or transmitted. In principle, each of these processes depends in turn on the wavelength of the incoming

Table 3.2: Typical values of emission coefficients for thermal radiation.

Polished metal and metal foil	0.01–0.13
Clean metal	0.1–0.3
Oxidised metal	0.25–0.7
Ceramic materials	0.4–0.8
Sandy soil	0.7–0.8
Glass	0.8–0.95
Human skin	0.88–0.97
Surface water	0.88–0.97
Paint (white and black!)	0.90–0.95
Stone, concrete, asphalt	0.90–0.95
Snow	0.95

light, but here too, the grey approach is often used, in which none of these processes depends on the wavelength.

The proportion of the incoming radiation that is absorbed is represented by the **absorption coefficient** a. According to Kirchhoff's law (which is not being proved here), for a grey body $a = e$ always holds true. A completely black object absorbs all incoming radiation; for this, $a = e = 1$.

The proportion of the incoming radiation that is transmitted is represented by the **transmission coefficient** t. Radiation that is not absorbed or transmitted by a body is reflected. This proportion is represented by the **reflection coefficient** r. Obviously,

$$a + t + r = 1 \qquad (3.201)$$

An object (or a medium) is completely transparent if $t = 1$, and completely **opaque** if $t = 0$. In the case of a transparent material (such as glass, water, or air), the transmission coefficient depends on the thickness of the material. For a flat layer of thickness L of such a material, Beer's law applies

$$t = \exp(-aL) \qquad (3.202)$$

where a is the **absorptivity** (unit: m^{-1}). If $L \gg 1/a$, then the layer is 'optically thick', and $t \approx 0$. If $L \ll 1/a$, the layer is 'optically thin', and $t \approx 1$. Here, too, there is often a strong dependency on the wavelength.

Window glass, for example, has an absorptivity of $a = 30 m^{-1}$ for visible light and $a \approx 10^6 m^{-1}$ for infrared radiation. This means that window glass 5 mm thick is optically thin for visible light and almost opaque for infrared radiation (verify this!). Something similar applies to the atmosphere: for solar radiation, the atmosphere has a transmission coefficient of t_{atm} that is roughly equal to 1, while for infrared radiation (emitted by the Earth) we can regard the atmosphere as a grey body with an emission coefficient of e_{atm} and an absorption coefficient of a_{atm} that both equal 0.8.

Example 3.17. The temperature of the Earth.

The Earth receives radiation energy from the Sun, part of which is reflected by the atmosphere of the Earth. This reflection coefficient, often referred to as the **albedo** of the Earth, has a value of around $r = 0.3$. The Earth itself also emits radiation. The Earth–Sun system is steady. The maximum in the solar spectrum occurs at $\lambda_{max} = 500$ nm.

Consider the Sun and the Earth as completely black bodies. According to this data, what is the steady-state temperature T of the Earth surface?

It can be assumed that the Earth surface is at a uniform temperature. Convective heat transport may be neglected.

Data: the radius of the Sun $R_s = 6.92 \times 10^8$ m

the radius of the Earth $R = 6.37 \times 10^6$ m

the distance between the Earth and the Sun $s = 1.49 \times 10^{11}$ m

Solution 1

The surface temperature of the Earth under steady-state conditions follows from a heat balance for the Earth that equates the incoming and outgoing heat flows to each other. In order to determine how great the incoming energy flow to the Earth is, it is first necessary to establish how much energy the Sun radiates and then the proportion of that radiation that reaches the Earth. With the help of Wien's displacement law, it follows for the surface temperature T_s of the Sun that

$$T_s = \frac{2.898 \times 10^{-3}}{\lambda_{max}} = 5,800 \ K \qquad (3.203)$$

The Sun therefore emits an energy flow ϕ_s of magnitude

$$\phi_s = 4\pi R_s^2 \sigma T_s^4 = 3.86 \times 10^{26} \ W \qquad (3.204)$$

Of this, only a small fraction reaches the Earth, because solar radiation is emitted in every direction. At a distance s from the Sun, all the emitted radiation is spread over a sphere whose surface measures $4\pi s^2$, of which the Earth takes up a surface area of πR^2. Of the radiation that reaches the Earth's atmosphere, fraction r is reflected by the atmosphere, while the remainder is absorbed by the Earth. The amount of solar radiation absorbed by the surface of the Earth is therefore

$$\phi_{in} = (1-r)\frac{\pi R^2}{4\pi s^2} 4\pi R_s^2 \sigma T_s^4 \qquad (3.205)$$

The amount of radiation emitted by the surface of the Earth is equal to

$$\phi_{out} = 4\pi R^2 \sigma T^4 \qquad (3.206)$$

From this, from the steady-state heat balance for the Earth, $0 = \phi_{in} - \phi_{out}$, it follows that

$$T = \left((1-r)\frac{R_s^2}{4s^2} \right)^{1/4} T_s = 256 \ K = -17 \ °C \qquad (3.207)$$

It should be noted first of all that this expression does not include the radius of the Earth! Also, this outcome is clearly incorrect: in reality, 'the' temperature of the surface of the Earth is around 15 °C! The error we have made is not to have taken account of the so-called greenhouse effect of the atmosphere.

Solution 2

For solar radiation, the atmosphere has a transmission coefficient t_{atm} of around 1, but a fraction a_{atm} of around 0.8 of the infrared radiation emitted by the Earth is absorbed by the atmosphere and then emitted in two directions – to the surface of the Earth and into space.

A steady-state heat balance for the surface of the Earth then gives

$$0 = (1-r)\frac{\pi R^2}{4\pi s^2} 4\pi R_s^2\, \sigma T_s^4 + 4\pi R^2\, \sigma e_{atm} T_{atm}^4 - 4\pi R^2\, \sigma T^4 \tag{3.208}$$

With $e_{atm} = a_{atm}$, this balance can then be rewritten as

$$T^4 - a_{atm} T_{atm}^4 = (1-r)\frac{R_s^2}{4s^2} T_s^4 \tag{3.209}$$

In the meantime, another unknown, namely T_{atm}, has been introduced and this asks for another (balance) equation. A steady-state energy balance for the atmosphere gives

$$0 = a_{atm} 4\pi R^2 \sigma T^4 - 2 \times 4\pi R^2 e_{atm} \sigma T_{atm}^4 \tag{3.210}$$

from which it follows that

$$T_{atm} = \left(\frac{1}{2}\right)^{1/4} T \tag{3.211}$$

Substituting equation (3.211) into equation (3.209) results in

$$T = \left(\frac{(1-r)\frac{R_s^2}{4s^2}}{1 - \frac{1}{2}a_{atm}}\right)^{1/4} \cdot T_s = 290.5 \text{ K} = 17.3 \text{ °C} \tag{3.212}$$

This outcome is much more in keeping with the reality.

! Summary

Heat can be transported by radiation (photons), as well as by molecules (convectively or through conduction). For a completely 'black' body, the flux of the radiation that is emitted can be described using Stefan–Boltzmann's law. It is also the case that a completely black body absorbs all incoming radiation.

In practice, no completely black radiators exist, and emission and absorption coefficients are used. In general, absorption and emission coefficients are numerically equal. As a rule,

$$\phi_{q,rad}'' = e\sigma T^4$$

is used and we refer to a grey body when $0 < e < 1$ and also the emission coefficient does not depend on the wavelength of the emitted radiation.

Whether a medium is opaque or transparent, and whether this medium can be regarded as a grey body, depends on the wavelength-dependent absorptivity of the material.

3.7.3 Heat transfer through radiation

Because every object emits heat radiation continuously, the heat transfer between two objects, 1 and 2, is the net result of the heat radiation $\phi_{1\rightarrow 2}$ from object 1 to object 2, and the heat radiation $\phi_{2\rightarrow 1}$ from object 2 to 1. If both objects are black radiators, then all the radiation from object 1 that falls on object 2 will be absorbed by object 2, and vice versa. This gives

$$\phi_{\text{net},1\to 2} = \phi_{1\to 2} - \phi_{2\to 1} \tag{3.213}$$

Usually, just a fraction of the radiation emitted by object 1 will fall on object 2, and vice versa. These fractions are denoted by **visibility factors** $F_{1\to 2}$ and $F_{2\to 1}$. For black bodies, this then leads to

$$\phi_{\text{net},1\to 2} = F_{1\to 2}A_1\sigma T_1^4 - F_{2\to 1}A_2\sigma T_2^4 \tag{3.214}$$

Visibility factors $F_{1\to 2}$ and $F_{2\to 1}$ are determined solely by the geometry of the two objects and their position in relation to each other. If the two objects have the same temperature, in other words if $T_1 = T_2$, then of course they do not exchange any net radiation, and it follows from equation (3.214) that

$$\frac{F_{1\to 2}}{F_{2\to 1}} = \frac{A_2}{A_1} \tag{3.215}$$

This is known as the **reciprocity relation** for the visibility factors. Although this relation has been derived here for the case of two black radiators at the same temperature, it is generally valid, as visibility factors depend only on the geometry of the objects,

For two parallel disks of equal size with a radius of R and separated by distance b, where $b \ll R$, the visibility factors are $F_{1\to 2} = F_{2\to 1} = 1$. For two concentric spheres or two infinitely long concentric cylinders of radii R_1 and R_2, where $R_1 < R_2$, the visibility factors are $F_{1\to 2} = 1$ (because all the radiation emitted by object 1 reaches object 2) and, due to equation (3.215), $F_{2\to 1} = R_1^2/R_2^2$ (because part of the radiation emitted by object 2 lands elsewhere on object 2).

Example 3.18. Radiant heat losses from an oven.
A spherical-shaped oven (diameter $D = 0.5$ m) with an exterior wall temperature $T_1 = 700$ °C is located in a factory. All the other objects in the factory, as well as the floors, ceilings, and walls, have a temperature $T_a = 20$ °C. The oven and all the other objects, floors, ceilings, and walls may be regarded as black radiators. Calculate the net loss of heat from the oven through radiation.

Solution
By interpreting the factory as a concentric sphere around the oven, with a diameter of $D_2 > D$, then it follows from equation (3.214), with $F_{1\to 2} = 1$ and $F_{2\to 1} \to 0$, that the net heat flux from oven to environment $\phi_{\text{net}} = A_1\sigma(T_1^4 - T_a^4)$. [The same result is also found by treating the remaining objects as separate black radiators.] The numerical answer that was asked for then is 3.3 kW.

For two non-transparent ($t = 0$) grey radiators, it is more difficult to calculate the net heat transport from object 1 to object 2 than is the case with two black radiators, because some of the radiation emitted by object 1 is reflected by object 2 back to object 1. For both objects ($i = 1, 2$), non-transparent and grey, obviously $a_i = e_i = 1 - r_i$. If there is no radiant heat transfer to other objects, it can be derived for two grey radiators with emissivities e_1 and e_2, surfaces A_1 and A_2, and mutual visibility factors $F_{1\to2}$ and $F_{2\to1}$, that:

$$\phi_{net,1\to2} = A_1\, e_{eff}\, \sigma\left(T_1^4 - T_2^4\right) \tag{3.216}$$

where

$$e_{eff} = \left(\frac{1-e_1}{e_1} + \frac{A_1}{A_2}\frac{1-e_2}{e_2} + \frac{1}{F_{1\to2}}\right)^{-1} \tag{3.217}$$

stands for the **effective emission coefficient** or **effective emissivity**.

For engineers, three frequently useful applications of equation (3.217) are:

– Radiant heat transfer between two infinitely large, parallel grey plates with emission coefficients of e_1 and e_2: when $F_{1\to2}=1$ (in other words, all the radiation from plate 1 falls on plate 2) and $A_1/A_2=1$, equation (3.217) results in

$$e_{eff} = \frac{e_1 e_2}{e_1 + e_2 - e_1 e_2} \tag{3.218}$$

– Radiant heat transfer between two concentric, grey spheres (or infinitely long cylinders) with emission coefficients of e_1 and e_2 and with diameters of D_1 and D_2: when $F_{1\to2}=1$ (all the radiation from the inner sphere or cylinder 1 falls on the outer sphere or cylinder 2), equation (3.217) now gives

$$e_{eff} = \left(\frac{1}{e_1} + \frac{D_1^2}{D_2^2}\frac{1-e_2}{e_2}\right)^{-1} \tag{3.219}$$

– Radiant heat transfer between a convex grey object and its grey environment. If we regard a convex grey object 1 as enclosed by a grey environment 2, where $A_2 \gg A_1$ and where $F_{1\to2}=1$ (because all the radiation from the object falls on the environment), then equation (3.219) is simplified to $e_{eff}=e_1$. Where $e_{eff}=e_1=1$ (i.e. for a black radiator in a 'large' environment), equation (3.216) is reduced to the equation used in Example 3.14.

This model is also excellently suited for calculating the net radiation losses from an object to the atmosphere, and the emissivity of the atmosphere does not therefore play a role here.

! Summary
Visibility factors play a major role in the case of heat transfer through radiation between two black or grey objects; they are determined solely by the geometry of the two objects and their position in relation to each other.

Where every grey object emits, reflects, and absorbs, the interest is always in the net heat transport:

$$\phi_{net,1 \to 2} = A_1\, e_{eff}\, \sigma \left(T_1^4 - T_2^4 \right)$$

where the effective emission coefficient, or effective emissivity, stands for

$$e_{eff} = \left(\frac{1-e_1}{e_1} + \frac{A_1}{A_2} \frac{1-e_2}{e_2} + \frac{1}{F_{1 \to 2}} \right)^{-1}$$

and is a function of the emission coefficients of both objects, of their surfaces, and of visibility factor $F_{1 \to 2}$. This relation can be simplified for some typical situations. If a 'small' object is enclosed by a 'large' environment, the emissivity of the environment plays no role.

3.7.4 Radiant heat transfer coefficient

It is often a good idea to write a heat flux through radiation in terms of a heat transfer coefficient and a driving force ΔT. This radiant heat transfer coefficient is denoted by h_r and can be obtained on the basis of equation (3.216). To this end, $T_1^4 - T_2^4$ is rewritten by linearising T^4:

$$T_1^4 = T_2^4 + 4\bar{T}^3 \cdot (T_1 - T_2) \tag{3.220}$$

since $dT^4/dT = 4T^3$ and using $\bar{T} = \frac{1}{2}(T_1 + T_2)$. In an alternative derivation, $T_1^4 - T_2^4$ is decomposed into factors:

$$T_1^4 - T_2^4 = \left(T_1^2 + T_2^2 \right)\left(T_1^2 - T_2^2 \right) =$$
$$= \left\{ (T_1 + T_2)^2 - 2T_1 T_2 \right\}(T_1 + T_2)(T_1 - T_2) \tag{3.221}$$

Using $\bar{T} = \frac{1}{2}(T_1 + T_2)$ and $T_1 T_2 \approx \bar{T}^2$, equation (3.220) is also obtained.

Combining equation (3.220) with equation (3.216) results in an expression similar to Newton's law of cooling (Section 3.2.1):

$$\phi_{net,1 \to 2} = h_r A_1 (T_1 - T_2) \tag{3.222}$$

in which

$$h_r \approx 4 e_{eff} \sigma \bar{T}^3 \tag{3.223}$$

It is clear that the coefficient h_r not only depends on radiation constant σ but also on emission coefficients e_1 and e_2 (and therefore on the materials of which both objects are made or with which they are coated) and on temperatures T_1 and T_2.

Verify that the above calculation in terms of $\bar{T}$ for $T_1 = 300$ K and $T_2 = 400$ K leads to an error of just 2%, and that for $T_1 = 300$ K and $T_2 = 1{,}000$ K, the error amounts to around 20%. Also verify that, for average temperatures in the vicinity of room temperature, h_r is of the order of 1–10 W/m²K, and therefore cannot be neglected in comparison with free convection, for example. With higher temperatures, h_r increases considerably, and therefore so does the contribution of radiation to the overall heat transport.

Example 3.19. A thermometer in a cold room.
In the middle of a room in which the temperature of the air is T_a hangs a thermometer. If the walls have uniform, lower temperature T_w, what temperature T_1 does the thermometer show?
   Carry out the calculation for $\bar{T} = 300$ K and $T_1 - T_w = 10$ K, on the basis of an emission coefficient of $e = 0.05$ for the thermometer. Consider the situation as steady.

Solution
The idea is that the thermometer shows the air temperature T_a. The problem here is that the thermometer 'sees' the colder walls. This means that radiation will cause heat to 'flow away' from the thermometer to the wall and the temperature T_1 of the thermometer will be lower than T_a. As a result of the difference in temperature ($T_a - T_1$), heat will then be transferred from the air to the thermometer. This transport will be maintained by free convection (why?). In a steady-state situation, the heat discharge through radiation will form an equilibrium with the heat supplied through free convection:

$$e\sigma(T_1^4 - T_w^4) = h(T_a - T_1) \tag{3.224}$$

from which it follows, approximately, that

$$T_a - T_1 = \frac{4e\sigma}{h}\bar{T}^3(T_1 - T_w) \tag{3.225}$$

If a normal value of 6 W/m² K is taken for heat transfer coefficient h for free convection, it follows that $T_a - T_1 \approx 0.5$ K.

Example 3.20. Solar radiation onto a flat roof.
The Sun shines onto a horizontal flat roof, and there is no wind. Heat flux ϕ_s'' through the solar radiation to the roof is 300 W/m². All the radiation is absorbed by the roof, while heat loss on the underside of the roof is negligible due to good insulation there. Calculate, for a steady-state situation, the surface temperature T_w of the roof if the ambient temperature T_a is 20 °C.

Solution
As a result of the heat radiation from the Sun, the temperature of the roof rises. When there is no wind, two heat transport mechanisms are possible, apart from heat conduction: free convection is caused because the roof heats up the air immediately above it, and the consequent lower air density induces free convection. The warm roof will of course start to radiate, but the heat radiation depends on the still unknown roof temperature, but can be calculated with the help of equation (3.216) in which $e_{eff} = e_1 = a_1 = 1$:

$$\phi_{net,1\to2}'' = \sigma(T_w^4 - T_a^4) \tag{3.226}$$

The heat transfer coefficient h for the heat discharge through free convection follows from equation (3.162). If, for the sake of simplicity, air can be regarded as an ideal gas, so that

$$\frac{\Delta \rho}{\rho_w} = \frac{T_w - T_a}{T_a} \tag{3.227}$$

(where ρ_w represents the density of air at $T = T_w$), then it follows that

$$\phi_{\text{free convection}}'' = 0.17 \,\lambda \left(\frac{g}{\upsilon a T_a}\right)^{1/3} (T_w - T_a)^{4/3} \tag{3.228}$$

In steady-state conditions, the fluxes (3.216) and (3.228) then reach equilibrium with the solar influx ϕ_s'':

$$\phi_s'' = \sigma\left(T_w^4 - T_a^4\right) + 0.17 \,\lambda \left(\frac{g}{\upsilon a T_a}\right)^{1/3} (T_w - T_a)^{4/3} \tag{3.229}$$

From this, T_w should be calculated iteratively or estimated with the help of an equation like (3.223), for example. The roof temperature that was asked for is around 44 °C. Neglecting the contribution of radiation to the heat discharge would have led to a roof temperature of around 63 °C. The contribution by radiation to free convection would have been of the same order of magnitude.

Summary
It is possible to use a heat transfer coefficient in the case of heat transfer through radiation. In general, this h_r depends on emission coefficients, temperatures, and surfaces and the radiating bodies:

$$h_r \approx 4\, e_{\text{eff}} \,\sigma \bar{T}^3$$

where e_{eff} represents the effective emission coefficient or effective emissivity.

Heat transfer through radiation and heat transfer through free convection frequently occur simultaneously, and with ambient temperatures they are often of the same order of magnitude. As temperatures rise, the importance of radiation increases.

4 Mass transport

4.1 Analogy between mass transport and heat transport

It emerged in Chapter 1 that **convective transport** of heat and convective transport of mass occur and can be described analogously. And indeed, the expressions for **convective transport** of heat and species look the same:

$$\phi_{qc} = \phi_v \cdot \rho\, c_p T \tag{4.1}$$

$$\phi_{mc} = \phi_v \cdot c \tag{4.2}$$

Both equations take the form of volumetric flow rate times the relevant concentration. The fluxes are given by the product of velocity and the relevant concentration.

Chapter 2 looked in detail at **molecular transport**. During their temperature movements, molecules transport their own mass and their own critical energy. In a phenomenological approach, the resulting diffusive net fluxes of mass and heat are described in the same manner: the mass and heat fluxes are proportional to the gradients in the mass or species concentration and thermal energy concentration respectively. The proportionality constant (or respectively, the diffusion coefficient ID and the thermal diffusivity coefficient a) both have the same dimension, that is m^2/s. Both coefficients are physical properties that reflect the mobility of the individual molecules and their capacity to pass on thermal energy. The phenomenological expressions for the fluxes (i.e. the flow rates per unit of cross-sectional area) are as follows:

$$\phi_q'' = -\lambda \frac{dT}{dx} = -a \frac{d(\rho c_p T)}{dx} \quad \text{Fourier's law} \tag{4.3}$$

$$\phi_m'' = -\text{ID} \frac{dc}{dx} \quad \text{Fick's law} \tag{4.4}$$

As noted in Chapter 2, the second version of Fourier's law is only correct if the product of $\rho \cdot c_p$ is constant; however, writing it in this way shows more clearly the similarity with Fick's law for species transport and explicitly links the transport of 'something' to the concentration gradient in that same 'something'.

With Fick's law, it should be restated here that it is of only limited validity: 'Fick' describes mass transport resulting from differences in concentration (and disregards other driving forces for mass transport), it applies to diluted systems, is limited to binary systems and gives a moderate or poor description of the diffusion of complex (i.e.

https://doi.org/10.1515/9783111246574-004

non-spherical and polar) molecules.[14] The analogy with heat transport therefore does not hold under every condition: this should be carefully borne in mind at all times!

The species concentration and the species flow in Fick's law can be expressed in two different ways: on the basis of mass (i.e. in kg/m³ and kg/m² s) and in terms of moles (i.e. in kmol/m³ and kmol/m² s). The latter is often the more obvious one to use in the case of chemical reactions. With Fick's law, the diffusion coefficient is defined in such a way that the net molecular transport of two species A and B through a surface in space is zero. A distinction should be made here according to whether 'kg' or 'kmol' is the basis:

a) **The net mass flow (in kg) is zero**:
In this **barycentric** approach, the following applies in every plane in the domain

$$\phi''_{m,A,x} + \phi''_{m,B,x} = 0 \tag{4.5}$$

where – with mass concentrations c_i in kg/m³ – the following applies

$$\phi''_{m,A,x} = -ID_{AB}\frac{dc_A}{dx} \tag{4.6}$$

and

$$\phi''_{m,B,x} = -ID_{BA}\frac{dc_B}{dx} \tag{4.7}$$

while

$$c_A + c_B = c = \text{constant} \tag{4.8}$$

so that

$$ID_{AB} = ID_{BA} \tag{4.9}$$

The definitions have been set in such a way that the diffusion coefficients of A in B and of B in A have the same numerical value. These definitions are generally used with diffusion in solids and liquids because when concentrations of A in B, for example, are not too high, the density can be assumed to be constant.

b) **The net mole flow is zero**:
The same equations as above apply here as well, but now with the concentrations c_i in kmol/m³ and with the fluxes ϕ''_{mol} in kmol/m²s. It is also true here that the diffusion coefficients of A in B and of B in A have the same numerical value. This choice is very

---

[14] It falls outside the scope of this introduction on transport phenomena to deal with the generally applicable Stefan-Maxwell equations for mass transport. Readers are advised to consult e.g. R. Taylor and R. Krishna, 'Multicomponent mass transfer, Wiley, 1993 or J.A. Wesselingh and R. Krishna, The Stefan-Maxwell approach to mass transfer, Chem. Eng. Sci. 52 (1997) 861–911.

suitable for describing diffusion in gases, as equation (4.8) certainly applies to ideal gases when pressure and temperature are constant.

In the rest of this chapter, we will work both in terms of kg and in terms of 'moles', depending on the issue of interest. In the case of species balances with chemical reactions, working in terms of moles is the preferred approach.

Thanks to the above analogies between heat and species transport, it will be possible to convert much of what was discussed and derived in Chapter 3 from heat transport to species transport terms without much difficulty. The technique of drawing up balances, which usually involves the terms that feature in equations (4.1)–(4.4), and deriving and solving differential equations from them with due regard to the boundary conditions, works in exactly the same way with questions relating to mass transport and heat transport.

However, there are two important differences:

a) Depending on the boundary conditions, the diffusion of mass of species A leaves species B various options for movement or transport which, in turn, influence the transport from species A.

b) If two media are in thermal equilibrium, then the temperature of the two is identical; however, if two different media (or phases) that contain dissolved A are in equilibrium with each other, then in general the concentrations of A in both media (or phases) will be **different**!

Before dealing with these exceptional situations in depth in Sections 4.3 and 4.6, we will first look in Section 4.2 in more detail at the analogies in the ways of describing molecular transport. In Sections 4.2 and 4.3, processes with a diffusive character will be examined, while convective mass transport in different conditions will be covered in Section 4.5, based on the analogy with convective heat transport. Mass transport in the most general sense will feature in Section 4.4. Finally, Section 4.6 will discuss mass transport from one phase to another phase, while Section 4.7 will deal with the simultaneous occurrence of heat and mass transfer.

! **Summary**

This section has referred to how mass transport is, in principle, entirely analogous to heat transport. It has pointed out limits of the validity of Fick's law in particular, and therefore of the analogy. In addition, two types of situation have been highlighted that merit separate attention, as the analogy does not hold for them. One can work in terms of kg as well of 'moles'.

4.2 Mutual diffusion based on the analogy with heat transport

This section deals with binary systems (i.e. systems with just two species, such as substance (or species) A in a pure solvent B) that involve only pure diffusion and in which no convective transport occurs. If the two types of molecule, A and B, exchange places and net, then Fick's law is applied. This means there is **mutual diffusion**. In this case, diffusion is entirely analogous to conduction. We will again deal with the three basic geometries.

4.2.1 Steady-state diffusion in Cartesian coordinates

Consider two parallel plates (distance between the two is D) between which two gaseous (or two liquid) substances A and B are located (see Figure 4.1). Suppose that there is a difference in concentration Δc_A between the plates. Macroscopically, the situation is un-changeable over time (steady state). The overall concentration c is constant (i.e. not a function of x): $c = c_A + c_B =$ constant. This means therefore that a difference in concentration for substance B of $\Delta c_B = -\Delta c_A$ exists between the slices. The result of this is that there are two equally sized but opposing fluxes, ϕ_A'' and ϕ_B''. This is easy to understand: draw up a mass balance for substance A for a thin slice between x and $x + dx$ (see Figure 4.1):

$$0 = -ID\frac{dc_A}{dx}\bigg|_x - \left(-ID\frac{dc_A}{dx}\bigg|_{x+dx}\right) \tag{4.10}$$

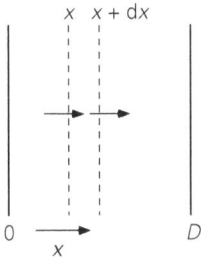

x $x + dx$

0 D
  x

Figure 4.1

Solving equation (4.10) with boundary conditions: $x = 0 \rightarrow c_A = c_{A,0}$ and $x = D \rightarrow c_A = c_{A,D}$ produces

$$c_A(x) = c_{A,0} + (c_{A,D} - c_{A,0})\frac{x}{D} \tag{4.11}$$

Equation (4.11) is entirely equivalent to equation (3.5) for heat transport. Now, too, the link between the driving force Δc_A and the mass flux can be determined:

$$\Delta c_A = \phi_{m,A}''\frac{D}{ID} \tag{4.12}$$

This is again completely analogous to Ohm's law: the driving force Δc_A and the mass flux are linearly linked by a resistance. Compare equation (4.12) with equation (3.17).

Precisely this logic can be used for the transport of substance B. The result is

$$\Delta c_B = \phi''_{m,B} \frac{D}{ID} \tag{4.13}$$

Because $\Delta c_A = -\Delta c_B$, it follows directly from equations (4.12) and (4.13) that both mass fluxes are indeed the same size but in opposing directions. The resistance to transport is of the same magnitude for both substances.

4.2.2 Steady-state diffusion in cylindrical coordinates

The mutual diffusion of two substances A and B between two concentric cylinders in a steady-state situation is shown in Figure 4.2. The overall concentration $c = c_A + c_B$ is again constant, that is, it is not a function of place. The concentrations on both cylinder surfaces have a fixed value: $r = R_1 \rightarrow c_A = c_{A1}$, $c_B = c_{B1}$ and $r = R_2 \rightarrow c_A = c_{A2}$, $c_B = c_{B2}$. Completely analogous to the cylinder geometry in the case of heat transport, it is possible to determine the concentration profile of, say, substance A from a mass balance for substance A for a volume between the cylinders of between r and $r + dr$ (and the same length L as both cylinders). This balance is as follows (remember that there is only transport from A as a result of mutual diffusion):

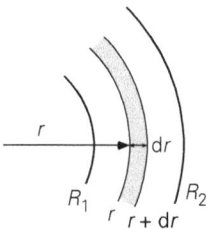

Figure 4.2

$$0 = 2\pi L \cdot \left(-ID\, r \frac{dc_A}{dr}\right)\Big|_r - 2\pi L \cdot \left(-ID\, r \frac{dc_A}{dr}\right)\Big|_{r+dr} \tag{4.14}$$

After simplification and for constant ID, equation (4.14) produces a second-order differential equation:

$$\frac{d}{dr}\left(r \frac{dc}{dr}\right) = 0 \tag{4.15}$$

Equation (4.15) is entirely analogous to equation (3.36); integrating once shows that $r \cdot dc/dr$ is constant, after which the solution proceeds further, as presented after equation (3.26). For the link between the driving force Δc_A and mass flow $\phi_{m,A}$, the following expression results:

$$\Delta c_A = \frac{\ln(R_2/R_1)}{2\pi \, ID \, L} \phi_{m,A} \tag{4.16}$$

This is entirely analogous to equation (3.30). Here, too, the distance $R_2 - R_1$ (to which the difference in concentration relates) does not appear as such, although it implicitly does in $\ln(R_2/R_1) = \ln R_2 - \ln R_1$! This is therefore a result of the cylinder geometry.

Example 4.1. Transport of oxygen in human tissue.
In a study on oxygen transport in human tissue, Nobel Prize laureate August Krogh (Denmark, 1874–1949) considered a cylindrical vein surrounded by an annular ring of tissue. The metabolism through which in the tissue oxygen is converted into carbon dioxide, was modelled by a zeroth-order reaction with reaction rate k_r (in mol/m^{3}s). Transport in the tissue takes place through diffusion with diffusion coefficient ID.

On the inner side of the tissue ring, adjacent to the vein, at $r = R_1$, an oxygen concentration $c = c_0$ was assumed, while on the outer side, at $r = R_2$ (with $R_2 > R_1$), no oxygen transport was supposed, that is $dc/dr = 0$. In Krogh's study, a steady-state case was considered.

Derive, on the basis of a proper mass balance, expressions for the radial concentration profile in the tissue as well as for the transport rate per unit of length at which oxygen is transferred from the vein into the tissue.

Solution
Actually, the mass balance (in terms of mol/s) looked after is an extension of equation (4.14): because of the chemical reaction, a consumption term should be added to the balance equation, namely $- k_r \cdot 2\pi r L dr$. Dividing all terms of the equation through $2\pi r \cdot L \cdot ID \, dr$ results in the second-order differential equation

$$\frac{d}{dr}\left(r \frac{dc}{dr}\right) = \frac{k_r}{ID} r \tag{4.17}$$

Integrating twice gives

$$c = A \, \ln r + \frac{k_r}{4\,ID} r^2 + B \tag{4.18}$$

With the help of the two boundary conditions provided, the radial concentration profile in the tissue becomes

$$(c - c_0)\frac{2\,ID}{k_r R_2^2} = -\ln\left(\frac{r}{R_1}\right) + \frac{r^2 - R_1^2}{2R_2^2} \tag{4.19}$$

and the oxygen supply rate (per unit of length of the vein) from vein into tissue

$$\frac{\phi_m}{L} = -ID \left.\frac{dc}{dr}\right|_{r=R_1} 2\pi R_1 = \pi k_r \left(R_2^2 - R_1^2\right) \tag{4.20}$$

The latter result could also have been found from a steady-state overall mass balance for an annular ring of tissue: supply rate = consumption rate in the metabolism.

4.2.3 Steady-state diffusion in spherical coordinates

Consider finally two concentric spheres with radii of R_1 and R_2 ($R_2 > R_1$). In the space between them are two substances, A and B, which are mutually diffusing. The situation is steady. The fixed boundary conditions

$$r = R_1 \rightarrow c_A = c_{A1}, c_B = c_{B1} \text{ and } r = R_2 \rightarrow c_A = c_{A2}, c_B = c_{B2}$$

apply at the edges. The link between the driving force and mass flow can of course be found by solving the concentration profile. The result of this calculation is

$$\Delta c_A = c_{A1} - c_{A2} = \frac{1}{4\pi ID} \left(\frac{1}{R_1} - \frac{1}{R_2} \right) \phi_{m,A} \tag{4.21}$$

For $R_2 \rightarrow \infty$, this results in an expression that is again entirely analogous to equation (3.45) for heat conduction from the surface of a sphere into an infinite motionless medium.

The above observations on mutual diffusion in a binary mixture also apply if diffusion occurs by just one substance through a motionless medium (which is actually mutual diffusion of that one substance and 'empty places'). This occurs in the case of diffusion of a gas through a solid, for example.

Example 4.2. Gas diffusion through a solid wall.
Pure helium (gas) is contained in a glass sphere (internal diameter $D_1 = 20$ cm, external diameter $D_2 = 21$ cm). The sphere is surrounded by air, in which the concentration of helium is negligible. Helium can diffuse through the glass, while the air is not able to do so (helium molecules are much smaller than 'air' molecules). This, then, is a case of diffusion by one substance (helium) through a motionless medium (glass).

The diffusion coefficient ID of helium through glass amounts to 0.2×10^{-11} m²/s. The solubility c^* of helium in glass (in kg/m³) is just 0.84% of the density (also in kg/m³) of the helium gas the glass is (long-term) exposed to. The temperature of the sphere and air is 20 °C.

If the pressure in the sphere is 2 bar, how great is the flow of mass through the glass?

Solution
Given that the diffusion coefficient is very small, the flow and therefore the decrease in helium mass in the sphere will be very small. This means that the transport of the helium through the glass can be assumed to be quasi-steady: the change in the boundary condition on the inside of the glass is so slight that the effect of the change on the transport can be disregarded.

The first step, then, is to draw up a mass balance for a concentric shell in the glass between r and $r + dr$, in the same way as in Section 3.1.4 for heat conduction in a spherical geometry. The solution to the resulting differential equation – which has the same form as equation (3.42) – produces the profile of the concentration of the helium in the glass. From this it follows, via equation (4.21), for the mass flow of helium, that:

$$\phi_{m,He} = \frac{4\pi\, ID}{\frac{1}{R_1} - \frac{1}{R_2}} (c^* - 0) \tag{4.22}$$

where c^* represents the helium concentration in the glass at the inside wall of the sphere, and the helium concentration in the glass on the outer wall is taken to be equal to 0. Equation (4.22), after all, is the result of a mass balance for a shell *inside* the glass and so therefore the helium concentrations at the inside and outside edges *inside* the glass must also be filled in.

Of course, even at the inside edge, the helium concentration in the glass is – and this includes equilibrium situations – much smaller than the helium concentration in the gas phase inside the glass sphere. It is given that the **solubility** c^* of helium in glass depends on the helium density (or concentration) c in the adjacent gas; here, it is sufficiently accurate to assume that helium behaves like an ideal gas:

$$c_{He} = \frac{mp}{RT} = 0.33 \, \text{kg/m}^3 \tag{4.23}$$

where m stands for the molar mass of helium. Filling in the various numerical values in equation (4.22) produces: $\phi_{m,He} = 1.5 \times 10^{-13}$ kg/s.

Example 4.3. A spherical fungus flake.
An aerated bioreactor is used to grow fungus flakes which for convenience sake may be conceived as spheres. Fore its growth, such a fungus needs oxygen which is consumed according to a zeroth-order reaction. As this growth takes place uniformly distributed across the flake, oxygen has to diffuse into the flake. The need of oxygen therefore increases as the flake grows. As soon as oxygen can no longer penetrate to the core of the flake, anaerobic (oxygen-less) reactions start producing toxic substances. The growth process has therefore to be interrupted before the core of the flake becomes poor in oxygen.

Consider now a quasi-steady state of a flake with radius R while ignoring the growth for the time being. Derive from an oxygen balance for such a flake the differential equation for the oxygen concentration c. Solve this differential equation, specify the boundary condition(s) used, find the concentration profile in the flake, and calculate the maximum flake diameter when using the following data:

Diffusion coefficient of oxygen in a flake ID = 5×10^{-9} m²/s; reaction constant for oxygen consumption $k_r = 3.6 \times 10^{-4}$ mol/m³s; and oxygen concentration c_0 at the flake surface inside the flake $c^* = 0.3$ mol/m³.

Solution
Again, this problem has to be solved by drawing up a mass balance for oxygen over a spherical shell inside a flake between r and $r + dr$, in the same way as in the preceding Example. The difference is that the balance now also comprises a negative production term owing to the zeroth-order chemical reaction. The balance for the steady-state condition suggested runs as

$$0 = \left(-4\pi r^2 \, \text{ID}\frac{dc}{dr}\right)\bigg|_r - \left(-4\pi r^2 \, \text{ID}\frac{dc}{dr}\right)\bigg|_{r+dr} - k_r \, 4\pi r^2 dr \tag{4.24}$$

Combining the two diffusion terms on the right-hand side and dividing all terms by $4\pi \, dr$ give

$$0 = \text{ID}\frac{d}{dr}\left(r^2 \frac{dc}{dr}\right) - k_r \, r^2 \tag{4.25}$$

Integrating once results in

$$r^2 \frac{dc}{dr} = \frac{k_r}{\text{ID}}\frac{1}{3}r^3 + a \tag{4.26}$$

The boundary condition $r = 0 \rightarrow dc/dr = 0$ makes $a = 0$ and therefore

$$\frac{dc}{dr} = \frac{k_r}{3\,\text{ID}}\, r \tag{4.27}$$

Integrating once more with the second boundary condition $r = R \rightarrow c = c^*$ leads to the concentration profile

$$c = c^* - \frac{k_r}{6\,\text{ID}}\left(R^2 - r^2\right) \tag{4.28}$$

The condition that $c = 0$ at $r = 0$ has to be avoided results in the conclusion that the maximum flake radius R_{max} is given by

$$R_{max}^2 = \frac{6\,\text{ID}\, c^*}{k_r} \tag{4.29}$$

Substituting the numerical data provided leads to a maximum flake diameter D_{max} of 1 cm.

Summary
The section above has looked at steady-state mutual diffusion in binary systems, analogous to heat transport. The link between driving force Δc_A and mass flow $\phi_{m,A}$ can be written analogously to Ohm's law. Driving force and mass flow are directly proportional; the proportionality constant can be taken as resistance, in this case to mass transport. Expressions have been found for flat and non-flat geometries – with the help of micro-balances for thin slices, rings and shells – which are entirely similar to those for heat transport.

4.2.4 Mass transfer coefficient and Sherwood number

It was stated in the previous section that the mass flow and the driving force are proportionate to each other. This finding can be used for linking driving force and mass flow to each other according to the recipe of Newton's law of cooling. For mass transport, the following applies:

$$\phi_{m,A} = k\,A\,\Delta c_A \tag{4.30}$$

where the constant k is called the **mass transfer coefficient** (SI unit: m/s). Entirely analogous with heat transport, $1/k$ is now interpreted as the resistance to mass transport. The results from the previous sections on steady-state diffusion in different geometries can now be shown in terms of a k value for every situation:

Mass transfer coefficient	Between two flat plates	For annular space between two cylinders (related to outer surface)	For sphere in an infinite medium
k	$\dfrac{\text{ID}}{D}$	$\dfrac{2\text{ID}}{D_2 \ln D_2/D_1}$	$\dfrac{2\,\text{ID}}{D}$

These relations too can now be put into a non-dimensional form. The non-dimensional number that goes with mass transfer is the **Sherwood number**, abbreviated to Sh and defined as

$$\text{Sh} = \frac{\text{resistance to mass transfer due to diffusion}}{\text{actual resistance to mass transfer}} = \frac{D/MD}{1/k} = \frac{kD}{\text{ID}} \qquad (4.31)$$

For mutual diffusion from a spherical surface in an infinite medium, Sh = 2, just as Nu = 2 applies to the cooling down of a sphere through conduction in an infinite medium.

The term **overall mass transfer coefficient** cannot yet be dealt with analogously to the term 'overall heat transfer coefficient'. Because this involves diffusion across the interface of two different media in which even in a state of equilibrium the concentrations will generally not be equal, discussion of the topic will have to wait until Section 4.6.

Summary

There is also a kind of 'cooling law' in the event that mass transport is determined by driving force (a difference in concentration in this case), the surface through which the mass flow goes, and the mass transfer coefficient k:

$$\phi_{m,A} = k A \Delta c_A$$

This mass transfer coefficient k is a key variable in a new non-dimensional number: the Sherwood number Sh = k D/ID. Specific expressions for the Sherwood number apply to the various basic geometries in the way that they do for the Nusselt number in relation to conductive heat transport.

4.2.5 Unsteady-state diffusion: penetration theory

As with unsteady-state heat conduction, we will only deal with flat basic geometry below. Consider a medium (see Figure 4.3) that is bordered on one side, at $x = 0$, by a flat wall, while on the other it stretches out into 'infinity'. The dimensions W and L in the two remaining directions are also very large. The medium consists of a species A and a species B. The situation is in equilibrium, which means that the concentrations of both species, c_{A0} and c_{B0}, respectively, are constant.

At a certain point in time, $t = 0$, the concentration of species A at position $x = 0$ is suddenly raised, Δc_A, while at the same time the concentration of species B at $x = 0$ is reduced by $\Delta c_B = -\Delta c_A$ [this way, requirement (4.10) is met: $c_A + c_B = c$]. These new boundary conditions then remain in place. What will happen?

These new concentrations will start to penetrate the medium, analogous to the penetration of heat (see Figure 3.11 and Section 3.3.1), where it is again assumed that mutual diffusion will be involved. [Notice that the boundary conditions have been chosen with this in mind.] Below, however, only species A will be considered.

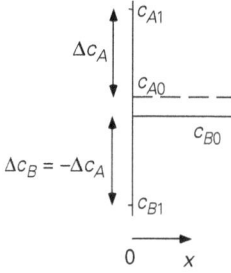

Figure 4.3

In order to be able to determine how the concentration of, say, species A progresses, it is necessary to draw up a mass balance for species A for a slice of the medium between x and $x + dx$. This balance is as follows:

$$\frac{\partial(WL\, dx\, c_A)}{\partial t} = WL \cdot \left\{ -\text{ID} \frac{\partial c_A}{\partial x}\Big|_x \right\} - WL \cdot \left\{ -\text{ID} \frac{\partial c_A}{\partial x}\Big|_{x+dx} \right\} \tag{4.32}$$

Equation (4.32) can be simplified to

$$\frac{\partial c_A}{\partial t} = \text{ID} \frac{\partial^2 c_A}{\partial x^2} \tag{4.33}$$

This result corresponds completely with equation (3.72), in relation to heat transport. Because the boundary conditions are also the same, the whole solution is the same: all the results in Section 3.3.1 can be replicated unconditionally. The solution to equation (4.33) is therefore once again the **error function**. This means that here too we are working with the **penetration theory**.

The **penetration depth** is now defined as $x = \sqrt{\pi \text{ID} t}$. This penetration depth has the same meaning as it does with heat transport: the mass flux that flows through the interface $(x = 0)$ into the medium is given by

$$\phi''_{m,A} = -\text{ID} \frac{\partial c_A}{\partial x}\Big|_{x=0} = \text{ID} \frac{c_{A1} - c_{A0}}{\sqrt{\pi \text{ID} t}} =$$

$$= \sqrt{\frac{\text{ID}}{\pi t}} (c_{A1} - c_{A0}) = k \cdot (c_{A1} - c_{A0}) \tag{4.34}$$

This means that equation (4.34) is entirely analogous to equation (3.78). Here, too, the driving force $\Delta c_A = c_{A1} - c_{A0}$ is a constant, while the mass transfer coefficient k is a function of time.

Of course, the validity of the penetration theory is again limited to a flat geometry and to short periods of time. Whether the time interval in question is indeed sufficiently short to be able to apply the penetration theory can be estimated by comparing the penetration depth with the length scale (e.g. the thickness D of the layer into which the heat penetrates from one side), or with the help of the Fourier number. The following must apply:

$$\sqrt{\pi \mathrm{ID}t} < 0.6\,D, \quad \text{or} \quad \mathrm{Fo} = \frac{\mathrm{ID}t}{D^2} < 0.1 \tag{4.35}$$

This condition is completely identical to condition (3.79) for applying short-term penetration theory in the case of transient conduction. The Fourier number is, apart from non-dimensional constants, also now the square of the ratio of two scales of length (or the ratio of two scales of time): see the discussion after equation (3.80).

For the mass transfer coefficient k, it is found that, similar to equation (3.91),

$$k(t) = \sqrt{\frac{\mathrm{ID}}{\pi t}} \tag{4.36}$$

From this follows the mean mass flux through plane $x = 0$ that enters the medium for t_e seconds (starting from $t = 0$), the mean transfer coefficient for this interval, and the Sherwood relation that goes with $k(t)$:

$$\overline{\phi''_{m,\mathrm{A}}} = \frac{1}{t_e} \int_0^{t_e} k(t)\,\Delta c_{\mathrm{A}}dt = 2\sqrt{\frac{\mathrm{ID}}{\pi t_e}}\,\Delta c_{\mathrm{A}} \tag{4.37}$$

$$\overline{k} = 2\sqrt{\frac{\mathrm{ID}}{\pi t_e}} = 2\,k(t_e) \tag{4.38}$$

$$\mathrm{Sh} = \frac{kD}{\mathrm{ID}} = 0.564\,\mathrm{Fo}^{-1/2} \tag{4.39}$$

Example 4.4. A contaminated landfill I.

Ten years ago, contaminated soil was deposited at a landfill in location L. During this period, the substance has slowly penetrated the clean soil underneath (caused by the rainwater seeping down, for example). This process can be described in terms of a diffusion or dispersion process with an effective diffusion coefficient $\mathrm{ID} = 2 \times 10^{-10}$ m^2/s. It can be assumed that the concentration of the contamination on the surface of the clean soil has a constant value of $c^* = 2$ kg/m^3.

To decide on the thickness of the soil that has to be dug up, and on what to do with it once it has been dug up, first a number of questions have to be answered:

a) How deeply has this substance diffused into the soil over the 10-year period?
b) How much of the substance has ended up in the underlying soil?

Solution

Question (a) is answered by calculating the penetration depth:

$$x_e = \sqrt{\pi\,\mathrm{ID}\,t_e} = 45 \text{ cm} \tag{4.40}$$

As it can be assumed that the clear layer of soil is much thicker than 45 cm, the penetration theory may be applied here. The penetration depth gives immediately a reasonable estimate of how far the contamination has penetrated.

The answer to question (b) is determined first by calculating the mean mass flow rate through the interface of the clean and contaminated soil:

$$\overline{\phi''_m} = \overline{k} \cdot (c^* - 0) = 2\sqrt{\frac{ID}{\pi t_e}}\,c^* \tag{4.41}$$

From this, it is easy to determine the mass that has penetrated the soil per unit of surface area during the 10 years:

$$M'' = \overline{\phi''_m} \cdot t_e = \frac{2}{\pi}\sqrt{\pi\,ID\,t_e}\cdot c^* = 0.57\,\mathrm{kg/m^2} \tag{4.42}$$

4.2.6 Unsteady-state long-term diffusion

Unsteady-state long-term diffusion, too, is entirely analogous to unsteady-state long-term heat conduction (Sections 3.3.5 and 3.3.6). This is nicely illustrated by the Fourier number: for heat transport it has been defined as $Fo = a \cdot t/D^2$ while for mass transport $Fo = ID \cdot t/D^2$ is used. Just like with long-term conduction, we speak of long-term diffusion

– when $Fo > 0.1$ for diffusion processes (at least at rather low concentration levels, see Section 4.3 further on) of a species from (or towards) one side into (or out of) a layer of a stagnant medium or a solid material; and

– when $Fo > 0.03$ in cases the diffusion is double-sided – from (or towards) the two sides of a stagnant or solid slab of material – or all round from (or towards) the surface of a solid or stagnant cylindrical or spherical body.

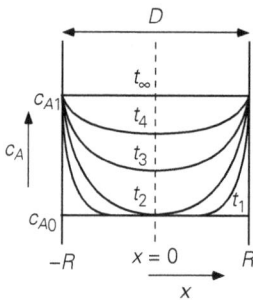

Figure 4.4

For long-term two-sided diffusion, the concentration profiles (see Figure 4.4) still have to obey the unsteady-state diffusion equation:

$$\frac{\partial c_A}{\partial t} = ID\,\frac{\partial^2 c_A}{\partial x^2} \tag{4.43}$$

but again – see also the conditions of equation (3.98) – with the following initial and boundary conditions:

$$c_A(t=0) = c_{A0} \quad \text{for } |x| \le R$$

$$c_A(x=R) = c_{A1} \quad \text{for } t > 0 \tag{4.44}$$

$$\frac{\partial c_A}{\partial x} = 0 \qquad \text{for } x = 0 \text{ and } t > 0$$

Here, too, it is the case that for sufficiently long times (Fo > 0.03), the profiles become spatially similar (providing that the boundary conditions remain the same the whole time, of course). The results in Sections 3.3.5 and 3.3.6 can be translated directly and Figures 3.17 and 3.18 can be used again, providing that T is replaced by c_A (and therefore also $\langle T \rangle$ and T_c by $\langle c_A \rangle$ and c_{Ac}, respectively) and, as remarked before, Fo is defined using ID instead of a!

For the slab, the cylinder, and the sphere, the same values for the Sherwood number apply as in the case of heat transport for the Nusselt number. Remember that it involves mass transport into a body where the resistance to the mass transport is located inside the body itself (the boundary conditions apply and are constant over time!).

So for long periods of time:

Slab	Sh = 4.93
Cylinder	Sh = 5.8
Sphere	Sh = 6.6

For two-sided and all round heating and cooling cases, the findings for penetration theory and long-term diffusion can be summarised as

Penetration theory	Fo < 0.03	$k = k(t)$	Δc_A = constant
Long-term diffusion	Fo > 0.03	k = constant	$\Delta c_A = \Delta c_A(t)$

In fact, this summary is identical to the one at the end of Section 3.3.5.

Example 4.5. A contaminated landfill II.
Consider now what happens with the landfill in location L (see Example 4.4) if no action is taken. The thickness of the layer of clean undersoil is 1 m. Below this is a layer of clay, which is (practically) impermeable as far as the contamination is concerned. The concentration of the contaminated soil right on the surface of the clean soil was $c^* = 2$ kg/m^3, and the diffusion coefficient ID = 2×10^{-10} m^2/s.

How long does it take before the mean concentration of the contamination amounts to 1 kg/m^3 in this 1 m layer of soil?

Solution
This problem can be solved most quickly with the help of Figure 3.17. It is first necessary to determine the parameter along the vertical axis, from the following:

$$\frac{c_1 - \langle c \rangle}{c_1 - c_0} = \frac{2-1}{2-0} = 0.5 \tag{4.45}$$

Verifying this with Figure 3.17 produces Fo = 0.05. From this it follows that (remember that for D it is not the thickness of the layer that should be taken, but twice the thickness: see also Example 3.10):

$$t = 1.0 \times 10^9 \text{ s} \approx 32 \text{ years}$$

! Summary
The analysis of unsteady-state mutual diffusion runs entirely analogously to that of unsteady-state heat conduction. Here, too, a distinction is made between the penetration theory (during short periods of time) and long-term diffusion. Again, the following summary applies to diffusion from or towards the two planes of a flat layer and to the cases of a cylinder and a sphere:

Penetration theory	Fo < 0.03	$k = k\,(t)$	Δc_A = constant
Long-term diffusion	Fo > 0.03	k = constant	$\Delta c_A = \Delta\, c_A(t)$

The same theory and principles apply here as in the case of heat transport. Even Figures 3.16 and 3.17 can be used again. The Fourier number is defined differently (with the help of the diffusion coefficient) and the place of the Nusselt number is taken by the Sherwood number to which in any case the same numerical values apply for the various basic geometries (slab, cylinder, and sphere).

4.3 Diffusion and drift flux

So far, we have very deliberately only discussed situations in which the transport of mass has been described using Fick's law: mutual diffusion and diffusion of just one species in a solid. With mutual diffusion, it is on average the case that the molecules of the two substances only exchange position within an enclosed space. In that case, in gases for example, there is no change in the overall concentration of the molecules (regardless of what kind they are) at a given location.

In general, diffusion problems will be more complex. It is very simple to generate another form of diffusion phenomena in a mixture of substances A and B: there is a net transport of substance A, while substance B is not transported, net. An example of this is the **Winkelman experiment** (see Figure 4.5): in a test tube there is a layer of volatile[15]

---

15 Example 2.2 looked at the case of a substance of a very slight level of volatility.

substance A. This substance evaporates constantly and diffuses to the edge of the test tube. Pure air is blown over the edge so that the concentration of substance A at the edge of the tube is (essentially) zero.

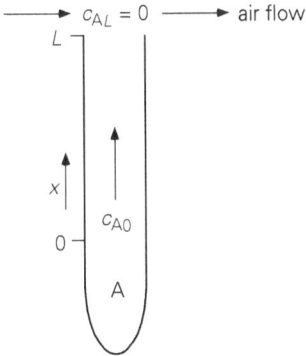

Figure 4.5

At first sight, it appears that this could be described using Fick's law. After all, there is a driving difference in concentration: the concentration of substance A just above the surface of the liquid minus the concentration at the edge of the tube. As a result of this, substance will diffuse to the exit of the test tube.

This, however, is not the full story. This can be understood by looking at the 'air' molecules. The concentration of the air molecules (shown as B) is not constant in the tube either: at the edge of the tube, the concentration c_B is equal to that of the outside air ($c_{BL} = p/RT$, in mol/m^3), while the air concentration at the surface of the liquid is much lower. The total concentration $c = c_A + c_B$ is constant because the pressure in the tube is the same throughout. If this were not the case, a large convective mass flow would occur immediately, because a difference in pressure (apart from just a hydrostatic pressure difference) always causes flow. This means that the concentration of the air just above the liquid surface is equal to $c_{B0} = c - c_{A0} = (p - p_{A0})/RT$ (where p_A is the vapour pressure of substance A, which for the sake of convenience is regarded as an ideal gas). Nonetheless, in the steady-state situation,[16] there is no net transport of air into the test tube from top to bottom. After all, where would this air go?

The diffusion flow of the air, which is a result of the difference in concentration, is apparently compensated by a convective flow: the total concentration c is not exactly constant, but is slightly higher at the liquid surface than at the edge of the tube.

---

16 In fact, the evaporation of the liquid in this test tube is of course an unsteady-state process. However, the height of the liquid (actually a boundary condition for the transport through the gas phase) changes so slowly (given the difference in density between the liquid and gas phase) that the situation can be regarded as quasi-steady for the transport through the gas phase.

However, this difference is so slight that the drop in pressure over the tube ($p_0 > p_L$) is so small that there is only a very slight convective flow, exactly enough to ensure that the net flow of the air molecules is zero!

This convective flow, caused by a very small difference in pressure, makes no distinction between substance A or B. Both are transported: there is therefore also a convective flow of substance A. This convective flow is called the **drift transport** (shown as ϕ in mol/s). The net mass fluxes for substances A and B in the test tube (with the concentrations in mol/m³) are now:

$$\phi_A'' = -ID\frac{dc_A}{dx} + \phi'' \frac{c_A}{c} \tag{4.46}$$

$$\phi_B'' = -ID\frac{dc_B}{dx} + \phi'' \frac{c_B}{c} \tag{4.47}$$

Equations (4.46) and (4.47) are still very general. In each of the two equations, the first term on the right-hand side represents the diffusion flux (according to Fick's law), while the second term shows that **drift flux** carries both species.

Because the total concentration is (almost) constant: $c = c_A + c_B$, for the concentration gradients, $dc_B/dx = -dc_A/dx$. Adding up equations (4.46) and (4.47) now produces:

$$\phi'' = \phi_A'' + \phi_B'' \tag{4.48}$$

In the Winkelman experiment, $\phi_B'' = 0$. It now follows therefore that $\phi'' = \phi_A''$ (see above). Entering this into equation (4.46) results in

$$\phi_A'' = -ID\frac{c}{c - c_A}\frac{dc_A}{dx} \tag{4.49}$$

Equation (4.49) is known as **Stefan's law** and describes a transport process denoted as **unilateral diffusion**.

If a mole balance for species A is now drawn up for a thin slice between x and $x + dx$ from the tube, then the following will be found (the situation is still quasi steady):

$$0 = \phi_A''|_x - \phi_A''|_{x+dx} \tag{4.50}$$

Combining equations (4.49) and (4.50) produces

$$\frac{c}{c - c_A}\frac{dc_A}{dx} = \text{constant} \tag{4.51}$$

With $c = $ constant, the solution to this differential equation, with the two boundary conditions $x = 0 \rightarrow c_A = c_{A0}$ and $x = L \rightarrow c_A = c_{AL}$, is

$$\frac{c - c_A(x)}{c - c_{A0}} = \left(\frac{c - c_{AL}}{c - c_{A0}}\right)^{x/L} \tag{4.52}$$

Now that the concentration profile is known, the flux of substance A can be calculated using Stefan's law, equation (4.49):

$$\phi_A'' = \frac{IDc}{L} \cdot \ln\left(\frac{c - c_{AL}}{c - c_{A0}}\right) \tag{4.53}$$

An analysis based solely on Fick's law produces the following erroneous result:

$$\phi_A'' = -ID\frac{c_{AL} - c_{A0}}{L} \tag{4.54}$$

The last two equations differ by one factor that is known as **Stefan's correction factor**:

$$f_D = \frac{c}{c_{A0} - c_{AL}} \cdot \ln\left(\frac{c - c_{AL}}{c - c_{A0}}\right) \tag{4.55}$$

A correction factor of this kind also occurs with unilateral diffusion from the surface of a sphere, for example, into 'infinity', such as in the case of an **evaporating droplet** or a **dissolving particle**. This correction factor is also at play of course in processes in the opposite direction, such as condensation or crystallisation. For Sherwood, this gives $Sh = 2f_D$ (instead of $Sh = 2$).

For c_{AL} and c_{A0} small in relation to c, Stefan's correction factor f_D can incidentally be written as

$$f_D = 1 + \frac{1}{2}\frac{c_{A0} + c_{AL}}{c} + \cdots \tag{4.56}$$

and it is therefore possible, if $c_{A0} + c_{AL} \ll c$, to assume that mutual diffusion will occur. This tallies with the limit situation that for $c_A \ll c$, Stefan's law, equation (4.49), transfers to Fick's law (see Example 2.2).

Both the basic equations (4.46) and (4.47) are of general validity for binary diffusion (i.e. without externally imposed flow) and therefore also cover mutual diffusion. In that case, the drift transport is equal to zero and Stefan's law transfers to Fick's law.

Remember that the above theory is formulated in mole flows and mole balances. This can also be done in mass flows and mass balances, but the way they are described is a little more complicated.

Other forms of mutual and unilateral diffusion are described by equations (4.46) and (4.47): see Example 4.6.

Example 4.6. Nickel carbonyl.
Consider a process in which carbon monoxide reacts with nickel (present in the form of a thin slice) to form gaseous nickel carbonyl:

$$Ni(s) + 4CO(g) \rightarrow Ni(CO)_4(g) \tag{4.57}$$

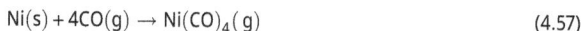

The chemical reaction (see Figure 4.6) occurs on the surface of the solid nickel, which is located in an atmosphere that initially consists entirely of CO. Because of the reaction, a layer is created in the vicinity of the nickel surface where there is less CO present and more nickel carbonyl. The reaction velocity is determined by the flux of CO molecules through this interface (thickness δ) to the nickel surface. [Referring to the film theory discussed in Section 3.5.2 may be useful.]

All CO molecules that reach the nickel surface react instantaneously, so that the CO concentration there is zero. At the same time, the nickel carbonyl diffuses away from the surface to the bulk of the gas phase, where the nickel carbonyl concentration can be assumed to be zero. Consider this process in the steady-state situation.

The question is: how large is the CO flux through the interface?

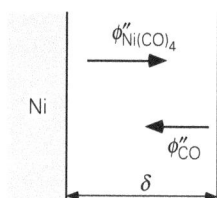

Figure 4.6

Solution
This is an example of a diffusion process that involves a drift flux: the drift flux is the sum of the nickel carbonyl flux and the CO flux and can be derived from the reaction equation. After all, four CO molecules have to be supplied for every nickel carbonyl molecule that is formed:

$$\phi''_{Ni(CO)_4} = -\frac{1}{4}\phi''_{CO} \tag{4.58}$$

The minus sign shows that both fluxes (in mol/s) are in opposing directions.

In this situation, the total concentration c can be taken to be practically constant: $c = c_1 + c_2 =$ constant (where $c_1 =$ concentration CO and $c_2 =$ concentration $Ni(CO)_4$, both in mol/m³). This means therefore that the gradients are equal in size but have opposing signs: $dc_1/dx = -dc_2/dx$. This means equations (4.46) and (4.47) become

$$\phi''_{CO} = -ID\frac{dc_1}{dx} + \frac{c_1}{c}\left(\phi''_{CO} + \phi''_{Ni(CO)_4}\right) \tag{4.59}$$

$$\phi''_{Ni(CO)_4} = -ID\frac{dc_2}{dx} + \frac{c_2}{c}\left(\phi''_{CO} + \phi''_{Ni(CO)_4}\right) \tag{4.60}$$

In both equations, the drift flux is written as the sum of both net fluxes, as derived in equation (4.48).

If the nickel carbonyl flux is now eliminated from equation (4.59) with the help of equation (4.58), the following applies:

$$\phi''_{CO} = -ID\frac{4c}{4c - 3c_1}\frac{dc_1}{dx} \tag{4.61}$$

A CO mole balance for a slice between x and $x + dx$ in the interface shows that the CO flux is constant, that is, that it is not dependent on x. The CO concentration profile can be determined from equation (4.61), with boundary conditions $x = 0 \rightarrow c_1 = 0$ and $x = \delta \rightarrow c_1 = c$:

$$c_1(x) = \frac{4}{3}c\left\{1 - \left(\frac{1}{4}\right)^{x/\delta}\right\}$$ (4.62)

From this, the CO flux can be calculated via equation (4.61):

$$\phi''_{CO} = -\left(\frac{4}{3}\ln 4\right)\frac{ID c}{\delta}$$ (4.63)

This example does not involve mutual diffusion, but it does not involve unilateral diffusion either. It involves two species that do not diffuse into each other one-to-one, but in a ratio that is imposed by the chemical reaction. It is therefore the chemical reaction that determines the strength of and the expression for the drift flux.

Summary

It is often the case that net transport involves one species passing through a medium (unilateral diffusion) or of multiple species diffusing in each other in a ratio other than one-to-one. It is only in the case of two species completely exchanging with each other (mutual diffusion) that the analogy with molecular heat transport (conduction) is a perfect one. The Winkelman experiment with an evaporating substance shows simply that account has to be taken of the so-called drift flux, ϕ''. In the case of unilateral diffusion, this has the effect that flows can arise as a result of differences in pressure, leading to the occurrence of convective transport. The general expression for mole fluxes in a binary system is

$$\phi''_A = -ID\frac{dc_A}{dx} + \phi''\frac{c_A}{c}$$

$$\phi''_B = -ID\frac{dc_B}{dx} + \phi''\frac{c_B}{c}$$

where

$$\phi'' = \phi''_A + \phi''_B$$

In the special case of unilateral diffusion (i.e. where the flow from one of the two species equals zero), the transport of the diffusing species can be described with the help of Stefan's law. Sometimes it is a phase change process, sometimes a chemical reaction that imposes the drift flux towards or from a boundary plane. Fick's law will suffice with low concentrations.

4.4 The general micro-balance for mass transport

This chapter on mass transport has so far looked only at diffusion (molecular transport), and then only in one direction, resulting from a concentration gradient in that direction. In general – both in industrial applications (in processing machines) and in other cases (in the home, in the environment) – time-dependent, three-dimensional concentration distributions will be at play: the concentration is a function of all three space coordinates and of time. As well as diffusion resulting from concentration gradients, the convective transport of species is also generally involved, as a result of flow (generally a three-dimensional velocity field).

This general description is nothing more than that given under heat transport (Section 3.4). This situation is therefore described entirely analogously with the help of a general micro-balance, the derivative of which is made on the basis of a small control volume, $dxdydz$, as shown in Figure 3.22 (**cubic volume element method**).

This derivation involves the basic form of a balance, and it should cover both the convective transport and the diffusion on the six planes. Each convective transport term takes the form $v \cdot c$. This means that a flow

$$[v_x c]_{x,y,z} \; dydz \tag{4.64}$$

passes into the cube through the left-hand side, and

$$[v_x c]_{x+dx,y,z} \; dydz \tag{4.65}$$

leaves again through the right-hand side. This makes the net contribution by the transport in the direction of x

$$-\frac{\partial}{\partial x}(v_x c) \; dxdydz \tag{4.66}$$

Similar expressions apply to the net transport in the two other directions. If the diffusion obeys Fick's law, then

$$-\left[ID\frac{\partial c}{\partial x}\right]_{x,y,z} \; dydz \tag{4.67}$$

will enter the cube on the left-hand side, because of diffusion, and

$$-\left[ID\frac{\partial c}{\partial x}\right]_{x+dx,y,z} \; dydz \tag{4.68}$$

will leave the cube on the right-hand side, because of diffusion. Net, these two terms give

$$\frac{\partial}{\partial x}\left[ID\frac{\partial c}{\partial x}\right] \; dxdydz \tag{4.69}$$

Similar expressions apply to the other edges of the cube. Finally, the accumulation term takes the form

$$\frac{\partial c}{\partial t} \; dxdydz \tag{4.70}$$

and the production term is

$$r \, dxdydz \tag{4.71}$$

where r represents the reaction rate (in mol/m^{3}s).

The general micro-balance is then created through the merging of the three net contributions of convective transport, of the three net diffusion terms (now with constant ID), of the accumulation term and of the production term, and by dividing all three by $dxdydz$

$$\frac{\partial c}{\partial t} = -\frac{\partial}{\partial x}(v_x c) - \frac{\partial}{\partial y}(v_y c) - \frac{\partial}{\partial z}(v_z c) +$$

$$+ \text{ID}\left\{\frac{\partial^2 c}{\partial x^2} + \frac{\partial^2 c}{\partial y^2} + \frac{\partial^2 c}{\partial z^2}\right\} + r \tag{4.72}$$

This species transport equation is entirely analogous to equation (3.135) for heat transport. Solving the concentration field also requires solving the velocity field, but this will not be dealt with until Chapter 5.

However, for many applications, no knowledge of the whole velocity and concentration fields are needed; all that is required are mass transfer models in which mass transfer coefficients for mass transport at the boundary or boundaries of the field of flow play a major part. The rest of this chapter is therefore devoted completely to this phenomenological approach.

Summary

The generally valid micro-balance for mass transport has been derived with the help of the cubic volume element method. This analyse has again taken account of both convective and molecular mass transport. It has been assumed that Fick's law applies.

However, the micro-balance can only be used to calculate the concentration distribution if the velocity field is known.

4.5 Mass transfer coefficients during convection

4.5.1 General observations

As with heat transport in Section 3.5, convective mass transport can be conveniently described using the analogue of Newton's law of cooling as a starting point:

$$\phi_m = k A \, \Delta c \tag{4.73}$$

This equation certainly applies in a single phase, but with a suitable choice of driving force can also be used for mass transfer across phase interfaces, providing the term **overall mass transfer coefficient** is used (see Section 4.6 for more details).

The transfer coefficient k is generally dependent on ρ, $\langle v \rangle$, ID, μ, D, and so on:

$$k = k(\rho, \langle v \rangle, \text{ID}, \mu, D, \ldots) \tag{4.74}$$

Here, too, a far-reaching analogy with convective heat transport can be expected. The same considerations as those discussed in Section 3.5 also apply to the choice of independent variables that could influence the mass transfer coefficient – except that now, quantities that are typical of those involved in heat transport must be converted into those that are typically associated with mass transport, such as $a \rightarrow$ ID. With the help of dimensional analysis, equation (4.74) can then be reduced to

$$\text{Sh} = \text{Sh}(\text{Re}, \text{Sc}, \ldots) \tag{4.75}$$

The Schmidt number Sc, defined as υ/ID, replaces the Prandtl number and is related to the ratio of the hydrodynamic boundary layer thickness and the mass transfer boundary layer thickness. See also the discussion about the role of boundary layers in convective heat transport after Figure 3.24. The Reynolds number now also shows the ratio of convective to molecular transport of momentum in the flow.

Expressions for the Sherwood number analogous to those for the Nusselt number for heat transport exist for various situations. These are discussed in some more detail below.

4.5.2 Mass transfer in the case of forced convection

Entirely analogously with convective heat transport, the following Sherwood correlations apply to various flow configurations:

Laminar pipe flow
Providing the velocity profile of the flow is fully developed, the following applies to mass transfer from a given point, $x = 0$:

$$\text{Sh} = 1.08 \left(\frac{\text{ID}x}{\langle v \rangle D^2} \right)^{-1/3} \quad \text{if} \quad \frac{\text{ID}x}{\langle v \rangle D^2} < 0.05 \tag{4.76}$$

$$\langle \text{Sh} \rangle = 1.62 \left(\frac{\text{ID}L}{\langle v \rangle D^2} \right)^{-1/3} \quad \text{if} \quad \frac{\text{ID}L}{\langle v \rangle D^2} < 0.05 \tag{4.77}$$

$$\text{Sh} = 3.66 \quad \text{if} \quad \frac{\text{ID}x}{\langle v \rangle D^2} > 0.1 \tag{4.78}$$

The same considerations apply to these expressions as those mentioned under heat transport (see Section 3.5.2).

Turbulent pipe flow
Film theory (see also Section 3.5.2, Figure 3.24, equation (3.146)) can also be used here in order to model the mass transfer, leading to $k = ID/\delta$; the following correlation comprising the Reynolds number and the Schmidt number can be used for calculation purposes:

$$Sh = 0.027\ Re^{0.8}\ Sc^{0.33} \quad \text{providing that}\ Re > 10^4\ \text{and}\ Sc \geq 0.7 \tag{4.79}$$

Forced flow around a sphere
The following correlation, analogous to equation (3.151), applies here:

$$Sh = 2.0 + 0.66\ Re^{0.50}\ Sc^{0.33} \quad \text{providing that}\ 10 < Re < 10^4\ \text{and}\ Sc \geq 0.7 \tag{4.80}$$

where it is assumed that no drift flux occurs (see Section 4.3).

Example 4.7. A wetted wall column.
Air containing a low concentration c_0 of SO_2 has to be stripped of this SO_2. This is done in a **wetted wall column** in which the air is flowing upward at a flow rate ϕ_v while an alkaline solution flows downward as a thin liquid film on the inside of the cylindrical column – reason why this piece of equipment is also denoted as a **film contactor**. The thickness of this liquid film is very small with respect to the column diameter D.

In this countercurrently operated process, the SO_2 has first to be transported from the bulk of the air flow to the interface with the liquid film; this gas phase transport is facilitated by operating in the turbulent regime such that the bulk of the gas flow is radially well mixed rendering the concentration c a function of x only, with the mass transfer resistance being concentrated in a thin air film adjacent to the liquid film. At the air–liquid interface, the SO_2 is transferred to the liquid where it reacts immediately with the alkaline to sulphite resulting in a zero SO_2 concentration at the film interface ($c_i = 0$).

The question is to derive the expression describing how during a steady-state operation of the process the SO_2 concentration c in the air decreases as a function of x, which is the upward (downstream) coordinate. It is also asked how the total amount $\dot{M}$ of SO_2 transferred per unit of time depends on column height L.

Solution
The question with respect to the SO_2 concentration profile $c(x)$ asks for drawing up a species (SO_2) mass balance over a thin slice of thickness dx at any position x somewhere in the column for a steady-state situation, with the result:

$$0 = \phi_v\ c|_x - \phi_v\ c|_{x+dx} - k \cdot \pi D \cdot (c - c_i)\ dx \tag{4.81}$$

This species balance leads to the differential equation

$$\frac{\pi}{4} D^2\ v\ \frac{dc}{dx} = -k \cdot \pi D \cdot c \tag{4.82}$$

with the boundary condition $c(0) = c_0$. The mass transfer coefficient k is to be found via the Sherwood number and with the help of equation (4.79), and is a constant (does not depend on x) – providing the two criteria as to the values of Re and Sc are met. The integration then straightforwardly results in

$$\frac{c}{c_0} = \exp\left(-\frac{4k}{vD}x\right) = \exp\left(-\frac{4\mathrm{ID}}{vD^2}\,\mathrm{Sh}\,x\right) \tag{4.83}$$

The second question relates to $\dot{M}$ which can be found either by integrating the third term of the right-hand side of equation (4.81) – which stands for the local mass transfer rate – from 0 to L:

$$\dot{M} = \int_0^L k \cdot \pi D \cdot c(x)\, dx = \phi_v\, c_0 \left\{1 - \exp\left(-\frac{4kL}{vD}\right)\right\} \tag{4.84}$$

or from the expression $\phi_v \cdot \{c_0 - c(L)\}$ which, due to balance (4.81), gives the same result. In both cases, result (4.83) is to be substituted for $c(x)$ and $c(L)$.

Example 4.8. An evaporating sphere.
A sphere made from pure substance X (with $\rho = 1{,}950$ kg/m^3) is hung from a thread in an air flow. The air flow being supplied does not contain species X has a temperature of 20 °C, a uniform velocity v of 5 m/s, and a kinematic viscosity v of 1.5×10^{-5} m^2/s. The diameter of the sphere at the start of the experiment is $D_0 = 25$ mm. Temperature effects due to the evaporation may be ignored.
The following applies to substance X: molar mass $m = 120$ g/mol, vapour pressure in the prevailing conditions $p^* = 0.01$ bar, and diffusion coefficient ID $= 0.5 \times 10^{-5}$ m^2/s.
Determine the point in time at which the diameter of the sphere is exactly 20 mm.

Solution
The mass balance for the evaporating sphere is

$$\rho \frac{dV}{dt} = \rho \frac{\pi}{2} D^2 \frac{dD}{dt} = -\pi D^2 k \cdot \left(m\frac{p^*}{RT} - 0\right) \tag{4.85}$$

Additional information here is that the concentration of X in the air far away from the sphere is zero (and remains so). Also, the volume of the sphere $\pi D^3/6$ is differentiated according to time.
Equation (4.85) is the differential equation that describes the change of the diameter of the sphere as a function of time. Before integrating this equation, it is first necessary to find the relationship between k and D. This relationship generally follows from equation (4.80) but, with a view to the integration of equation (4.85), this has an awkward form. From equation (4.80), it follows that Sh has a minimal value of 80 during the evaporation. With pinpoint accuracy, the relationship between k and D is simplified to

$$k = \frac{\mathrm{Sh}\,\mathrm{ID}}{D} = 0.66 \left(\frac{v}{v}\right)^{1/2}\left(\frac{v}{\mathrm{ID}}\right)^{1/3} D^{-1/2} \tag{4.86}$$

Entering this into equation (4.85) leads to

$$D^{1/2}\frac{dD}{dt} = -C, \text{ where } C = 1.32\frac{m}{\rho}\frac{p^*}{RT}\mathrm{ID}^{2/3}v^{-1/6}v^{1/2} \tag{4.87}$$

Thanks to boundary condition $D(0) = D_0$, the solution to this differential equation is

$$\frac{2}{3}\left(D^{3/2} - D_0^{3/2}\right) = -C \cdot t \tag{4.88}$$

Substituting the numerical data shows that $D = 20$ mm at $t = 5.5 \cdot 10^3$ s, which is around 1.5 h.

Chilton and Colburn

The above analogy between convective heat and mass transport implies that k can be determined from knowledge of h (and vice versa). In particular, for Re values that are not too small, it is generally possible to write the following:

$$Nu = C\ Re^m Pr^n \tag{4.89}$$

and

$$Sh = C\ Re^m\ Sc^n \tag{4.90}$$

In each of the cases given above, n is equal to 1/3 – related to the ratio of the respective boundary layer thicknesses (see the discussion after Figure 3.24). Chilton and Colburn used this in order to introduce two new non-dimensional transfer numbers: one for heat transfer (j_H) and one for mass transfer (j_D):

$$j_H = \frac{Nu}{Re\ Pr^{1/3}} = \frac{h}{\rho c_p v}\left(\frac{v}{a}\right)^{2/3} \tag{4.91}$$

$$j_D = \frac{Sh}{Re\ Sc^{1/3}} = \frac{k}{v}\left(\frac{v}{D}\right)^{2/3} \tag{4.92}$$

These numbers relate the rate at which heat and mass, respectively, are transferred (in the direction normal to the main flow) to the convective transport of the main flow (see also Figure 3.23). This makes these numbers a measure for the effectiveness of the transfer process. In this sense, j_H and j_D are comparable to the Reynolds number: see equation (2.49).

Combining the four above equations now reveals – where the Reynolds number is not too small – that both these new numbers for geometrically similar situations are equal to each other and dependent only on the Reynolds number:

$$j_H = j_D = C\ Re^{m-1} \tag{4.93}$$

The discovery that j_H and j_D are equal to each other can be rewritten as a link between transfer coefficients k and h with the help of the definitions of j_H and j_D. Owing to definitions (4.91) and (4.92), the convective velocity drops out of the relation between k and h:

$$k = \frac{h}{\rho c_p} \cdot \left(\frac{\text{ID}}{a}\right)^{2/3} = \frac{h}{\rho c_p} \text{Le}^{-2/3} \qquad (4.94)$$

The Lewis number (Le) appears in equation (4.94): for the definition; see equation (2.38). It therefore seems that k can be determined by measuring h and by looking for a number of mass constants. In practice, this is highly beneficial, as only relatively simple temperature measurements have to be carried out for h, while determining k often requires very time-consuming and difficult mass transfer experiments, in many cases for different species. With the help of equation (4.94), therefore, only knowledge of h and some physical properties are needed. The relationship does not contain the Reynolds number, and not explicitly the geometry either.

Example 4.9. Dissolving incrustation.
Over time, a solid has become encrusted on the bottom of an ideally stirred tank (diameter $D = 3$ m). In order to remove this, a pure solvent is allowed to flow through the tank. The encrusted matter slowly dissolves in this solvent. The flow rate of this solvent is $\phi_v = 0.5 \times 10^{-3}$ m³/s.

Heat transfer measurements on the tank have been carried out which have shown that
a) the heat transfer coefficient h on the bottom depends on the number of revolutions of the stirrer according to $h \propto N^{2/3}$,
b) for a certain level of revolutions, $h = 3$ kW/m²K, and
c) the power P that is supplied to the stirrer varies by N^3.

Data:

Diffusion coefficient of the crust in the solvent	ID $= 10^{-9}$ m²/s
Density of solvent	ρ = 1,200 kg/m³
Specific heat of solvent	c_p = 3 kJ/kg K
Thermal conductivity coefficient of solvent	λ = 0.5 W/mK
Solubility of the crust in the solvent	c^* = 50 kg/m³

The question is this: during this slow dissolution process, what is the constant dissolution velocity in the event that $h = 3$ kW/m²K, and how does this dissolution velocity change if twice as much power is supplied to the stirrer?

Solution
A dissolution process of the kind described means that mass has to be transferred from a thinly saturated layer just above the crust to the bulk of the stirred tank, which as a result of this dissolution and flow of solvent contains a constant (in terms of time) concentration of dissolved mass. In this steady-state situation, the mass balance for the crust that is dissolved in the solvent (c = concentration in the bulk) is

$$0 = -\phi_v c + k \frac{\pi}{4} D^2 (c^* - c) \qquad (4.95)$$

The second term on the right-hand side is the dissolution velocity that was asked for, but knowledge of the bulk concentration is required in order to be able to calculate this. To that end, it is necessary to solve equation (4.95), and for that, a value for k is needed. Because the heat transfer experiment was carried out in the same conditions in the same geometry, it is possible to use expression (4.94) to work out that

$k = 3.1 \times 10^{-5}$ m/s. From this, it follows that the concentration in the bulk is equal to $c = 15.2$ kg/m^3. This means that the following applies to the dissolution velocity:

$$\phi_m = k \frac{\pi}{4} D^2 (c^* - c) = 7.6 \times 10^{-3} \text{ kg/s} \qquad (4.96)$$

When doubling the level of power supplied, the following applies:

$$P_n = 2P \rightarrow N_n = 2^{1/3} N \rightarrow$$
$$h_n = 2^{2/9} h \rightarrow k_n = 2^{2/9} k = 3.6 \times 10^{-5} \text{ m/s} \qquad (4.97)$$

and therefore

$$c_n = 16.9 \text{ kg/m}^3 \rightarrow \phi_{m,n} = 8.5 \times 10^{-3} \text{ kg/s} \qquad (4.98)$$

4.5.3 Mass transfer in the case of free convection

How heat is transported if free convection is involved was discussed in Section 3.5.3. This mechanism is also possible with mass transfer. The analogy here is, if anything, even greater than is the case with forced convection. After all, it does not matter whether the difference in density that is needed for free convection comes from expansion as a result of a change of temperature or is caused by dissolving salt, for example, leading to a local increase in density. An example in which mass transport through free convection is involved concerns a vertical wall that consists of salt and which is in contact with pure water. The water at a large distance from the slice is pure, and has a density of ρ_∞. The layer of water immediately adjacent to the salt slice has a higher density as salt has dissolved into it. Free convection will now therefore occur of its own accord.

The effect of free convection on the mass transfer is accounted for by a Grashof number, exactly as in the case of heat transfer. The Grashof number Gr that is used in describing free convection has previously been written, not without reason, in equation (3.157), in terms of the difference in density that occurs:

$$\text{Gr} = \frac{x^3 g}{v^2} \frac{\Delta \rho}{\langle \rho \rangle} \qquad (4.99)$$

In the relationships involving the Sherwood number and the Grashof number, the Schmidt number must of course appear where, in the case of heat transfer, the Prandtl number figured in the relationships involving the Nusselt number and the Grashof number. Mass transfer coefficient k, in the case of free convection, can therefore be obtained from

$$\text{Sh} = f(\text{Gr} \cdot \text{Sc}) \qquad (4.100)$$

The same relations apply to mass transfer resulting from free convection as they do to heat transfer resulting from free convection (where the geometry is equal) providing that the following 'translations' are made: $\text{Pr} \to \text{Sc}$ and $\text{Nu} \to \text{Sh}$.

! **Summary**
Mass transfer with forced and free convection is entirely analogous to heat transport with forced and free convection. Translations only have to be made as follows:

$$\text{Nu} \leftrightarrow \text{Sh}$$
$$\text{Pr} \leftrightarrow \text{Sc}$$

Chilton and Colburn have used this in order to link k and h to each other exclusively via physical properties, where the values of the Reynolds number are sufficiently large.

4.6 Mass transfer across a phase interface

4.6.1 Introduction on dealing with two phases

In such processes as evaporating or dissolving **pure** substances, it is sufficient to consider the transport in only one phase. The other phase (the evaporating droplet, the dissolving particle) then only produces the boundary condition for the concentration field. In Examples 2.2 and 4.8, the (temperature-dependent) **equilibrium vapour pressure** p^* is used, and the (temperature dependent) **solubility** c^* in Examples 4.2, 4.3, and 4.9.

Actually, this is the second important difference between heat transport and mass transport that was already mentioned in Section 4.1 (in addition to unilateral diffusion). If two substances are in thermal equilibrium, they have the same temperature. However, if two substances containing a dissolved species A are in equilibrium with each other, then in general the concentrations c_A in both substances will be different. In Section 4.6.2, it will be explained that the chemical potentials of species A in the two phases will be equal rather than the concentrations of A.

Something similar to this was dealt with in Example 4.2 with the solubility of helium in glass; there, the concept of **solubility** was introduced for the first time to instantaneously produce and steadily maintain the equilibrium concentration c^* **in** the glass at the interface with the gas. This c^* is different from but depends on the density (in the case of a pure gas) or concentration (in the case of gas mixture) in the adjacent gas phase.

Many relevant processes do not deal with pure substances indeed, but with mixtures or solutions. In many gas–liquid cases and for equilibrium conditions only it is then possible to use **Henry's law**

$$p_A = H_A \cdot y_A \tag{4.101}$$

in which the **Henry coefficient** H_A denotes the proportionality coefficient between the mole fraction y_A of dissolved gas A in a liquid and the partial pressure p_A of the gas above the liquid. Henry's law may best be used for dilute solutions.

Example 4.10. Mass transfer to a rain drop.
A rigid, spherical rain drop 5 mm in diameter D is falling at its terminal velocity v through a layer of air (thickness $z_0 = 250$ m) that contains a high ammonia concentration which is expressed in terms of a (constant) partial vapour pressure p_0. The ammonia concentration in the rain drop right before it enters the air/ammonia blanket is zero. The diffusion coefficient ID of ammonia in water amounts to 1.49×10^{-9} m²/s (at the prevailing temperature of 20 °C) and the value of the Henry coefficient of ammonia is 2.8×10^5 Pa.

The rain drop retains its spherical shape[17] all the time and remains rigid (no internal flows), the air with the ammonia behaves as an ideal gas, and evaporation of water from the drop and the related heat effect may be ignored.

The eventual question is about the ratio of the average ammonia concentration in the rain drop at the moment it leaves the air/ammonia blanket, to the ammonia concentration c_0 in the air.

Solution
The mass transfer of the ammonia from the air into the rain drop is a transient process: the amount of ammonia in the rain drop may increase in time as long as the rain drop is in contact with the air/ammonia mixture (and saturation conditions have not been attained). It should further be realised that the mass transport comprises two steps: the external transport of ammonia from the bulk of the air towards the phase interface and the internal transport from the phase interface into the interior of the drop.

The external mass transport is of the convective type, owing to the terminal velocity of the drop through the air/ammonia mixture, while the internal mass transport in the water (where the molecules are much closer together than in the gas phase) is due to diffusion, individual ammonia molecules find their way in between the water molecules. Although it therefore seems reasonable to submit that the internal, diffusive mass transport is rate limiting – the internal resistance to mass transport may be the largest – this supposition should be verified quantitatively.

At the phase interface, equilibrium is attained instantaneously, exactly as in Example 3.5, but this does not imply that at either side of the phase interface the concentrations are equal: rather, Henry's law applies here. In such a two-phase mass transfer process, this law does not apply to the average variables in the bulk of the phases involved as long as equilibrium has not been attained yet.

Obviously, the internal mass transport resistance is both molecular and transient. Then the question is relevant whether either penetration theory at short times (Section 4.2.5) applies or the unsteady-state long-term diffusion solution (Section 4.2.6). To answer this question, the contact time is needed: the time interval τ during which the rain drop is in contact with the air/ammonia mixture. This contact time can be calculated from the terminal velocity of the drop which in its turn follows from a force balance over the drop: see Section 2.3.3. The same terminal velocity is also needed to calculate the external mass transfer coefficient with the help of the relevant Sh relation from Section 4.5.2.

From a force balance, the terminal velocity is found: $v = 11.7$ m/s; the contact time then follows: $\tau = z_0/v = 21.4$ s. This terminal velocity leads to Re = 3900; and equation (4.80), along with Sc = 0.64, then results in Sh = 37.5 and $k_e = 0.21$ m/s.

---

17 Actually, a rain drop of 5 mm diameter will not be spherical. We use that here for simplicity.

After 21.4 s, the penetration depth – see for example, equation (4.40) – is just 0.32 mm: pretty much smaller than the drop diameter. This makes this case an obvious example of short-time penetration theory; this value of 0.32 mm illustrates how slow diffusion in the liquid phase is and – in other words – how large the internal resistance to mass transport is! Although at $t = 0$ the internal mass transfer coefficient k_i is infinitely large according to equation (4.36), this k_i decreases sharply to a value of 4×10^{-6} m/s at $t = 21.4$ s according to the same equation and very fast becomes very much smaller than k_e.

Further on in this Section 4.6, we will see that in the case of mass transfer across a phase interface, the answer to the question in which phase the resistance to mass transport is largest does not depend on just the ratio k_e/k_i. However, in this case, the internal resistance to mass transport is rate limiting, indeed.

The external resistance to mass transport may therefore be ignored; as a result, the ammonia concentration at the air side of the phase interface is taken equal to the bulk concentration in the air, while the ammonia concentration c_g at the water side of the phase interface is calculated by using Henry's law: $c_g = 0.0034 \cdot p_0$. Finally, the average ammonia concentration $\langle c_i \rangle$ in the rain drop is found with the help of similar expressions as equations (4.37) and (4.42): $\langle c_i \rangle = 119\, c_0$. The latter result – a much higher average concentration inside the drop than outside – illustrates once more that in the case of mass transfer from one phase to another, the concentration difference is NOT the driving force for mass transport.

In the case of such multi-component processes as distillation, extraction, absorption, and stripping in particular, the concentrations in both phases vary both in time and in space. In many of those cases, transport limitations in both phases, at either side of the interface, have to be considered explicitly.

4.6.2 The partition coefficient

By way of example, a layer of benzene has been added to a layer of water in Figure 4.7a. If this system is in equilibrium, then the temperature in the water and in the benzene will be the same: $T^W = T^B$. Of course, the temperature T_i of the interface between the two phases is at the same value (see Figure 4.7b). The driving force for heat transport is therefore zero: $\Delta T = T^W - T^B = 0$.

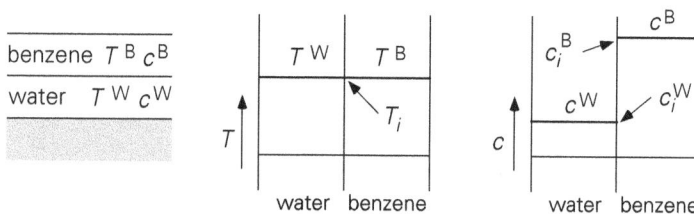

Figure 4.7: a/b/c (from left to right).

Figure 4.7c shows the equilibrium situation for a layer of benzene on a layer of water when acetic acid has been dissolved in both. In the **equilibrium** situation shown, the concentration is uniform in every layer (otherwise transport would occur, and then we would not speak of an equilibrium!). In general, it is now not the case that c^B and c^W are equal. This is the result of differences in the relationship between the **chemical potential** (which accounts for the interactions between the molecules of the dissolved species and the solvent) and the concentration: there is only equilibrium if a dissolved substance in both solvents experiences the same chemical potential, and this is generally the case with different concentrations.

Nonetheless, the equilibrium situation requires that the driving force for mass transport is equal to zero. Clearly, the driving force is not: $\Delta c = c^B - c^W$, as even with equilibrium this is not equal to zero. This can be solved by introducing the **partition coefficient** m defined such that the driving force, at *equilibrium*, is zero:

$$0 = c^B - m \cdot c^W \Leftrightarrow m = \left[\frac{c^B}{c^W} \right]_{\text{equilibrium}} \tag{4.102}$$

The partition coefficient m is a property of the three substances that are at play here (in this case, benzene, water, and acetic acid). In the case of the system acetic acid–benzene–water, the value of m is greater than unity (see Figure 4.7c).

Example 4.11. Aerating water.
A small quantity of water at 20 °C is saturated with oxygen. This is done by putting the water into contact with a large quantity of air at 20 °C and with a pressure of 1 bar. The mass fraction of oxygen in the air is about 20%. After a long time, the water is saturated and the O_2 concentration c_w in the water amounts to 8 mg/L. The water is now brought into contact with air of the same composition and temperature, but at a pressure of 0.5 bar. In the equilibrium situation, what will be the O_2 concentration in the water?

Solution
In order to be able to answer the question, it is first necessary to calculate the partition coefficient m. For this, the O_2 concentration in the 1 bar air has to be established. The density of air can be calculated with the ideal gas law, from the air composition and with the molar masses of N_2 and O_2: $\rho = 1.2$ kg/m^3. The oxygen concentration in air is therefore $c_{\text{air}} = 0.2 \times 1.2 = 0.24$ kg/m^3. This value is much higher than the saturation concentration in water: $c_w = 0.008$ kg/m^3. With the partition coefficient at equilibrium defined as $m = c_{\text{air}}/c_w$, the value for m is found to be 0.24/0.008 = 30.

The concentration of oxygen in the air at 0.5 bar is $c_{\text{air}} = 0.12$ kg/m^3 and so the new concentration in the water at equilibrium (after a long time) will be $c_w = 4$ mg/L. Starting from water with 8 mg/L oxygen dissolved, oxygen is transported from the water to the air at the lower pressure – although the oxygen concentration in the air is higher than that in the water – until the oxygen concentration in the water will have fallen to 4 mg/L.

4.6.3 Mass transfer across a phase interface

If there is no equilibrium between the two layers in terms of the chemical potential, mass transfer will occur across the phase interface. An example of such a non-equilibrium situation is shown in Figure 4.8: in this example for the acetic acid-in-water/benzene system $c_b^W < c_b^B$ but also $m\,c_b^W > c_b^B$. The driving force for the acetic acid transport from the bulk (denoted by subscript b) of the water to the bulk of the benzene is then $\Delta c = m\,c_b^W - c_b^B$, because the definition of equation (4.102) is not complied with in relation to the bulk of both phases. At the interface, however, the equilibrium *does* instantaneously adapt:

$$c_i^B = mc_i^W \tag{4.103}$$

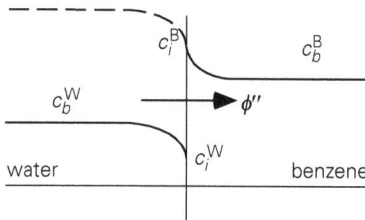

water   benzene

Figure 4.8

This implies that no equilibrium any longer prevails in the bulk of either of the two liquids, and that in the bulk of either phase a transport is created towards and away from the interface, respectively. As can be seen from the profiles in Figure 4.8, the acetic acid moves from the bulk of the water to the benzene-water interface, across the interface, and then from the benzene–water interface into the bulk of the benzene. Essentially, this transport is analogous to the heat transport in Figure 3.10.

The acetic acid flux will be affected by the mass transfer coefficients in both phases: k_B and k_W. Analogous to Example 3.5, the fluxes in the two layers separately run as

$$\phi_B'' = k_B\left(c_i^B - c_b^B\right) \tag{4.104}$$

$$\phi_W'' = k_W\left(c_b^W - c_i^W\right) \tag{4.105}$$

As a matter of fact – an interface cannot host mass – both fluxes are equal, and therefore the subscript with the flux will be omitted from now on.

With the help of equation (4.103), an expression can be found for c_i^W:

$$c_i^W = \frac{k_W\,c_b^W + k_B\,c_b^B}{k_W + mk_B} \tag{4.106}$$

Apart from m in the denominator, equation (4.106) strongly resembles equation (3.85). Combining equations (4.105) and (4.106) results in

$$\phi'' = \left[\frac{1}{k_B} + \frac{m}{k_W}\right]^{-1}\left(mc_b^W - c_b^B\right) \tag{4.107}$$

entirely analogously to equation (3.62) for heat transfer across an interface. It can be seen that the partition coefficient m appears not just in the driving force but also in the **overall mass transfer coefficient** K:

$$\frac{1}{K} = \frac{1}{k_B} + \frac{m}{k_W} \tag{4.108}$$

which expresses – on the analogy of equation (3.59) and Ohm's law – that the total resistance to mass transport consists of two partial resistances.

The (4.107)–(4.108) system can also be written differently (although it is of equal validity):

$$\phi'' = \left[\frac{1}{mk_B} + \frac{1}{k_W}\right]^{-1}\left(c_b^W - \frac{1}{m}c_b^B\right) \tag{4.109}$$

$$\frac{1}{K*} = \frac{1}{mk_B} + \frac{1}{k_W} \tag{4.110}$$

Exactly as with heat transport, also with mass transport it is common practice to ignore the smallest of the two partial resistances; to decide on this, the ratio of the two partial resistances should be inspected. It follows from the above derivation that not only k_W and k_B should be considered for calculating this ratio but also the partition coefficient m should be involved:

$$\frac{\text{resistance in benzene}}{\text{resistance in water}} = \frac{1/(mk_B)}{1/k_W} \tag{4.111}$$

In this respect, mass transfer across a phase interface deviates from heat transfer from one medium to another. Yet, this ratio of partial mass transport resistances can be interpreted as a **Biot** number.

Example 4.12. H_2S stripper.
In a reactor (see Figure 4.9), H_2S is produced in oil. In order to remove this dissolved H_2S from the oil (or to 'strip' it), hydrogen is passed through the oil continuously (at a flow rate ϕ_v). The H_2S is transferred to the gas, which allows it to flow out of the reactor. The increase in the gas flow rate that results may be disregarded. The reactor is stirred intensively, so that gas phase and liquid phase can be regarded as ideally mixed.

Because of the intensive stirring, the gas stream is broken up into very small bubbles, with a large phase interface A available for mass exchange; as a result, it may be assumed that the liquid phase and the gas phase are practically in equilibrium such that $c_l = mc_g$ where c_l and c_g denote the concentrations of H_2S in the oil (liquid) and gas, respectively. The volume of the gas phase in the reactor is V_g, and the volume of the liquid phase is V_l. Finally, assume that A is constant during this process.

From a certain point in time ($t = 0$), no more H_2S is produced in the reactor. The concentration of H_2S in the oil at $t = 0$ is denoted by c_{l0}.

The question now is to determine c_l as a function of time.

Solution
Firstly, two mass balances will have to be drawn up for the H_2S: one for the oil phase and one for the gas phase. For the driving force for the mass transfer to the gas phase, the following is chosen: $\Delta c = c_l - mc_g$.

This means that the overall mass transfer coefficient K is based on the liquid phase. The two balances are now as follows:

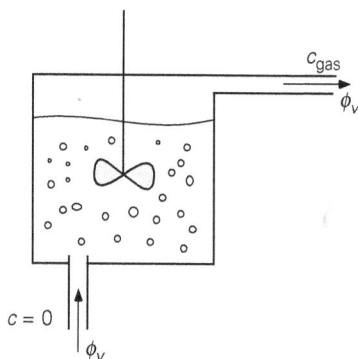

Figure 4.9

$$V_l \frac{dc_l}{dt} = -KA(c_l - mc_g) \tag{4.112}$$

$$V_g \frac{dc_g}{dt} = -\phi_v c_g + KA(c_l - mc_g) \tag{4.113}$$

In equation (4.113), the incoming gas was assumed not to contain any H_2S.

In a process like this, the resistance to mass transfer is often in the liquid phase: then $K \approx k_l$. In such cases, one often talks about the quantity $k_l a$ with the specific interfacial area a defined as A/V_l.

There is now a problem with equations (4.112) and (4.113): it has been stated that the liquid phase and the gas phase are virtually in equilibrium. This would mean that the driving force $\Delta c = c_l - mc_g$ is zero and that there is therefore **no** H_2S transfer from the oil to the gas. However, this is not a true representation of the facts. What is actually happening is that the driving force has become almost zero precisely because the product of KA is very large (because of the intensive stirring). The transfer from oil to gas is therefore the product of '∞' and '0'. This problem is elegantly circumvented by adding up both equations: the transfer from oil to gas then disappears:

$$V_l \frac{dc_l}{dt} + V_g \frac{dc_g}{dt} = -\phi_v c_g \tag{4.114}$$

From this, c_g can be eliminated given the fact that virtual equilibrium between the two phases predominates in the reactor: $c_g = c_l/m$. This produces

$$\frac{dc_l}{dt} = -\frac{\phi_v}{mV_l + V_g} c_l \tag{4.115}$$

and with the boundary condition at $t = 0 \to c_l = c_{l0}$ it then follows that

$$c_l(t) = c_{l0} \exp\left(-\frac{\phi_v}{mV_l + V_g} t\right) \tag{4.116}$$

4.6.4 Penetration theory at a phase interface

The derivation in Section 4.6.3 is also applicable to non-steady-state situations. Here, too, the fluxes that go to and from the interface are taken as starting points. Of course, these two are equal at all times, as an interface cannot absorb mass.

In the early stage of a process of contacting two layers of immiscible liquids with some species being susceptible to exchange by just diffusion, penetration theory applies – just like with unsteady-state heat conduction (Section 3.3). In either phase, the mass transfer coefficient k is then given by

$$k = \sqrt{\frac{ID}{\pi t}} \qquad (4.117)$$

which is completely analogous to the model description presented in Sections 3.3.2 and 3.3.3.

Expression (4.117) can then be substituted into the expression for the mass flux ϕ_m'' to render equations like (4.104) and (4.105). Of course, penetration theory and its related expressions are valid as long as the penetration depth $\sqrt{\pi IDt}$ is sufficiently smaller than the thickness of the layer into which diffusion takes place – that is, as long as the driving force for diffusion is constant and still equal to the value $|c_{b0} - c_i|$ at the time $t = 0$ when the layers were brought into contact. [For longer times, long-term relations may be required analogous to those treated in Section 3.3.5.]

Example 4.13. Acetic acid–benzene–water.
A layer of benzene (thickness 1 cm) is laid carefully (without creating any flow) on a layer of pure water (also 1 cm thick). Acetic acid has been dissolved in the benzene with a concentration of 10^{-2} mol/L. The partition coefficient m, defined as the ratio between the acetic acid concentrations in benzene and water at equilibrium, is 10. The diffusion coefficients of acetic acid in benzene and in water are 0.5×10^{-9} m²/s and 1.5×10^{-9} m²/s, respectively.

There are a number of questions that need to be answered about the situation in these two layers, ten minutes after the layers come into contact with each other:
a)   Determine the penetration depths in both layers.
b)   Calculate the interface concentrations.
c)   How great are the concentration gradients on both sides of the interface?

Solution
Superscripts B and W again denote benzene and water, respectively, while subscripts b and i serve to represent a value in the bulk and at the interface, respectively.
   Penetration depth of acetic acid in water after 10 min: $\sqrt{\pi ID_W t} = 1.7$ mm.
   Penetration depth of acetic acid in benzene after 10 min: $\sqrt{\pi ID_B t} = 0.97$ mm.

Both penetration depths are clearly smaller than the thickness of the layers, so application of the penetration theory is allowed. This means that the fluxes on both sides of the interface are known; moreover, they are equal to each other, as all the acetic acid that comes out of the benzene goes into the water. In formula form:

$$\phi_m'' = \sqrt{\frac{ID_B}{\pi t}}(c_b^B - c_i^B) = \sqrt{\frac{ID_W}{\pi t}}(c_i^W - c_b^W) \tag{4.118}$$

Because both fluxes must be equal at any point in time, a link between the four concentrations at stake follows from equation (4.118). In addition, the ratio of the interface concentrations is known as m; that means $c_i^B = 10 c_i^W$. With data, $c_b^B = 10^{-2}$ mol/L and $c_b^W = 0$, it now follows that:

$$c_i^B = 8.5 \, \text{mmol/L}, \quad c_i^W = 0.85 \, \text{mmol/L}$$

The gradients at the interface now follow from the concentrations and the penetration depths:

$$\left.\frac{dc^W}{dx}\right|_i = -\frac{c_i^W - c_b^W}{\sqrt{\pi ID_W t}} = -0.5 \, \text{mol/L}^{-1} \, \text{m}^{-1} \tag{4.119}$$

$$\left.\frac{dc^B}{dx}\right|_i = -\frac{c_b^B - c_i^B}{\sqrt{\pi ID_B t}} = -1.55 \, \text{mol/L}^{-1} \, \text{m}^{-1} \tag{4.120}$$

Comparing this example with Example 3.5 (bitumen on oak) shows that these expressions are essentially identical. The only difference is the role of the partition coefficient, m. If this were now to be introduced artificially for heat transport (with $m = 1$ at equilibrium), then both expressions would be completely identical.

❗ Summary

For mass transfer across a phase interface, there is an important difference with heat transport: in equilibrium, the concentrations of species X in two phases A and B that are in contact with each other, generally, are **not** equal. They are related by the so-called partition coefficient m, which is a property of the three substances involved: A, B, and X.

This has consequences for the driving force for the mass exchange between the two phases: this is $\Delta c = c^A - m \cdot c^B$, where m is defined by requiring that $\Delta c = 0$ for the equilibrium situation in this relation. This also means that the expression for the overall mass transfer coefficient includes m, as well as k_A and k_B. For the flux ϕ_m'' of X, the following applies:

$$\phi_m'' = \left[\frac{1}{k_A} + \frac{m}{k_B}\right]^{-1}(c_B - m c_B)$$

where $\left[\dfrac{1}{k_A} + \dfrac{m}{k_B}\right]^{-1}$ stands for the total mass transfer coefficient K.

4.7 Simultaneous transport of heat and mass: the wet bulb temperature

In this section, we will finally analyse what happens when flow is forced to run along a damp surface. It is to be expected that the liquid will evaporate in such a situation: a question of mass transport. Given the heat that is required for evaporation, however, this evaporation results in the surface cooling down, which therefore leads to a heat flow from the surroundings to the damp body. We will look at this process with the help of the following example.

Consider a spherical liquid droplet in a gas flow, which is not saturated with the vapour of the liquid. At first, the temperature of the liquid is the same as that of the gas. Because of the evaporation of the liquid, the droplet cools off and a steady-state situation is eventually reached (see Figure 4.10). Assume in the analysis that the total quantity of liquid that evaporates is so slight that the diameter of the sphere remains constant.

So much heat has to be supplied from the air in this quasi-steady-state situation that it actually compensates for the heat of evaporation associated with the mass flow rate:

$$\phi_q = \phi_m \cdot \Delta h_v \tag{4.121}$$

where Δh_v represents the **enthalpy of evaporation** of the liquid.

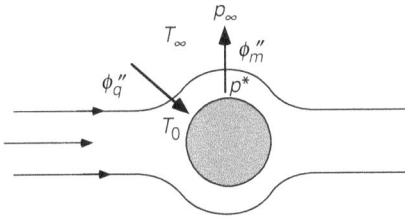

Figure 4.10

The mass flow rate depends on the driving force Δc in the gas phase and the mass transfer coefficient k in the gas phase. For the heat flow, this depends on ΔT and h, entirely analogously:

$$\phi_m = \pi D^2 k \, \frac{\mathfrak{m}}{RT} (p^* - p_\infty) \tag{4.122}$$

$$\phi_q = \pi D^2 h \, (T_\infty - T_0) \tag{4.123}$$

In these two equations, $\mathfrak{m}$ stands for the molar mass of the liquid, p^* for the vapour pressure of the substance at temperature T_0, T_∞ for the temperature of the incoming gas, p_∞ for the vapour pressure of the incoming gas (which of course has to be less than p^*, otherwise no evaporation could occur), and T_0 for the **wet bulb temperature** which is the temperature that the droplet assumes in the equilibrium situation; for T in equation (4.122), the usual choice is the mean of T_∞ and T_0.

Substituting equations (4.122) and (4.123) into equation (4.121) gives

$$\frac{p_\infty - p^*}{T_\infty - T_0} = -\frac{RT}{m \, \Delta h_v} \frac{h}{k} \tag{4.124}$$

For the two unknowns T_0 and p^*, one needs two equations: equation (4.124) and an expression for the saturation pressure $p^*(T)$. For a graphical solution technique, the **humidity chart** of Figure 4.11 is used (for air–water at 1 bar). Because of the minus sign on the right-hand side of equation (4.124), the straight lines in Figure 4.11 (shown with the term 'adiabatic saturation line') all have a negative slope.

If Re is not too small, then the following applies to h/k, thanks to **Chilton and Colburn**:

$$\frac{h}{k} = \rho \, c_p \, \mathrm{Le}^{2/3} \tag{4.125}$$

where Le stands for the **Lewis number**: see also equation (2.38). This means that the following applies, where Re is not too small:

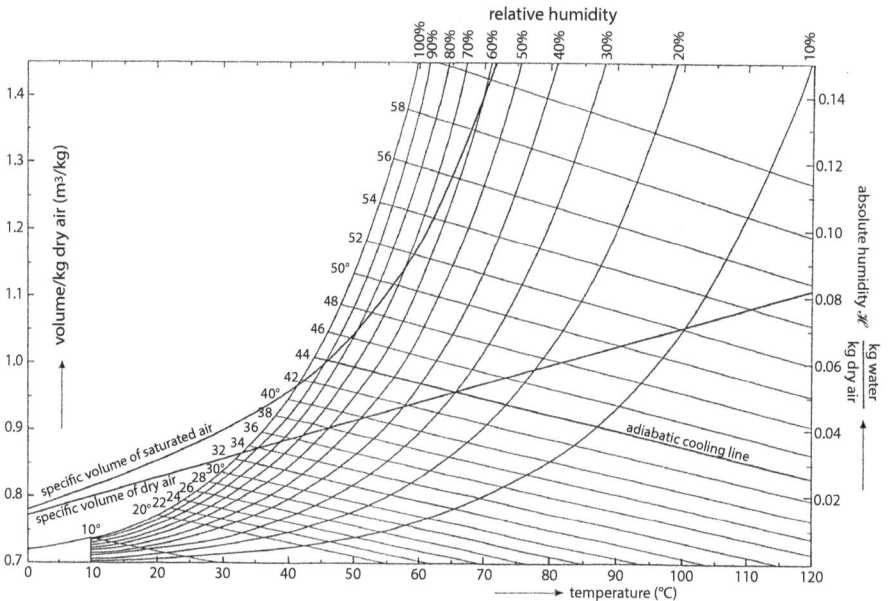

Figure 4.11

$$\frac{p_\infty - p^*}{T_\infty - T_0} = -\frac{RT\rho c_p}{\eta \, \Delta h_v} \mathrm{Le}^{2/3} \tag{4.126}$$

Equation (4.126) is actually an equation for T_0, because the **equilibrium vapour pressure** p^* is a function of the temperature of the liquid, so for p^* the value at the **vapour pressure line** at T_0 has to be taken. Notice that in equation (4.126) neither the Reynolds number nor the diameter of the sphere occurs! For conditions of exclusively molecular transport around the sphere (shown by Nu = 2 and Sh = 2), relation (4.126) applies as well.

The principle of the wet bulb temperature is used among other things to help determine the humidity of a gas flow. Many drying and humidity-related fields often involve – rather than vapour pressures – the use of the **absolute humidity** which is defined as the mass (in kg) of vapour in a gas per kg of dry gas. Other terms also feature, such as **relative humidity** – the pressure of the vapour in a gas mixture divided by the saturation pressure of vapour at the same temperature – and **dew point** – the temperature at which condensation of vapour begins when humid gas is cooled down: in other words, the temperature at which humid gas is saturated with vapour.

Some examples in which the humidity chart is used, are given below.

Example 4.14. Determining the dew point.
What are the absolute humidity and the relative humidity of air at 70 °C with a wet bulb temperature of 55 °C? What is the dew point?

Solution
The intersection of the adiabatic saturation line of this air with the saturation line is at 55 °C. We will now move along this adiabatic saturation line to the right to 70 °C and see that the absolute humidity is 0.105 kg water/kg dry air. We also see that the relative humidity is 39%.

Cooling the air increases the relative humidity; the absolute humidity does not change in the process. In the humidity chart, we can see that air with an absolute humidity of 0.105 kg water/kg dry air is saturated at 53.7 °C. This is the dew point we were looking for.

Example 4.15. Condensing through cooling.
The same air is cooled to 30 °C.
How much vapour will condense for every m³ of saturated air at 30 °C?

Solution
The humidity chart tells us that saturated air at 30 °C contains 0.027 kg water/kg dry air. Therefore, 0.105−0.027 = 0.078 kg water/kg dry air has condensed. We also see from the humidity chart that the saturated volume for air at 30 °C is 0.88 m³/kg dry air. The quantity of condensed vapour is therefore

$$\frac{0.078}{0.88} = 0.089 \text{ kg water/m}^3 \text{ saturated air at 30 °C.}$$

Incidentally, this principle of simultaneous transport of mass and heat is not restricted to liquid–gas systems. When **dissolving** crystals, for example, in a liquid, this combined effect of mass and heat transfer occurs as well.

! **Summary**

One of the best illustrations of the analogy between heat and mass transfer is found with the wet bulb, where heat transport supplies the heat of evaporation that is needed for the evaporation process (the mass transport). In some circumstances, the wet bulb temperature is independent of flow conditions and sphere dimensions.

The wet bulb also illustrates that in some cases mass transport is limited by heat transfer. In any case, heat transport plays a role in processes that involve a phase transition.

5 Fluid mechanics

5.1 Introduction

Chapter 1 saw the introduction of balances, which proved to be very useful in that and subsequent chapters when solving all kinds of transport problems. The form of these balances is always the same:

$$\frac{d}{dt} = \text{in} - \text{out} + \text{production} \tag{5.1}$$

At the start of this chapter on fluid mechanics, it seems a good idea to demonstrate again, using balances, how flow is obtained through the supply of momentum (unit: Ns) and how a flow obeys and is described by mass, momentum, and energy balances.

Example 5.1. A wastewater pump.
A basin contains water, in which a highly corrosive substance is dissolved. The water has to be pumped into tanks via a transport pipeline. However, the pump that is available may not come into contact with the corrosive substance, so a separate structure is created, as shown in Figure 5.1.

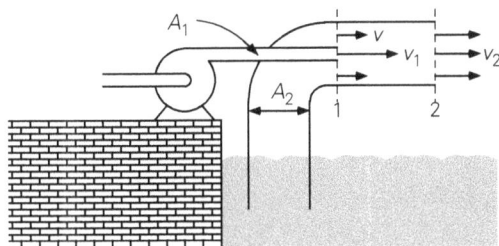

Figure 5.1

A smaller pipe with cross-sectional area $A_1 = 20$ cm^2 is connected to the discharge side of the pump and is then inserted into the pipeline with cross-sectional area $A_2 = 180$ cm^2) on the exterior of a bend. The pump pushes clean water at a uniform velocity of $v_1 = 10$ m/s into the pipeline. Part of the momentum of this water is transferred to the contaminated water in the pipeline. This way, water is sucked out of the basin and pumped to the tanks. At some distance (at point 2 in the diagram) downstream of the mixing point (point 1), the velocity at A_2 is again uniform and equal to $v_2 = 2$ m/s. This situation can be considered as steady state over time. Also, for the sake of simplicity, friction on the walls of the pipes may be disregarded, and it can be assumed that the density of the clean and contaminated water is equal (i.e. 10^3 kg/m^3). Derive, for the domain between planes 1 and 2, balance equations for mass, momentum in the horizontal direction, mechanical energy, and thermal energy.

Solution
First to follow is the velocity v of the contaminated water at point 1, thanks to a **mass balance** for the entire volume between 1 and 2, from the given velocities v_1 and v_2:

https://doi.org/10.1515/9783111246574-005

$$0 = \rho A_1 v_1 + \rho (A_2 - A_1) v - \rho A_2 v_2 \qquad (5.2)$$

Entering the values given shows that velocity v is equal to 1 m/s and that the flow rate of the contaminated water is 16 L/s.

Every force operating in the direction of flow has to be included in a balance for the **momentum** in the direction of flow. Given that friction on the walls can be disregarded, friction forces along the walls of the pipe are not relevant. And because gravity does not work in the direction of flow, the only force to remain is pressure forces at the entrance and exit planes at points 1 and 2. There is a uniform pressure p_1 in the whole of plane 1: after all, there is no radial flow there. This means that the momentum balance for the same control volume between planes 1 and 2 is

$$0 = \rho A_1 v_1 \cdot v_1 + \rho (A_2 - A_1) v \cdot v - \rho A_2 v_2 \cdot v_2 + A_2 p_1 - A_2 p_2 \qquad (5.3)$$

Eliminating v from equation (5.3) with the help of equation (5.2) produces for the pressure drop from 1 to 2:

$$p_1 - p_2 = -\rho \frac{A_1}{A_2 - A_1} (v_1 - v_2)^2 < 0 \qquad (5.4)$$

There is therefore no pressure drop, but an increase: $p_2 = p_1 + 8 \times 10^3$ Pa! The fact that at point 1 there is **under**pressure compared with at point 2 is due to the transfer of momentum from the discharge pipe of the pump. This input of momentum is now the driving force behind the flow: it sucks as it were the contaminated water from the basin. Where the velocities between clean and contaminated water are levelled out and the sucking action is over, pressure p_2 is higher precisely because of the input of momentum through plane 1, which is not yet expressed in (static) pressure p_1. From Bernoulli's law, we know that kinetic energy can be converted into 'pressure' energy and that deceleration can lead to an increase in pressure.

The transfer of momentum to the contaminated water is, it should be mentioned, accompanied by dissipation of mechanical energy. This follows directly from the **mechanical energy balance**:

$$0 = \rho A_1 v_1 \left(\frac{1}{2} v_1^2 + \frac{p_1}{\rho} \right) + \rho (A_2 - A_1) v \left(\frac{1}{2} v^2 + \frac{p_1}{\rho} \right) +$$
$$- \rho A_2 v_2 \left(\frac{1}{2} v_2^2 + \frac{p_2}{\rho} \right) - \phi_m e_{diss} \qquad (5.5)$$

Eliminating v with the help of the mass balance (5.3), eliminating $p_1 - p_2$ with the help of the momentum balance (5.4), and dividing by the mass flow rate produces, for the dissipation:

$$e_{diss} = \frac{A_1 v_1}{A_2 v_2} \frac{1}{2} v_1^2 + \frac{A_2 v_2 - A_1 v_1}{A_2 v_2} \frac{1}{2} \left(\frac{A_2 v_2 - A_1 v_1}{A_2 - A_1} \right)^2 +$$
$$- \frac{1}{2} v_2^2 - \frac{A_1}{A_2 - A_1} (v_1 - v_2)^2$$
$$= 18 \, J/kg \qquad (5.6)$$

Of course, this mechanical energy is lost in the heating up of the water.

The increase in temperature can be determined by drawing up a **thermal energy balance**. Assume here that the water at point 1 is at uniform temperature T_1. The internal energy concentration u of the contaminated water at point 1 is now therefore equal to that of the clean water (the same T and c_p). The thermal energy balance for the control volume is

$$0 = \rho A_1 v_1 \cdot u_1 + \rho (A_2 - A_1) v \cdot u_1 - \rho A_2 v_2 \cdot u_2 + \phi_m e_{diss} \qquad (5.7)$$

and with the help of mass balance (5.2) can be easily simplified to

$$u_2 - u_1 = e_{diss} \longrightarrow T_2 - T_1 = \frac{e_{diss}}{c_p} = 4.3 \cdot 10^{-3}\,\text{K} \tag{5.8}$$

It was possible to make statements about the flow in this example because information was known about the velocities v_1 and v_2 (and therefore about the driving force and resulting overall flow rate). In spite of the different balances, this information led us to values for the contaminated water flow rate, difference in pressure and increase in temperature. This is in fact a specific example, just as in Example 1.20 where the flow rate and the difference in pressure between two positions in a flow were given. In practice, this much information will not be available, especially at a design stage.

Usually just one fact is known and the remainder are needed: for a given difference in pressure (or, generally, a driving force), how great will the flow rate be; or: what magnitude of the difference in pressure is required in order to bring about a particular flow rate. In these cases, it will be necessary to know more about the associated energy dissipation. This will concern the e_{diss} term in the mechanical energy balance introduced in Chapter 1.

The emphasis in the first part of this chapter lies on **engineering fluid mechanics***: how energy dissipation can be calculated with the help of empirical friction or pressure drop coefficients. First, a number of different flow meters will be looked at (in Section 5.2), before the focus shifts onto the link between flow rate through and pressure drop over transport pipelines (Sections 5.3 and 5.4) and packed beds (Section 5.5). Balances play an important role here as well.

Laminar flows are covered in the second part of this chapter. The technique that is used in describing these flows entails drawing up force balances for typical volume elements somewhere in the flow field (micro-balances). In the case of laminar flows, the transport of momentum perpendicular to the direction of flow (usually expressed in terms of shear stresses: see Section 2.1.4) is solely maintained by the molecules; for this reason, a distinction has to be made between fluids that obey **Newton's viscosity law** concerning the link between shear stress and velocity gradient (i.e. concerning **molecular momentum transport**, the so-called **Newtonian fluids**) and fluids that do not do so (the so-called **non-Newtonian liquids**). This will be dealt with separately in Sections 5.6 and 5.7, respectively. At the end of the chapter (in Section 5.8), the Navier–Stokes equations will be introduced, which describe any flow in a multi-dimensional domain, including time-dependent, and compressible, or incompressible ones.

5.2 Flow meters

In Section 1.3.4, two instruments were introduced with which the flow velocities of liquids or gases can be determined, namely the Pitot tube and the Venturi tube. Three more

flow rate meters will be introduced in this section: the weir, the orifice meter, and the rotameter. The weir requires a somewhat more detailed examination based on the Bernoulli equation than the Pitot tube and the Venturi tube, while energy dissipation should certainly not be disregarded as far as the orifice meter and rotameter are concerned.

5.2.1 Weir

Figure 5.2 shows a **weir** in a stretch of water (a ditch, a canal, or a stream). The weir, which is effectively a wall in the water and thereby forms an obstruction to the flow, extends across the full width of the water. Let us assume for now that the crest of the weir is sharp and horizontal. The water level upstream of the weir is higher than the crest, while that downstream is lower. If the upstream water level is sufficiently higher than the crest, and providing that the supply of water is sufficiently great, a layer of water will pass over the sharp crest (because of the inertia of the water) and plunge into the water downstream.

This sheet of water, or **nappe**, which stretches across the full width of the stretch of water, passes over the crest without any friction at all (so $e_{diss} = 0$) and moves freely through the air so that the pressure throughout the sheet of water just after the weir is atmospheric. Figure 5.2a gives a side view, and shows that the water level starts to fall as it approaches the weir (because the water accelerates towards the weir). Because the supply of water occurs across the whole depth of the stretch of water, water that is lower than the crest of the weir will also flow over it. All told, then, the water velocities above the sharp edge of the weir are not purely horizontal.

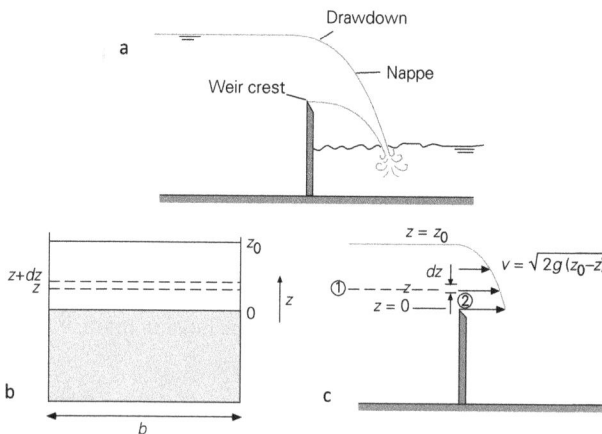

Figure 5.2

In a steady-state situation, the flow rate over the weir depends on the upstream water level, z_0 (measured from the crest of the weir). This means that the flow rate can be regulated by moving the weir vertically upwards or downwards to reduce or increase z_0. The question here is how the flow rate depends quantitatively on z_0.

To answer this question, we should realise that the flow rate in the stretch of water is equal to the product of velocity times cross-sectional area, both in a vertical plane 1 sufficiently far upstream of the weir and in the vertical plane 2 (see Figure 5.2c). In plane 1 nothing is known about the local velocity profile or the averaged velocity, while in plane 2 the flow rate depends on z_0. Figure 5.2b shows how the flow rate in plane 2 is made up of contributions from thin, horizontal strips of height dz, the average velocity in each strip being (potentially) a function of z. How the water exactly flows of the crest, depends on the local energy housekeeping as described in terms of the mechanical energy balance between the points 1 and 2, with point 1 sufficiently far upstream of the weir and with point 2 exactly above it (see Figure 5.2c).

To this end, it is useful and usual to simplify the situation somewhat: see Figure 5.2b and c. The fall of the water level, or drawdown, upstream of the weir is ignored, the velocity of the water above the crest of the weir is assumed to be horizontal, and the pressure is assumed to be atmospheric ($p = p_0$) already above the sharp crest (and not just in the nappe). Further, ρ is constant, $\phi_w = 0$, $\phi_q = 0$ and $e_{\text{diss}} = 0$. This means that the **Bernoulli equation** – equation (1.126) – can be used here:

$$\frac{1}{2}v_1^2 + g\,z_1 + \frac{p_1}{\rho} = \frac{1}{2}v_2^2 + g\,z_2 + \frac{p_2}{\rho} \tag{5.9}$$

For points 1 and 2 at the same height z (above the crest of the weir), the following applies:

Point 1: $z_1 = z$; $\quad v_1 = v(z)$; $\quad p_1 = p_0 + \rho g(z_0 - z)$
Point 2: $z_2 = z$; $\quad v_2 \gg v_1$; $\quad p_2 = p_0$

It should be pointed out that this latter assumption is only justified if z_0 is markedly smaller than the depth of the stream upstream of the weir (at point 1). It is evident that the unknown variable here is $v_2(z)$. Entering this into equation (5.9) gives

$$v_2 = \sqrt{2g(z_0 - z)} \tag{5.10}$$

Actually, v_2 depends on z. Equation (5.10) expresses that the velocity at height z depends on the thickness $z_0 - z$ of the liquid layer above height z that in a way 'pushes' the liquid at height z over the weir. Equation (5.10) further implies that at height z_0 (*i.e.* on the free water surface) the water velocity is zero. This result relates to the above simplification $v_2 \gg v_1$ that leads to $v_1(z) = 0$. Close to the water surface ($z = z_0$), this simplification is rather dubious.

Equation (5.10) can now be used to calculate the contribution made by a thin strip between z and $z + dz$ (see Figure 5.2b) to the flow rate over the weir:

$$d\phi_V = v_2(z) \cdot b \, dz = b\sqrt{2g(z_0 - z)}\,dz \tag{5.11}$$

The overall flow rate follows from integrating equation (5.11) while using $\zeta = z/z_0$:

$$\phi_V = b\sqrt{2g} \int_0^{z_0} \sqrt{z_0 - z}\, dz =$$

$$= b\sqrt{2g}\, z_0^{3/2} \int_0^1 \sqrt{1 - \zeta}\, d\zeta = \frac{2}{3}b\sqrt{2g\, z_0^3} \tag{5.12}$$

A new and important aspect of the above analysis is the use of the Bernoulli equation for calculating the velocity locally, that is, at height z – in line with the physics of the situation.

In fact, the situation at a weir is not essentially different to that of Example 1.21, except that in Example 1.21 the variation of the velocity over the 'height' of the hole could be disregarded, because the size of the hole was considered very small in comparison with the liquid height H in the container.

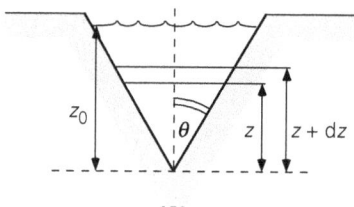

Figure 5.3

There are all kinds of other types of weir or overflow, such as those in the form of a triangle (see Figure 5.3). The analysis used in this case is entirely in parallel with that of a completely horizontal weir, with the only difference being in determining the flow rate. After all, the width b of the strip is also a function of z: $b = 2z \tan \theta$. The flow rate can now be calculated from

$$\phi_V = 2 \tan \theta \sqrt{2g} \int_0^{z_0} z\sqrt{z_0 - z}\, dz = \frac{8}{15} \tan \theta \sqrt{2g}\, z_0^{5/2} \tag{5.13}$$

Expressions (5.12) and (5.13) overestimate the actual flow rates. This is caused primarily by the simplifications that were made for Figure 5.2b. In fact, the lowering of the water level in the run towards the weir means that the size of the area that is available to flow is smaller. In addition, the velocity above the crest is not horizontal throughout. These effects are corrected with the help of a discharge coefficient C_d, with $C_d < 1$, to be entered into the equation for the flow rate; this means, for example, that equation (5.12) changes to

$$\phi_V = \frac{2}{3} C_d\, b \sqrt{2g\, z_0^3} \tag{5.14}$$

This method is entirely comparable to that in Example 1.21 for the outflow through a small hole: see equation (1.169). There, too, the area through which the flow passes, is, as a result of contraction, smaller than the through-flow area available from a geometric perspective, and consequently the flow rate is also smaller. In effect, C_d stands for the ratio of the actual area through which the flow passes to the size of the area available for flow but not entirely utilised by the fluid.

If the water does not 'shoot' over the crest, but simply runs slowly over it, or if the weir does not have a sharp edge, then energy dissipation will play a role, and the flow rates such as those shown in equations (5.12)–(5.14) will have to be corrected accordingly as well. The above derivative is not entirely correct either in the event that the water level downstream of the weir is not particularly low so that there is no real 'free fall' of water passing over the weir.

Weirs are also used with trays in distillation towers (in oil refineries, for example) for separating (or fractionating) multicomponent mixtures according to their boiling point. A distillation tower contains a large number of trays, two of which are shown in Figure 5.4. Each tray is kept at a certain temperature, with the tray at the bottom of the tower being the hottest, and the tray at the top, the least hot. The aim with each tray is to maintain a balance with regard to the composition of vapour and liquid, for which rising vapour and falling liquid flows have to be brought into close contact. This is achieved by bubbling vapour through a layer of liquid on each tray. The height of the liquid on each tray, and therefore the duration of the contact between the vapour and the liquid, depends very much on the weir height.

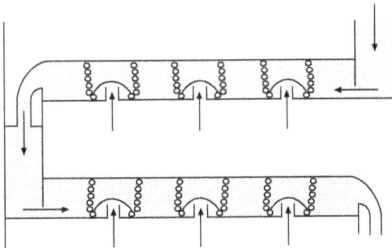

Figure 5.4

5.2.2 Orifice meter

The **Venturi tube** was discussed in Section 1.3.4 – this is used for measuring the flow rate through a pipe. A Venturi tube includes both a gradual narrowing of the diameter and a very gradual widening that prevents energy dissipation. This narrowing can also be made very 'sudden', in which case **energy dissipation** can no longer be disregarded! This is the case with the **orifice meter**.

This involves placing a disc in the pipe with a small hole, or orifice, in it (see Figure 5.5). As can be seen in the diagram, eddies occur primarily to the rear of the disc, which cause a considerable dissipation of mechanical energy. Contraction of the flow also occurs here as it passes the discharge opening. An analysis using Bernoulli now therefore has to be 'corrected' to allow for dissipation and contraction with the help of a **discharge coefficient** C that depends on the points at which pressures p_1 and p_2 are measured. The discharge coefficient is also a function of Re and of the ratio of the pipe diameter and the orifice diameter. For example, for an orifice meter with a sharp-edged hole, the following applies: $C = 0.62$ providing that Re $> 10^4$ – see also the discharge coefficient in Example 1.21.

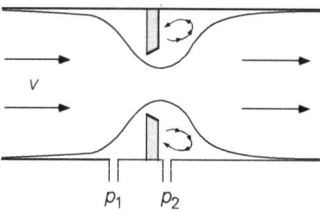

p_1　p_2　　　　　**Figure 5.5**

A significant disadvantage of using orifice meters is dissipation: after the orifice meter, the pressure does not return to the value that was present upstream, because mechanical energy has been destroyed. This means that a considerable pressure drop can occur over the orifice meter. In summary, the flow rate (for liquids) as a function of the pressure drop measured is as follows:

$$\phi_m = C\left(\text{Re}, \frac{D_2}{D_1}\right) \cdot \frac{A_2}{\sqrt{1 - \left(\frac{A_2^2}{A_1^2}\right)}} \cdot \sqrt{2\rho\,(p_1 - p_2)} \tag{5.15}$$

where D_1, A_1 and D_2, A_2 are the diameter and the cross-sectional area of the pipe and the orifice opening, respectively.

5.2.3 Rotameter

The **rotameter** is a smart variant of the orifice meter. Again, the principle is based on a change to the diameter in the pipe and measuring pressure difference that results from this. Instead of a fixed disc in the pipe, however, a float is fitted (see Figure 5.6). The rotameter should always be fitted exactly vertically. The reason for this is simple: the upward flow exerts a force on the float, which in a steady-state situation makes equilibrium with gravity and with the **buoyancy** of the float, which allows it to float.

If the tube of the rotameter were to have a constant diameter, there would be just one flow rate at which the float would be in equilibrium. This is impractical for a flow rate meter. For this reason, the rotameter has a variable diameter: the tube gets wider and wider towards the top. This means that with a certain flow rate, the float will remain floating at a particular height. It is precisely this property that makes the rotameter such a useful measuring instrument: the position of the float is a measure for the force of the upward flow on the float and therefore for the flow rate; by making the tube transparent, it is possible to see the position of the float and therefore determine the flow rate.

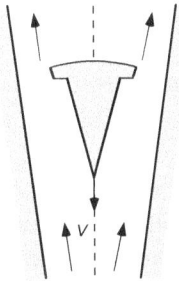

Figure 5.6

In principle, the force of the flow on the float (float volume V_o, A_o the area of the largest cross section of the float, ρ_o the density of the float material) can be modelled according to the recipe in Section 2.3. In Section 2.3, it was assumed that the object around which the flow was moving was in an 'infinite' medium, so that there were no influences from the sides. With the rotameter, the tube wall is very nearby. This is why the rotameter is modelled differently: analogous to the expression for the orifice meter. The pressure drop over the float as a result of the flow is replaced by the net gravity that affects the float, thanks to a force balance on the float. For the flow rate, this produces

$$\phi_m = C \cdot (A_{\text{tube}}(z) - A_o) \cdot \sqrt{2\rho_f \left(\rho_o - \rho_f\right)g \frac{V_o}{A_o}} \tag{5.16}$$

where ρ_f is the density of the liquid and $A_{\text{tube}}(z)$ the cross-sectional area of the vertical tube at height z. If Re (related to the annular gap between float and tube wall) is sufficiently large, then discharge coefficient C is constant.

Equation (5.16) expresses how the flow rate depends on the vertical position z in the tube that is taken by the float in response to the forces exerted on it. A calibration graph is usually used instead of equation (5.16) for the relation between position z of the float and the flow rate. Note that every type of float (shape, material) requires a different calibration graph. A rotameter is generally made in such a way that A_{tube} increases linearly by height z so that the calibration graph of the rotameter is virtually linear.

! Summary

Three flow meters have been analysed in this section: the weir, the orifice meter, and the rotameter. All three analyses have taken the Bernoulli equation as their starting point. A correction had to be made to the results of the analyses due to contraction of the liquid flow and because of energy dissipation. In principle, this so-called discharge coefficient depends on the geometries and the Reynolds number.

In the case of the weir, dissipation plays a secondary role. With the orifice meter and rotameter, however, there is a considerable degree of dissipation and therefore a striking pressure drop with the orifice meter or float, respectively.

5.3 Pressure drop along a straight pipeline

In Example 1.10, we looked at a horizontal straight round tube, through which water flowed as a result of a difference in pressure over the tube imposed by a pump. With the help of an energy balance, it was derived that the pressure energy (=mechanical energy) in the pipe was converted to internal energy (=thermal energy). This process is known as dissipation of mechanical energy. The underlying mechanism is friction. In this example, both pressure drop and flow rated were given. The link between pressure drop and flow rate will be discussed in more detail below so that one can be calculated if the other is known.

5.3.1 The concept of the friction factor

To this end, consider a straight horizontal pipe (length L, diameter D) through which water is flowing at a given mean velocity $\langle v \rangle$ – see Figure 5.7.

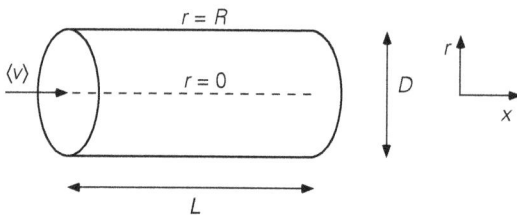

Figure 5.7

How large is the pressure drop that is needed in order to allow the water to flow through the pipe at mean velocity $\langle v \rangle$?

This pressure drop is needed in order to overcome the **friction** at and resulting from the pipe walls. Friction is the result of shear stress between individual fluid layers or between a fluid layer and a wall. The term 'shear stress' was introduced in Section 2.1.4 as an alternative description to molecular momentum transport. The

shear stress is a force per unit of surface area exerted by one layer of fluid on another (adjacent) layer or wall. This force is parallel to the direction of flow.

In order to be able to calculate the pressure drop in the pipe, it is therefore necessary to know what the shear stress at the wall is. The shear stress exerted by the wall
on the fluid can in principle be calculated from the following:

$$\tau_{w \to f} = -\tau_{f \to w} = -\left\{ -\mu \left[\frac{dv_x}{dr}\Big|_{r=R}\right] \right\} = \mu \left[\frac{dv_x}{dr}\Big|_{r=R}\right] \tag{5.17}$$

To calculate the shear stress at the wall, it would therefore be necessary to know the
velocity profile in the pipe so that the derivative from that could be used to determine
the velocity at the wall. In general, this is a difficult task, if not an impossible one.
This is why the technique of the dimensional analysis will again be used in order to
find out what the shear stress depends on.

The shear stress is a function of the mean fluid velocity $\langle v \rangle$ in the pipe, the diameter D of the pipe, the viscosity μ, and the density ρ of the fluid:

$$\tau_{w \to f} = f(\langle v \rangle, D, \mu, \rho) \tag{5.18}$$

Carrying out the dimensional analysis produces

$$\frac{\tau_{w \to f}}{\rho \langle v \rangle^2} = k \left(\frac{\mu}{\rho \langle v \rangle D}\right)^a = k \cdot Re^{-a} \tag{5.19}$$

The shear stress exerted by a wall on a fluid flow can be modelled in the same way as
the force exerted by a flow on an immersed body (see Section 2.3):

$$\tau_{w \to f} = -f \cdot \frac{1}{2} \rho \langle v \rangle^2 \text{ with } f = f(Re) \tag{5.20}$$

The factor f is known as the **friction factor**. The minus sign in equation (5.20) expresses the opposing character of the force exerted by the wall on the fluid.

With a view to determining the pressure drop over a pipe length L for a steady-
state flow through a straight horizontal pipe from Figure 5.8, a momentum balance
can be drawn up. The control volume is formed by the pipe wall and the two planes 1
and 2, separated by distance L.

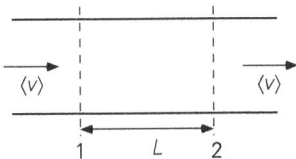

Figure 5.8

As long as the cross-sectional areas A_1 and A_2 of the tube at positions 1 and 2 are equal (=A), the x-momentum flow 'in' and the x-momentum flow 'out' will cancel each other out. This means the x-momentum balance is reduced to a force balance:

$$0 = p_1 A - p_2 A + \tau_{w \to f} \cdot SL \tag{5.21}$$

For a cylindrical tube, $A = \pi D^2/4$. The symbol S stands for the **wetted perimeter** (wetted by the 'liquid') of the tube; for an entirely filled cylindrical tube this is πD. Equation (5.21) is written in as general terms as possible, which means it also applies, for example, to an open ditch through which water is flowing. The term 'wetted perimeter' fully covers the cause of the pressure drop: friction on the wall.

The pressure drop can be determined easily from equation (5.21):

$$p_1 - p_2 = - \tau_{w \to f} \cdot \frac{SL}{A} \tag{5.22}$$

For the pressure drop resulting from energy dissipation due to wall friction (shear stress at the wall), combining equations (5.20) and (5.22) produces

$$p_1 - p_2 = f \cdot \frac{SL}{A} \cdot \frac{1}{2} \rho \langle v \rangle^2 \tag{5.23}$$

For a cylindrical tube (with diameter D) completely filled with fluid, it holds

$$\frac{SL}{A} = \frac{\pi DL}{\frac{\pi}{4} D^2} = 4 \frac{L}{D} \tag{5.24}$$

Entering this into equation (5.23) produces what is known as the **Fanning pressure drop equation** that for flow through cylindrical tubes reproduces the wall friction:

$$p_1 - p_2 = \Delta p = 4f \cdot \frac{L}{D} \cdot \frac{1}{2} \rho \langle v \rangle^2 \tag{5.25}$$

where the coefficient $4f$ is denoted as the **Fanning friction factor**[18] being a function of the Reynolds number: $4f = 4f\,(\text{Re})$.

For non-cylindrically shaped tubes, an equation that is entirely analogous to equation (5.25) is used. To that end, diameter D is replaced, not only in equation (5.25) but also in Re, by the so-called hydraulic diameter D_h, which is defined as

$$D_h \equiv \frac{4A}{S} \tag{5.26}$$

---

18 In the world of engineering, also the **Moody friction factor** f is widely used, being defined by equation (5.25) without the '4' – implying that the numerical values of the Moody f are equal to 4 times Fanning's f values. This textbook uses the Fanning friction factor just because of the ease of the hydraulic diameter.

Remember that A is the surface through which the flow passes. This is not necessarily the same as the cross-sectional area of the pipe. Moreover, S is the perimeter on which the shear stress is exerted (what is referred to as the **wetted perimeter**) and this in turn is not necessarily the same as the perimeter of the pipe.

Example 5.2. Flow in an open channel.
Water flows through a rectangular channel (height of walls is H, width is w). The height of the water level in the channel is $h < H$.
   What is the hydraulic diameter of this system (see Figure 5.9)?

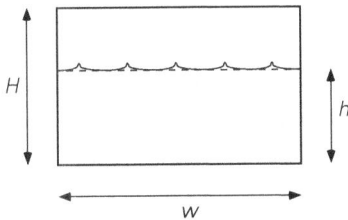

Figure 5.9

Solution
The cross-sectional area A through which the water flows is given by $w \cdot h$ – not by $w \cdot H$. Similarly, for the wetted perimeter, $S = 2h + w$ and not $2H + 2w$: after all, there is no wall exerting shear stress on the surface of the water (the effect of the air is negligible compared with the effect of a channel wall!). The use of H instead of h for the wetted perimeter is also erroneous: the height of the channel wall is not relevant here. For the hydraulic diameter, all this produces:

$$D_h = \frac{4A}{S} = \frac{4wh}{2h + w} \tag{5.27}$$

5.3.2 The use of the friction factor

As already mentioned, the Fanning friction factor is a function of the Reynolds number (in relation to the tube diameter). In Figure 5.10, $4f$ is plotted as a function of Re for tubes with a circular cross section. More or less analogously to how the resistance coefficient C_D for an object immersed in a flow depends on the Reynolds number related to the particle (see Figure 2.16), here too there are two separate regimes, namely that of **laminar** and of **turbulent pipe flow**.
   For the **laminar** regime, the following can be derived (see Section 5.6.4):

$$4f = 64/\text{Re} \quad \text{providing that } \text{Re} < 2{,}000 \tag{5.28}$$

while in the turbulent regime, the empirical formula of Blasius applies:

$$4f = 0.316\ \text{Re}^{-1/4} \quad \text{providing that } 4{,}000 < \text{Re} < 10^5 \tag{5.29}$$

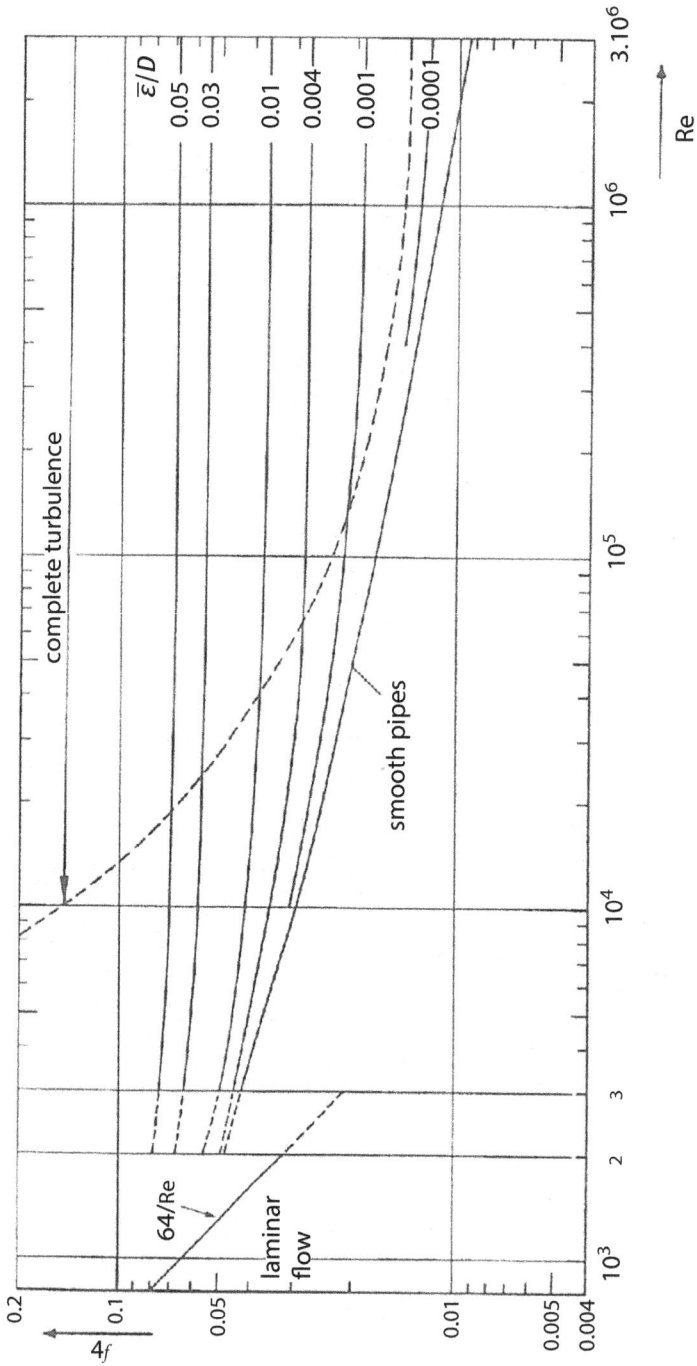

Figure 5.10

Remember that with C_D, the range in which $C_D \propto 1/Re$ is limited to Re < 1, while the laminar regime with tubular flow extends to Re ≈ 2,000! This once more illustrates that the critical value of the Reynolds number for the transition laminar–turbulent depends on the geometry.

In Figure 5.10, several lines have been drawn for $4f$. These show that the resistance that a fluid encounters in a tube also depends on the **wall roughness**. This is logical (see Figure 5.11), given that an extra eddy will occur behind any protrusion in the wall. Of course, this creates more dissipation of usable mechanical energy and therefore leads to more resistance. Or, in terms of shear stress, the extra eddy means that momentum will be transferred from the liquid to the wall more effectively. And shear stress is an alternative interpretation of molecular momentum transport.

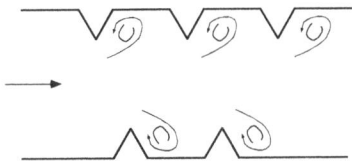

Figure 5.11

The roughness of the wall is usually described by the (non-dimensional) **relative roughness** $\bar{\varepsilon}/D$, where $\bar{\varepsilon}$ is the absolute roughness: the mean height of the 'protrusions' on the wall. As shown in the $4f$ versus Re plot, $4f$ becomes constant if $\bar{\varepsilon}/D$ or Re are sufficiently great. The larger the relative roughness, the more $4f$ will be constant as the value of Re gets smaller. This can be understood by using the image of the eddies transporting momentum. With a high level of relative wall roughness, the protrusions stick out relatively far, and the eddies effectively carry momentum from the bulk of the flow to the wall.

There now follow some examples for calculating pressure drops in straight pipelines.

Example 5.3. Pressure drop along a horizontal oil pipeline.
A viscous oil ($\mu = 70 \times 10^{-3}$ Ns/m^2, $\rho = 900$ kg/m^3) flows through a straight, horizontal, cylindrical tube (length 10 m, diameter 10 cm, smooth wall) in a steady state at a flow rate of 7.85×10^{-4} m^3/s.
    How great is the pressure drop Δp along this pipeline?

Solution
Equation (5.25) can be used directly for this calculation. From the flow rate, it follows that $\langle v \rangle = 0.1$ m/s, from which it follows that Re = 129. The flow is therefore laminar and thanks to equation (5.28), it follows that $4f = 0.5$. Entering this into equation (5.25) now produces $\Delta p = 225$ Pa.

Example 5.4. Velocity in a horizontal water pipeline.
Water flows through a long, horizontal, straight pipe (length 1 km, diameter 10 cm, relative roughness 0.001). Again, the situation is steady. The pressure drop along the pipe is 2 bar.
    What is the velocity of the water in the pipe?

Solution
Now, the velocity has to be solved with the help of equation (5.25):

$$\langle v \rangle = \sqrt{\frac{D}{L}\frac{2\,\Delta p}{\rho}\frac{1}{4f}}$$

(5.30)

This produces the same problem as when determining the uniform velocity of a body falling or rising through a fluid (see Section 2.3.3). Velocity $\langle v \rangle$ can only be calculated if $4f$ is known, while $4f$ itself is a function of $\langle v \rangle$. Here, too, an iterative solving procedure offers a way out: choose a velocity, say, of $\langle v \rangle = 2$ m/s; this gives Re = 2×10^5; looking this up on the $4f$ versus Re plot gives (reading it from the relative roughness line = 0.001!) $4f = 0.02$; entering this value into equation (5.30) produces a new $\langle v \rangle$: namely, 1.4 m/s. This then means Re = $1.4 \times 10^5 \rightarrow 4f = 0.02$: and that's it! In general, this calculation procedure will have to be repeated several times before $\langle v \rangle$ becomes constant.

Incidentally, a problem of the kind illustrated in Example 5.4 can also be solved in another way. To do this, equation (5.25) is rewritten as follows:

$$\frac{\rho D^3}{4\mu^2}\frac{\Delta p}{L} = \frac{1}{2}f \cdot \left(\frac{\rho \langle v \rangle D}{\mu}\right)^2 = \frac{1}{2}f \cdot \mathrm{Re}^2$$

(5.31)

The left-hand side of this equation generally consists solely of variables that are given; this is also the case in Example 5.4. This implies the value of $\frac{1}{2}f \cdot \mathrm{Re}^2$ is known. Re can then be directly read from a Re versus $\frac{1}{2}f \cdot \mathrm{Re}^2$ graph (see Figure 5.12), which has been constructed from Figure 5.10, from which $\langle v \rangle$ then follows.

In Example 5.4, this method produces

$$\frac{1}{2}f \cdot \mathrm{Re}^2 = \frac{\rho D^3}{4\mu^2}\frac{\Delta p}{L} = 5 \cdot 10^7$$

(5.32)

Looking up Re in Figure 5.12 produces Re = 1.2×10^5 and therefore $\langle v \rangle = 1.2$ m/s. This value is slightly different from the first result. Both solutions can be made more similar to each other by using more accurate plots and by taking the readings more meticulously.

Example 5.5. Velocity in a horizontal milk pipeline.
Milk ($\rho = 10^3$ kg/m^3, $\mu = 2.1 \times 10^{-3}$ Ns/m^2) flows under steady-state conditions through a long, horizontal, straight pipeline (diameter 4 cm, length 20 m, smooth wall) as a result of a pressure drop of 2 kPa.
What is the mean velocity in the pipeline?

Solution
This task can be solved using one of the two methods described above. However, there is a third way: using the Blasius equation as given in equation (5.29), providing that the requirement is met that the Reynolds number falls in the specified Re range. But of course, this can only be verified retrospectively. Entering this expression into equation (5.25) produces, for the pressure drop resulting from friction:

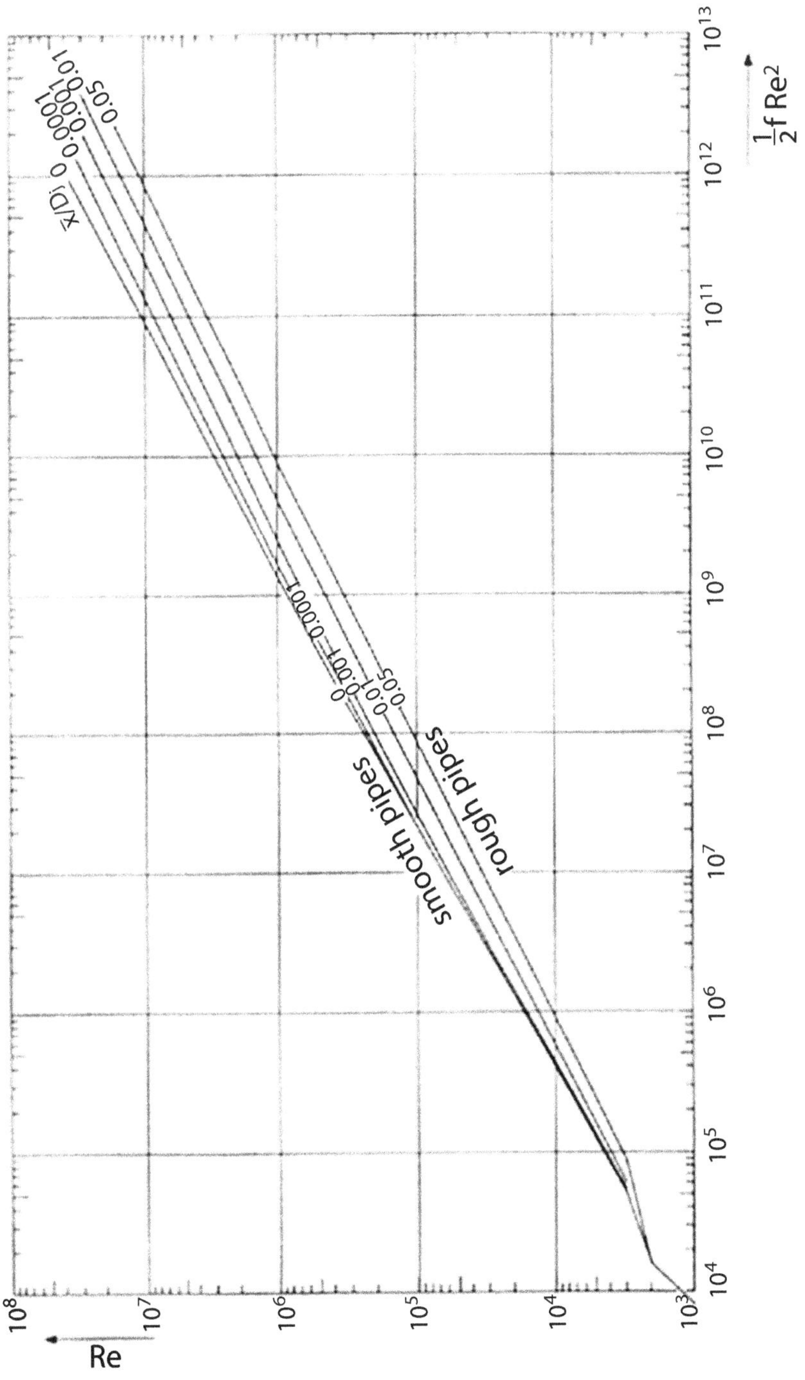

Figure 5.12

$$\Delta p = 0.316 \ Re^{-1/4} \cdot \frac{L}{D} \cdot \frac{1}{2} \rho \langle v \rangle^2 =$$

$$= 0.158 \ L \, D^{-5/4} \mu^{1/4} \rho^{3/4} \langle v \rangle^{7/4} \tag{5.33}$$

Entering this information gives $\langle v \rangle = 0.5$ m/s. For the Reynolds number, this produces $Re = 9{,}500$; Blasius may indeed be used, therefore. Remember that the Blasius equation is only valid for pipes with smooth walls.

The above examples all relate to cylindrical pipelines. In general, the relationship between the pressure drop and the velocity in the pipeline is modelled with the hydraulic diameter. For pipelines or flow channels with any given diameter (but constant throughout the entire length of the pipeline), the very same $4f$ (Re) can now be used **providing that** the flow is turbulent. The Reynolds number must then be determined with the help of the hydraulic diameter, of course. This agreement does not apply for the laminar flow area.

Example 5.6. Water transport through a rectangular pipe.
Water flows in steady state through a horizontal, rectangular pipe (height 3 cm, width 5 cm, length 100 m, smooth walls). The water velocity is 1 m/s. What is the pressure drop along the channel?

Solution
It is first necessary to calculate the hydraulic diameter:

$$D_h = \frac{4A}{S} = \frac{4 \, bh}{2h + 2b} = 3.75 \text{ cm} \tag{5.34}$$

from which it follows for the Reynolds number

$$Re_h = \frac{\rho \langle v \rangle D_h}{\mu} = 3.75 \cdot 10^4 \tag{5.35}$$

With the help of Figure 5.10, it again follows that $4f = 0.02$, and entering this value into equation (5.25) produces

$$\Delta p = 4f \cdot \frac{1}{2} \rho \langle v \rangle^2 = 2.7 \times 10^4 \text{ Pa} = 0.27 \text{ bar} \tag{5.36}$$

5.3.3 The analogy with heat and mass transfer

For the case of turbulent flow, the **film theory** for heat transfer was discussed in Section 3.5.2; and reference was also made in Section 4.5.2 to the film theory in the case of mass transfer. The subject of the **hydraulic film thickness** δ_h was raised as well. The film involves laminar momentum transport, for which Newton's viscosity law, equation (2.4), applies. Outside the film, the flow is turbulent and the velocity profile is therefore very flat, thanks to the momentum transport by the eddies in the turbulent field of flow. In the film theory, which models the **friction** at the wall, the following applies to the momentum transport from the bulk to the wall:

$$\tau_{f\to w} = f \cdot \frac{1}{2}\rho \langle v \rangle^2 = \mu \frac{\langle v \rangle}{\delta_h} \tag{5.37}$$

entirely analogously to the heat flux, expression (3.144), according to the film theory approach for turbulent flow. Equation (5.37) states that the **friction drag** lies entirely in the film.

It has already been demonstrated, in Section 3.5.2, that film thicknesses δ_q and δ_h depend on the degree of turbulence in the flow: δ_q and δ_h decrease as the Reynolds number increases. This means that heat transfer under turbulent conditions and therefore also the associated Nusselt number depend on the degree of turbulence and hence on δ_h as well. The following realignment is in accordance with this reasoning:

$$\text{Nu} = \frac{hD}{\lambda} = \frac{D}{\delta_q} = \frac{D}{\delta_h} \cdot \frac{\delta_h}{\delta_q} = \frac{f}{2} \cdot \frac{\rho\langle v \rangle D}{\mu} \cdot \frac{\delta_h}{\delta_q} \tag{5.38}$$

where equation (5.37) is used.

By also using the expression $\delta_h/\delta_q \propto \text{Pr}^{1/3}$ (discussed previously in relation to Figure 3.24 in Section 3.5.2), the following results:

$$\text{Nu} \propto \frac{1}{2}f \cdot \text{Re Pr}^{1/3} \tag{5.39}$$

or

$$\frac{1}{2}f \propto \frac{\text{Nu}}{\text{Re Pr}^{1/3}} \tag{5.40}$$

The result obtained is very similar to equations (4.91) and (4.92). Taking expressions (4.93) and (5.29) into account as well, the **Chilton–Colburn analogy** can be extended to

$$j_H = j_D = \frac{1}{2}f = C \cdot \text{Re}^{m-1} \tag{5.41}$$

The extended analogy of expression (5.41) only applies when the **friction factor** stands solely for **wall friction**, in accordance with the derivation. **Form drag**, which can be a significant part of the flow resistance of objects around which flow is passing (Section 2.3), is **not** associated with analogous heat or mass transfer effects.

Summary

This section has looked at the link between flow rate through and frictional pressure drop in a pipeline. Pressure drop, energy dissipation, wall friction, and shear stress have been discussed in terms of their mutual relationships. The Fanning pressure drop equation has been derived

$$p_1 - p_2 = 4f \cdot \frac{L}{D} \cdot \frac{1}{2}\rho \langle v \rangle^2$$

with a plot for $4f$ as a function of Reynolds and with two expressions for the friction factor $4f$ for a cylindrical pipe: one for the laminar regime ($4f = 64/Re$) and the Blasius equation for the turbulent regime.

The terms 'hydraulic diameter' and 'relative wall roughness' have also been introduced for use with turbulent pipe flows. Various solution strategies have been discussed for calculating the flow rate through a pipe when the pressure drop is known; one of these strategies is iterative, exactly the same as when calculating a uniform particle velocity from a force balance.

Also, the Chilton–Colburn analogy for turbulent flows has been extended to include the friction factor, providing the flow resistance is only determined by the wall friction.

5.4 Pressure drop in pipeline systems

In Section 5.3, we discussed how dissipation and the related frictional pressure drop for a straight pipeline are modelled with the help of the Fanning pressure drop equation. Because dissipation stands for the destruction of mechanical energy, it is obvious that the Fanning equation should be linked to the mechanical energy balance as introduced in Chapter 1 for steady-state conditions.

For a fluid of constant density flowing through a straight pipeline, this follows from equation (1.125):

$$0 = \phi_m \left(\frac{1}{2}v_1^2 + \frac{p_1}{\rho} + g\,z_1 - \frac{1}{2}v_2^2 - \frac{p_2}{\rho} - g z_2 \right) + \phi_w - \phi_m\, e_{\text{diss}} \tag{5.42}$$

For a horizontal, straight pipeline of constant cross-sectional area and in the absence of a pump, combining equations (5.25) and (5.42) expresses the quantity of energy that is dissipated per unit of mass:

$$e_{\text{diss}} = e_{\text{fr}} = 4f \cdot \frac{L}{D_h} \cdot \frac{1}{2} \langle v \rangle^2 \tag{5.43}$$

in which the use of the hydraulic diameter D_h – see equation (5.26) – renders the equation also valid for non-cylindrical conduits. With regard to the Fanning equation, combining equations (5.42) and (5.43) has the advantage that there is space for pressure changes other than those caused by friction at the wall of the conduit. Equation (5.43) expresses explicitly what the effect of **dissipation** is on the mechanical energy housekeeping in a pipeline.

Consider two identical tubes with the same diameter, one of which is placed horizontally, and the other vertically. An identical flow rate passes through both. The pressure drop over both pipes is a direct consequence of friction on the wall. This friction depends only on the shear stress on the wall, and that in turn is determined by the velocity profile just next to the wall. In the horizontal pipe there is a pressure drop that just compensates the frictional force. In the vertical pipe (in which the liquid is flowing from bottom to top), the pressure drop has to make up for both frictional force and gravity. However, the wall friction depends only on the velocity gradient on the wall and not on the orientation of the pipe. For this reason, the same expression for the dissipation can be used. Notice that the mechanical energy balance now states exactly that in the vertical pipe the pressure energy has to compensate the potential energy ('gravity') and the dissipation (**frictional energy losses**).

Of course, there are many pipelines that are not straight and do not have a constant diameter. As a rule, a system of pipes consists of all kinds of bends, valves or flaps, and narrow or wide sections (these are generally referred to as **appendages**, or **fittings**). All these types of 'obstacle' or impediment to the flow give rise to extra dissipation. In many cases, extra eddies occur in which mechanical energy is dissipated. For **turbulent** flow, all these appendages (fittings) can be easily modelled using an expression for the energy dissipation per unit of mass that is entirely analogous to equation (5.43):

$$e_L = K_L \cdot \frac{1}{2} \langle v \rangle^2 \qquad (5.44)$$

The convention is that here for $\langle v \rangle$, the average velocity of the flow downstream of the relevant appendage is used. K_L is the **loss coefficient** of the appendage (fitting) and is constant, that is, not dependent on Re, **providing that** the flow is sufficiently turbulent. The loss coefficients can be found in tables (see, e.g., Perry's Handbook[19]). A small number of loss coefficients are given below:

Fitting	K_L
Gate valve, open	0.2
Gate valve, 1/2 closed	≈ 6
Gate valve, 3/4 closed	≈ 24
90° bend, sharp angle	1.3
90° bend, long	0.5
Entrance loss pipe, sharp	0.5
Entrance loss pipe, rounded	0.05
Exit loss pipe	1

With a view to the use of the mechanical energy balance for pressure drop calculations for pipeline systems with fittings, it can be stated in summary that, providing that the flow is turbulent, the **specific energy dissipation** e_{diss} (in J/kg) is generally modelled as the sum of e_{fr} and e_L as given by equations (5.43) and (5.44), respectively:

---

19 Perry's Chemical Engineers' Handbook, 9th Ed., Eds. Don W. Green and Marylee Z. Southard, McGrawHill, 2019, Chapter 6.

$$e_{\text{diss}} = \sum_i \left(4f \frac{L}{D_h} \frac{1}{2} \langle v \rangle^2 \right)_i + \sum_j \left(K_L \cdot \frac{1}{2} \langle v \rangle^2 \right)_j \tag{5.45}$$

The first part of the right-hand term of this equation represents the sum of the friction losses in all straight pipe sections, while the second part covers the dissipation in all the appendages. Note that changes in the diameter in the system are correctly accounted for: for each fitting the downstream velocity $\langle v \rangle$ has to be used.

Example 5.7. Pumping upwards.
A watery liquid has to be pumped from a very large container (A) to another very large container (B), which is situated at a higher level. For this, a pump is added to the pipeline system illustrated in Figure 5.13. Both containers are open at the top. The pipeline is 50 m long and has a diameter of 5 cm. The walls of the pipeline are smooth. There are two sharp right-angled bends and an open gate valve in the pipeline. The entrance to the pipeline (from container A) is sharp. The height of the liquid in container A is 3 m, and in container B it is 13 m (measured from the ground).
   Calculations are needed for the following:
a)   the power that the pump has to supply in order to pump 2 L/s of water;
b)   the value of the greatest pressure that is present in the system.

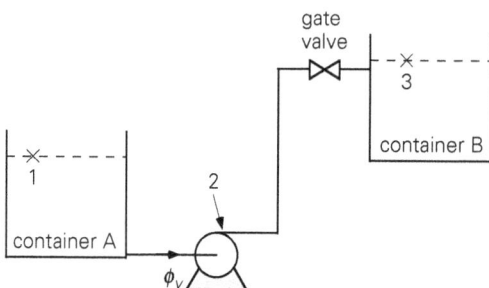

Figure 5.13

Solution
Both questions can be solved with the help of the mechanical energy balance. First of all, this balance is applied to the system between points 1 and 3 (see Figure 5.13): point 1 is located on the surface of the water in container A, where the following applies:
   $p_1 = p_0$, where p_0 is the ambient pressure; $v_1 \approx 0$; $z_1 = $ given.
   Point 3 is also chosen on the surface of the water in container B, so: $p_3 = p_0$; $v_3 \approx 0$; $z_3 = $ given.
   In general, it is advised to choose your reference points (inlet and outlet) at positions where you know the pressure.
The loss coefficients K_L of the various fittings have to be looked up (with a view to the dissipation in the system):

Two sharp right-angled bends	→	2 × 1.3
Open valve	→	0.2
Sharp entrance	→	0.5
Exit	→	1

The (average) velocity in the pipeline simply follows from the flow rate: $v = 1.0$ m/s. From this it follows that the Reynolds number for the flow is 5×10^4. The diameter of the pipeline is constant and so we can account for the total dissipation resulting from the wall friction and the fittings in one go – by using equation (5.45). Looking up $4f$ when $Re = 5 \times 10^4$ produces: $4f = 0.02$. All this gives:

$$e_{diss} = 4f \cdot \frac{L}{D} \cdot \frac{1}{2}v^2 + (2 \cdot 1.3 + 0.2 + 0.5 + 1) \cdot \frac{1}{2}v^2$$

$$= 12.15 \text{ J/kg} \tag{5.46}$$

For the pump capacity, it then follows that

$$\phi_w = \phi_m\{g(z_3 - z_1) + e_{diss}\} = 220 \text{ W} \tag{5.47}$$

The highest pressure occurs just behind the pump, of course (!). This pressure can be calculated by drawing up a mechanical energy balance between points 1 and 2 (see Figure 5.13). The following applies to point 2:

$$p_2 = ?; \ v = 1 \text{ m/s}; \ z_2 \approx 0$$

This means the mechanical energy balance between points 1 and 2 becomes

$$\frac{p_2 - p_1}{\rho} = g(z_1 - z_2) - \frac{1}{2}v^2 - e_{diss} + \frac{\phi_w}{\phi_m} \tag{5.48}$$

The dissipation per unit of mass e_{diss} now only relates to the inflow into the pipeline, that is,. $e_{diss} = e_L = 0.5 \cdot \frac{1}{2}v^2$ providing that the length of the pipe from container A to the pump is small enough to be ignored. With this, $p_2 = 2.4$ bar is found.

Example 5.8. A basin running empty.
Processed water flows through a cylindrical conduit (length 20 m, diameter 10 cm, wall roughness 1 mm) from a large open basin, situated high up, into an open channel. The water level in the basin is 4 m above that in the channel. The pipeline contains 2 sharp 60° bends (with $K_L = 1.86$) and a gate valve for controlling the water flow rate. The loss coefficients for the pipe entrance and exit are 0.2 and 1.0.
    The question is to calculate the water flow rate when the gate valve is half closed.

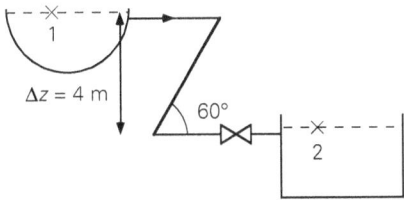

Figure 5.14

Solution
The solution to this problem is again found through a mechanical energy balance, between point 1 on the water surface in the basin and point 2 on the water surface in the channel (see Figure 5.14) – again two points on a water surface, because here statements can be made on quite some variables:
For point 1: $p_1 = p_0$, $v_1 \approx 0$, $z_1 = z_2 + \Delta z$, and for point 2: $p_2 = p_0$, $v_2 \approx 0$. Entering this into the mechanical energy balances produces

$$0 = -g \, \Delta z - e_{diss} \tag{5.49}$$

The total energy dissipation in the pipeline system at an (average) velocity v is given by

$$e_{diss} = \left(4f\frac{L}{D} + K_{L,tot}\right)\frac{1}{2}v^2 \tag{5.50}$$

where $K_{L,tot} = 2 \times 1.86 + 0.2 + 1.0 + 6 = 10.9$ takes care of the effects of all the fittings.
   Substituting equation (5.50) into equation (5.49) gives

$$v = \sqrt{\frac{2g\,\Delta z}{4f\frac{L}{D} + K_{L,\,tot}}} \tag{5.51}$$

From this, it is possible to obtain v with an iterative procedure because $4f$ is a function of v (which can be seen in Figure 5.10 with a relative roughness of 0.01).
   As an opening estimate, let $v = 1$ m/s, say. This gives $Re = 10^5$ and $4f = 0.035$. Using this value of $4f$ in equation (5.51) for calculating a better value for v gives $v = 2.1$ m/s and from this it follows that $Re = 2.1 \times 10^5$ and $4f = 0.035$. This means the iteration process need not be taken any further. The flow rate can now be easily calculated: $\phi_m = 16.5$ kg/s.

Notice that in both examples, the flow in the pipeline is turbulent so that the loss coefficients are indeed constant.

! **Summary**
The pressure drop in a pipeline system is not determined solely by friction. For that reason, it makes sense to use the mechanical energy balance, providing that expressions are available for the energy dissipation.
   For dissipation resulting from wall friction in a straight pipe, the following applies:

$$e_{fr} = 4f \cdot \frac{L}{D} \cdot \frac{1}{2}\langle v\rangle^2$$

If the Reynolds number of the flow is sufficiently large, the dissipation resulting from an appendage (or fitting) is modelled using a constant loss coefficient:

$$e_L = K_L \cdot \frac{1}{2}\langle v\rangle^2$$

where the downstream velocity should be selected for $\langle v\rangle$. In general, the total dissipation e_{diss} is the sum of the wall friction e_{fr} in all straight pipe sections and of the energy dissipation e_L in all fittings.

5.5 Pressure drop across a packed bed

Pipeline systems are not the only systems in which pressure drop and dissipation are important. In fact, pressure drop and dissipation are important quantities in any flow system. In this section, we will therefore look at another flow system, one that is found extensively in the process industry: **the packed bed**. This consists of a container filled with particles resting on each other. Liquid flows through the open space located between the particles. The question now is how the dissipation and pressure drop in this system depends on the velocity at which the flow passes.

Before tackling this problem, it is useful first to look at how the dissipation is modelled if the flow does not go through the 'inside' of a pipeline, but actually alongside the 'outside' of a body. Consider the simplest case of a spherical particle that is motionless in a flowing medium with a uniform approach velocity of v. The force exerted by the flow on the particle is (see Section 2.3)

$$F_D = C_D \cdot A_\perp \cdot \frac{1}{2}\rho v^2 \qquad (5.52)$$

The dissipation around this particle can now be calculated analogously to the reasoning in Section 5.4: there, the dissipation can be determined by using the mechanical energy balance. For a horizontal straight pipeline with a constant diameter, the following applies, based on equation (5.42):

$$0 = \phi_m \frac{p_1 - p_2}{\rho} - \phi_m \, e_{\text{diss}} \qquad (5.53)$$

However, this equation can be written differently by relating the difference in pressure $p_1 - p_2$ to the shear stress on the wall via a force balance, see equation (5.21). If the overall **frictional force** exerted by the wall on the liquid via the shear stresses is referred to as F_{fr}, this difference in pressure can be rewritten as

$$\phi_m \frac{p_1 - p_2}{\rho} = \rho \, Av \cdot \frac{p_1 - p_2}{\rho} = A(p_1 - p_2)v = F_{\text{fr}} \cdot v \qquad (5.54)$$

In an analogous approach, for the dissipation around a single submersed free particle around which flow is passing, the following applies:

$$\phi_m \, e_{\text{diss}} = F_D \cdot v \qquad (5.55)$$

and from this it then follows

$$\phi_m \, e_{\text{diss}} = C_D A_\perp \frac{1}{2}\rho v^2 \cdot v = C_D A_\perp \cdot \frac{1}{2}\rho v^3 \qquad (5.56)$$

In a uniform packed bed of more or less spherical, non-porous particles, the flow around the particles and therefore the drag and the drag coefficient on each of these particles are of course not the same as with a free particle. However, it is to be expected that the frictional force on an individual particle in the bed can be modelled on the same lines as in the case of the free particle. The overall energy dissipation in a bed of N particles then follows by multiplying the dissipation resulting from the friction along one particle in the bed by the number of particles in column N.

For the flow around a particle in the bed, the local flow velocity has to be taken: that is to say, the velocity $\langle v \rangle$ in the channels between the particles as they rest against and on top of each other. However, this is not a very good velocity to include in the calculation, because although the overall volumetric flow rate that goes through the packed bed is known, $\langle v \rangle$ is not. Nonetheless, this can be easily resolved.

The particles together occupy a volume, V_d, in the total volume, V_0, of the bed. Then, the volume fraction of the free space – known as the **bed porosity** ε – can be defined as $\varepsilon = (1 - V_d/V_0)$. The cross-sectional area of the empty container is A_0 and so εA_0 is the cross-sectional area in the packed bed that is available for the flow. Further, v_0 is the velocity that the liquid would have if it flowed through the empty container at the same flow rate – referred to as the **superficial velocity**.

The link between the volumetric flow rate and the flow velocity $\langle v \rangle$ is therefore:

$$\phi_v = A_0 v_0 = \varepsilon A_0 \langle v \rangle \rightarrow \langle v \rangle = \frac{v_0}{\varepsilon} \tag{5.57}$$

The volume occupied by the particles is $(1 - \varepsilon) A_0 L$ (where L is the height of the bed); this means the number of particles (with diameter d) in the bed is

$$N = \frac{(1-\varepsilon)A_0 L}{\frac{\pi}{6}d^3} \tag{5.58}$$

Eventually, we arrive at the energy dissipation in the packed bed:

$$\phi_m e_{\text{diss}} = \frac{(1-\varepsilon)A_0 L}{\frac{\pi}{6}d^3} \cdot C_D \cdot \frac{\pi}{4}d^2 \cdot \frac{1}{2}\rho \left(\frac{v_0}{\varepsilon}\right)^3 \tag{5.59}$$

Finally, eliminating $\phi_m = \rho A_0 v_0$ produces

$$e_{\text{diss}} = \frac{3}{2}C_D \cdot \frac{(1-\varepsilon)}{\varepsilon^3} \cdot \frac{L}{d} \cdot \frac{1}{2}v_0^2 \tag{5.60}$$

We are now left with the matter of determining C_D for the flow around the particles in a packed bed. It is to be expected that C_D will be a function of the Reynolds number. However, this is not the only question. What does this function look like? And, which Reynolds number describes the flow through the bed? The typical velocity is $\langle v \rangle$, but the typical length is not simply the diameter of the particles.

A better dimension would appear to be the hydraulic diameter of the gaps between the particles. This can be estimated on the basis of the bed porosity ε and the size of the particles, d. This relates to $4A/S$, where A represents the cross-sectional area of the gap that is available for the flow, and S the wetted perimeter. The gaps are of course not neat straight channels, which is why it is better to use a **hydraulic diameter** that is defined on the basis of the volume of a gap and the overall surface area of the gap:

$$D_h = \frac{4 \times \text{volume of gap}}{\text{surface area of gap wall}} \tag{5.61}$$

Next, both the numerator and the denominator in this expression are related to the bed volume. For the numerator, this produces ε, while the denominator acquires the wall surface area of all the gaps per unit of volume, which is referred to as the specific surface area a (in m^2/m^3). This can be determined by multiplying the surface area πd^2

of a single particle by the number of particles in the bed N and finally dividing it by the total bed volume.

With the help of equation (5.58) for N, we get the following for the specific surface area:

$$a = \frac{6(1-\varepsilon)A_0 L}{\pi d^3} \cdot \pi d^2 \cdot \frac{1}{A_0 L} = \frac{6(1-\varepsilon)}{d} \tag{5.62}$$

The Reynolds number is now:

$$Re_h = \frac{\rho \langle v \rangle D_h}{\mu} = \frac{\rho \frac{v_0}{\varepsilon} \frac{4\varepsilon}{6(1-\varepsilon)/d}}{\mu} = \frac{2}{3} \frac{\rho v_0 d}{(1-\varepsilon)\mu} \tag{5.63}$$

The implicit result of this is that the Reynolds number for a packed bed includes the superficial velocity as a typical velocity and the diameter of the particles as the typical length scale, and that the porosity occurs here.

If Re_h is sufficiently large it is found that C_D becomes constant, analogous to the flow around a single sphere; however, the value is different, namely 2.3 (instead of 0.43). This result holds where $Re_h > 2000$. Where $Re_h < 1$, it is seen that C_D is inversely proportional to Re_h, entirely in accordance with a single sphere around which a flow is passing; however, here too the proportionality constant is different: $C_D = 150/Re_h$ (for a single sphere the constant is 24). The change to C_D for a packed bed as a function of Re_h is shown in Figure 5.15.

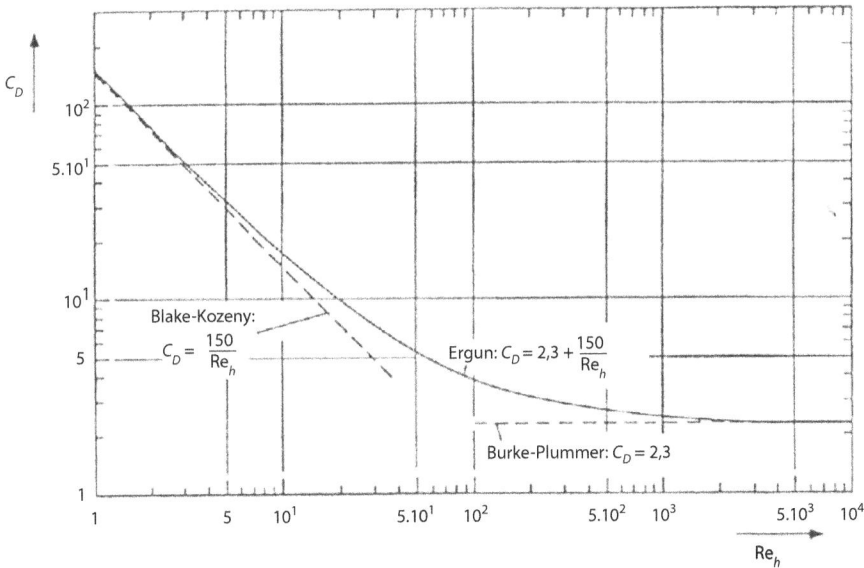

Figure 5.15

Ergun has drawn up an empirical equation for C_D, which gives a good description for the entire Re_h range in Figure 5.15. He simply added up the laminar and turbulent expressions

$$C_D = 2.3 + \frac{150}{Re_h} \tag{5.64}$$

Entering this **Ergun relation** into equation (5.60) for the specific energy dissipation produces:

$$\frac{p_1 - p_2}{\rho} = e_{diss} = \frac{1-\varepsilon}{\varepsilon^3}\left\{170\frac{\mu}{\rho v_0 d}(1-\varepsilon) + 1.75\right\}\frac{L}{d}v_0^2 \tag{5.65}$$

Example 5.9. Pressure drop over a packed bed.
A flow of water with a flow rate of 0.3 L/s has to have Ca^{2+} ions removed from it. This takes place in a cylindrical packed bed made up of spherical ion-exchanging particles (diameter $d = 2$ mm). The porosity of the packed bed ε is 0.4. The length of the column is 2 m, and the diameter D of the column is 0.25 m. The water flows through the column from the bottom to the top.
  How large is the pressure drop over the column?

Solution
The answer follows from the mechanical energy balance for the whole packed bed. It is then clear that the pressure drop consists of two contributions: the pressure drop resulting from gravity (the hydrostatic pressure drop) and the pressure drop resulting from friction. This second contribution follows from determining e_{fr} according to equation (5.65). For this, v_0 first has to be determined from the flow rate:

$$v_0 = \frac{\phi_V}{\frac{\pi}{4}D^2} = 6.1 \times 10^{-3}\,\text{m/s} \tag{5.66}$$

This means the specific dissipation is

$$e_{diss} = \frac{1-\varepsilon}{\varepsilon^3}\left(170\frac{\mu}{\rho v_0 d}(1-\varepsilon) + 1.75\right)\frac{L}{d}v_0^2 = 3.52\,\text{J/kg}$$

and the pressure drop is:

$$\Delta p = \rho g L + \rho e_{diss}$$
$$= 1.96 \times 10^4 + 3.52 \times 10^3 = 2.31 \times 10^4\,\text{Pa} \approx 0.23\,\text{bar} \tag{5.67}$$

Summary
In this section, we have derived how pressure drop across and energy dissipation in a packed bed depend on the superficial velocity through the bed. To this end, the bed is regarded as a collection of particles that encounter forces from the liquid flowing around them. Terms like porosity, specific surface area and hydraulic diameter of the gaps between the particles have been introduced. The drag coefficient C_D of a particle in such a bed depends on a Reynolds number defined with the superficial velocity v_0, the diameter of the particle, and the porosity. All of this leads to the Ergun relation

$$\Delta p = \frac{1-\varepsilon}{\varepsilon^3}\left(170\frac{\mu}{\rho v_0 d}(1-\varepsilon)+1.75\right)\frac{L}{d}\rho v_0^2$$

for the frictional pressure drop across a packed bed.

5.6 Laminar flow of Newtonian fluids

In Chapters 3 and 4, we discussed extensively how temperature and concentration profiles could be determined for molecular transport in a few simple geometries. The starting point was always an energy or mass balance for a thin slice of the material (for example, between x and $x + dx$). The flows 'in' and 'out' were always modelled using Fourier's or Fick's law (or Stefan's law). A similar analysis is also possible for determining the velocity profile in a fluid flowing in the laminar regime.

The treatment of laminar flow will remain restricted to steady-state one-dimensional flows in geometrically simple situations. Finally, this section will only look at fluids that comply with **Newton's viscosity law** (see equation (2.4)).

5.6.1 Laminar flow in Cartesian coordinates: due to drag by a moving wall

The first situation concerns a flat geometry: a liquid (a lubricating oil, for instance) between two very large horizontal plates, parallel to each other. The distance between the plates is D. The lower plate is motionless, and the upper one is moving horizontally in the direction of x at a constant velocity, v_0. The velocity is low enough – actually: the Reynolds number is low enough – for the liquid to be able to flow in 'layers' parallel to both plates. This means that the flow is **laminar** and that solely the individual molecules provide for the transport of x-momentum in the direction of y. This steady-state situation is illustrated in Figure 5.16.

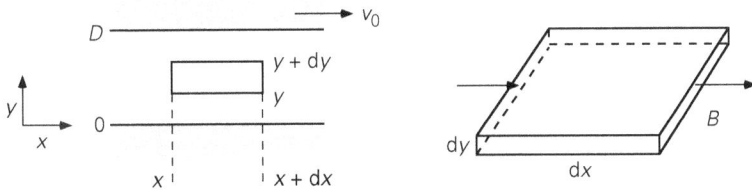

Figure 5.16

In order to determine the velocity profile, an x-momentum balance for a thin slice with thickness dy (between y and $y + dy$), with length dx (between x and $x + dx$) and width B (perpendicular to the surface of the drawing) has to be drawn up.

However, in the case of these parallel plates and in the steady-state conditions stated, as much convective momentum flows into the control volume (the thin slice) through the left-hand plane (at $x = x$) as convective momentum flows out through the right-hand plane (at $x = x + dx$). Further, because the flow is laminar and one-dimensional, no convective momentum transport is crossing the other surfaces of the control volume. If in addition, in view of the discussion in Section 2.1.4, the molecular (or diffusive) momentum transport is described in terms of shear stresses, then the momentum balance is reduced to a **force balance**. This force balance then contains information about the shear stress profile.

For the situation in Figure 5.16, it is only the force balance in the direction of x that is important. Only two forces are exerted on the control volume in the direction of x, viz. a shear stress on the lower plane and a shear stress on the upper plane. The force on the lower plane is $B dx \cdot \tau_{yx}|_y$: this expression represents the force that the layer flowing just underneath the control volume exerts on the layer just above it in the control volume. Remember that the shear stress is defined precisely in this way so that the layer with the smallest coordinate exerts $+\tau_{yx}$ on the layer with the greater coordinate (see Section 2.1.4 again). This means that on the top side of the control volume, the force $B dx \cdot \left[-\tau_{yx}|_{y+dy}\right]$ is exerted on the control volume by the fluid flowing over the control volume. The force balance is now therefore:

$$0 = B \, dx \cdot \tau_{yx}|_y + B \, dx \cdot \left[-\tau_{yx}|_{y+dy}\right] \tag{5.68}$$

This balance can be simply reduced to the differential equation that must be satisfied by the shear stress:

$$\frac{d}{dy}\tau_{yx} = 0 \tag{5.69}$$

The solution to this equation is

$$\tau_{yx} = \text{constant} \tag{5.70}$$

It is useful to realise that so far no account has been taken of whether the liquid complies with Newton's viscosity law:

$$\tau_{yx} = -\mu \frac{dv_x}{dy} \tag{5.71}$$

The **shear stress profile** that has been established for this situation is clearly independent of the properties of the liquid. This finding has general validity: the shear stress profile is only dependent on the force balance (or rather the momentum balance) and not on the type of liquid.

With a view to determining the **velocity profile**, however, it is necessary to use a link between shear stress and velocity gradient. Suppose **now** that the liquid between the slices is Newtonian. This means that inserting expression (5.71) into equation (5.70) leads to

$$- \mu \frac{d}{dy} v_x = \text{const} \;\rightarrow\; \frac{d}{dy} v_x = C_1 \tag{5.72}$$

Solving equation (5.72) gives

$$v_x(y) = C_1\, y + C_2 \tag{5.73}$$

Both integration constants can be found with the help of the two given boundary conditions: $y = 0 \rightarrow v_x = 0$ and $y = D \rightarrow v_x = v_0$. This produces the velocity profile:

$$v_x(y) = v_0 \cdot \frac{y}{D} \tag{5.74}$$

This finding is equivalent to equation (3.5) for heat conduction through a slab and to equation (4.11) for diffusion through a medium between two plates. Substituting equation (5.74) into Newton's viscosity law gives a constant value for the shear stress

$$\tau_{yx} = -\mu \frac{v_0}{D} \tag{5.75}$$

just as the fluxes ϕ_q'' and ϕ_m'' are also constants according to equations (3.6) and (4.12), respectively.

Both the shear stress profile and the velocity profile have been drawn in Figure 5.17. It can clearly be seen here that the shear stress is negative: every 'underlying' layer exerts a force in the negative x direction on the layer lying on top, and so resists the flow resulting from the moving upper plate.

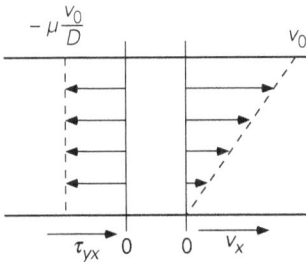

Figure 5.17

Finally, the above derivation clearly shows that in steady-state laminar flows there is no role for the fluid's density: the flow is dominated by viscosity. This applies to all flows in Section 5.6 – and supports remarks made in Chapter 2 with respect to incorporating or leaving out density and viscosity in dimensional analyses of flows and convective heat and mass transfer.

5.6.2 Laminar flow in Cartesian coordinates: due to a pressure gradient

The second situation again is a flat geometry and therefore Figure 5.16 still is a useful sketch, the difference however being that now the force driving the flow in the positive x-direction is not a moving plate or wall – as both plates do not move now – but an imposed pressure gradient $[-dp/dx]$. Here, the minus sign is introduced to arrive at a positive driving 'force' in the positive x-direction, because as a matter of fact dp/dx is negative for a flow in the positive x-direction which is currently the case. Rather than $[-dp/dx]$ also $\Delta p/L$ could have been written, where Δp stands for the positive pressure difference over the distance L between two positions upstream and downstream that drives the flow. The unit of the pressure gradient is Pa/m, which can better be interpreted as Ns/m^3s to express that a pressure gradient is a means to supply momentum (in Ns) to a unit of volume per unit of time. Where we now consider a one-dimensional flow between parallel plates, the pressure gradient is constant in the x-direction and may be denoted by a constant, given Γ for the sake of convenience.

Also in this situation the momentum balance reduces to a force balance, now with four forces acting in the x-direction on the slice of volume $Bdxdy$. Two of these forces are the same as in the preceding case: $B\,dx \cdot \tau_{yx}|_y$ and $B\,dx \cdot \left[-\tau_{yx}|_{y+dy}\right]$ which act along the bottom and top surfaces of the slice, respectively; the other two forces are the result of the pressure field and act normal to the two ends of the slice: $B\,dy \cdot p|_x$ and $B\,dy \cdot \left[-p|_{x+dx}\right]$, where the minus sign in the latter force expresses that the force exerted from outside on the slice at position $x + dx$ is in the negative direction. The force balance for the steady state says that the sum of the forces makes zero:

$$0 = B\,dx \cdot \tau_{yx}|_y + B\,dx \cdot \left[-\tau_{yx}|_{y+dy}\right] +$$
$$+ B\,dy \cdot p|_x + B\,dy \cdot \left[-p|_{x+dx}\right] \tag{5.76}$$

Dividing all terms by $Bdxdy$ and combining similar terms leads to the following differential equation:

$$\frac{d\tau_{yx}}{dy} = -\frac{dp}{dx} = \Gamma \tag{5.77}$$

with the solution:

$$\tau_{yx} = \Gamma y + C_1 \tag{5.78}$$

Again, the question whether the fluid is a Newtonian or a non-Newtonian liquid has not been considered so far. For any flow between two flat, stationary plates owing to a pressure gradient, the shear stress profile is therefore a linear function of the transverse coordinate y – irrespective of the properties of the fluid. Completely in agreement with the definition of the shear stress ('underlying' acting upon 'overhead': see again Section 2.1.4), the lower stationary plate at $y = 0$ slows the fluid down by exerting a force in the negative x-direction while the fluid flow exerts a force in the positive x-direction on the upper plate at $y = D$.

The integration constant, C_1, can be found with the help of the fact of life that in the case of two stationary plates the velocity halfway the two plates, in the plane of symmetry, is maximum. This implies that the derivative of the x-velocity versus y at that position is zero – and in its turn, at least for a Newtonian liquid, this results in the condition: $y = \frac{1}{2}D \rightarrow \tau_{yx} = 0$. This turns equation (5.78) into

$$\tau_{yx} = \Gamma \cdot \left(y - \frac{1}{2} D \right) = \Gamma \cdot D \cdot \left(\frac{y}{D} - \frac{1}{2} \right) \tag{5.79}$$

For a Newtonian liquid, equation (5.79) leads to

$$v_x = -\frac{\Gamma}{\mu} \cdot \left(\frac{1}{2} y^2 - \frac{1}{2} D y \right) + C_2 \tag{5.80}$$

This second integration constant, C_2, can be determined with the help of a boundary condition for the velocity, due to the symmetry, at either $y = 0$ or $y = D$, as $v_x = 0$ at both positions. As a result: $C_2 = 0$. The eventual velocity profile is then given by

$$v_x = \frac{\Gamma}{2\mu} \cdot y \cdot (D - y) \tag{5.81}$$

This equation shows that for a viscous or laminar flow the velocity profile does not depend on the fluid's density – confirming the remark in Chapter 2 that density may and should be ignored when carrying out a dimensional analysis for (horizontal) viscous flows.

Rather than by determining C_1 with the help of $\tau_{yx} = 0$ at $y = \frac{1}{2}D$, the same result would have been obtained – still for a Newtonian liquid – by deriving an expression for the velocity profile by substituting expression (5.71) into equation (5.78); the result then contains both C_1 and C_2 which can be found by the two above boundary conditions for the velocity at $y = 0$ and $y = D$. Using both boundary conditions implicitly implies that the plane $y = \frac{1}{2}D$ is a symmetry plane for the velocity profile where the velocity is maximum and the shear stress equals zero.

Note that for the shear stress profile the plane $y = \frac{1}{2}D$ is not a plane of symmetry! From the molecular interpretation of the shear stress τ_{yx} as a flux of x-momentum in the y-direction (see Section 2.1.4) we still can conclude that in this case $\tau_{yx} = 0$ at $y = \frac{1}{2}D$. We again use here a symmetry argument: seen from $y = \frac{1}{2}D$ the flow looks the same whether we look upwards towards $y = D$ or downwards towards $y = 0$. As a result, the transport of x-momentum in the y-direction at $y = \frac{1}{2}D$ must be zero.

Example 5.10. Slot coating II.[20]

We now return to the flow geometry discussed earlier in Example 2.7. Figure 5.18 again shows the cross-sectional view through the die of a specific coating machine. The very viscous Newtonian liquid is supplied via the vertical channel and is then entrained by the lower plate or belt (web) that moves to the right with velocity U. The result is that that this plate or belt gets a coating of thickness δ_∞ which is smaller than the gap height δ.

Restricting the analysis to the x–y plane suffices. Consider a steady-state situation in which the pressure P_1 is obviously higher than the ambient pressure P_0, as a result of which the liquid also extends over a fixed, steady-state distance L_2 in the 'wrong' direction. Velocities in the y-direction (among other things, in the area where the vertical flow is converted into flows in the x-direction) may be ignored. The situation may therefore be conceived as a steady-state laminar flow of a very viscous Newtonian liquid between two flat plates the lower one of which moves and the upper, stationary one consists of two parts.

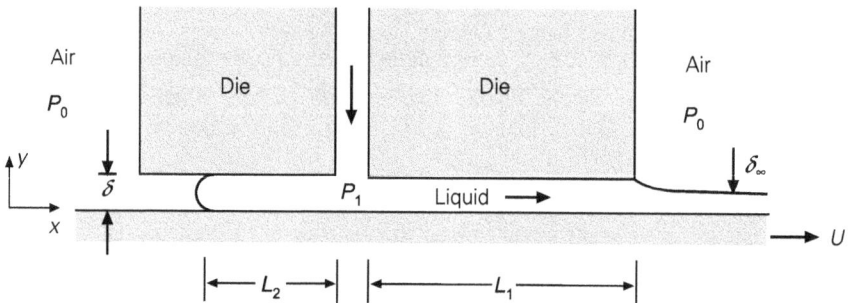

Figure 5.18

The following subquestions should be answered:
a) Derive expressions for the shear stress profile and the velocity profile in domain L_1 (sufficiently far from the vertical supply channel to allow for the presumption that the flow there is in the x-direction only). Make a sketch of the two profiles.
b) Derive expressions for the shear stress profile and the velocity profile in domain L_2 (sufficiently far from the vertical supply channel and from the convex end at $x = L_2$ to allow again for the presumption that the flow is in the x-direction only). Make also a sketch of these two profiles.
c) Derive an expression for the flow rate (per unit of depth) to the right and – in close connection – for the coating thickness δ_∞ outside the die (sufficiently far from the tip of the die).
d) Derive an expression for L_2 with the help of the observation that the flow to the left is zero. Compare your answer with the outcome of Example 2.7.

Solution

Starting point for answering the above questions is in equation (5.77). Substituting Newton's viscosity law, equation (5.71), into equation (5.77) results in

20 This problem has been derived from a problem in W.H. Deen, *Analysis of Transport Phenomena*, Oxford University Press, 2011.

$$\mu \frac{d^2 v_x}{dy^2} = \frac{dp}{dx} = -\Gamma \qquad (5.82)$$

with the solution

$$v_x = -\frac{\Gamma}{2\mu} y^2 + \frac{C_1}{\mu} y + C_2 \qquad (5.83)$$

The boundary conditions for both domains L_1 and L_2 are

$$\begin{aligned} y = 0 &: v_x = U \\ y = \delta &: v_x = 0 \end{aligned} \qquad (5.84)$$

and result – together with equation (5.83) – in the following expressions for shear stress and velocity profiles:

$$\tau_{yx} = -\frac{\mu U}{\delta} + \Gamma \cdot \left(y - \frac{1}{2}\delta \right) \qquad (5.85)$$

$$v_x = U \cdot \left(1 - \frac{y}{\delta} \right) + \frac{\Gamma}{2\mu} y\,(\delta - y) \qquad (5.86)$$

The difference between the two domains L_1 and L_2 is that in domain L_1

$$\Gamma = -\frac{dp}{dx} = \frac{P_1 - P_0}{L_1} > 0 \qquad (5.87)$$

while in domain L_2

$$\Gamma = -\frac{dp}{dx} = \frac{P_0 - P_1}{L_2} < 0 \qquad (5.88)$$

This implies that in domain L_1 the velocity profile is made up of two contributions both acting in the same positive x-direction: a drag term due to the moving wall plus a second term due to the pressure difference $P_1 - P_0$, while in domain L_2 the pressure term is negative and counteracts the drag term. The respective velocity profiles have been sketched in Figure 5.19.

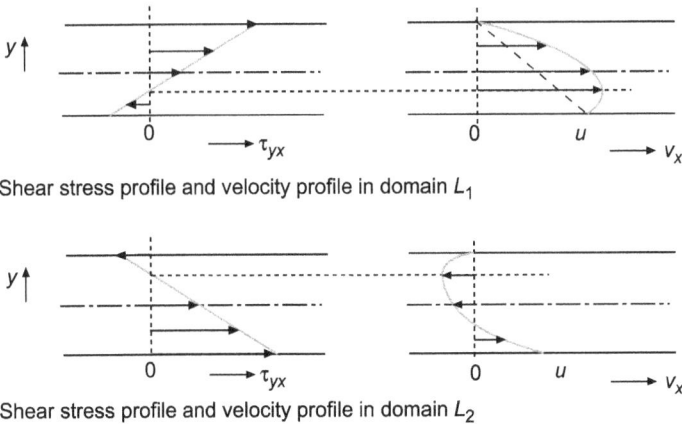

a. Shear stress profile and velocity profile in domain L_1

b. Shear stress profile and velocity profile in domain L_2

Figure 5.19

The flow rate (per unit of depth) in domain L_1 to the right, asked in subquestion c, is found by integrating the velocity profile:

$$\phi'_v = \int_0^\delta v_x(y)dy = \int_0^\delta \left[U \cdot \left(1 - \frac{y}{\delta}\right) + \frac{\Gamma}{2\mu} y\,(\delta - y)\right] dy =$$

$$= \frac{1}{2} U\delta + \frac{P_1 - P_0}{12\mu L_1}\, \delta^3$$

(5.89)

Outside the gap, all liquid is moving at speed U as the boundary condition of zero velocity at the upper surface of the liquid layer does no longer apply there – resulting in a new expression for the flow rate per unit of depth:

$$\phi'_v = U\,\delta_\infty$$

(5.90)

Owing to an overall mass balance, these two flow rates are equal, resulting in an expression for the eventual coating thickness δ_∞:

$$\frac{\delta_\infty}{\delta} = \frac{1}{2} + \frac{1}{12}\, \Gamma^*$$

(5.91)

in which the non-dimensional pressure gradient Γ^* denotes the ratio of pressure force to drag force:

$$\Gamma^* = \frac{(P_1 - P_0)\,\delta^2}{\mu U L_1}$$

(5.92)

The distance L_2 – the topic of subquestion d – can be found from the observation that in the steady-state considered there is no net flow to the right or to the left in domain L_2: the lower layers of the liquid are dragged to the right by the moving web, while the upper layers are pushed to the left by the pertinent pressure difference. These two effects just balance and determine the steady-state length L_2. By imposing the condition

$$\phi'_v = \int_0^\delta v_x(y)\, dy = 0$$

(5.93)

and by substituting the equations (5.86) and (5.88) into equation (5.93), an expression for L_2 is obtained:

$$L_2 = \frac{(P_1 - P_0)\,\delta^2}{6\,\mu U}$$

(5.94)

which shows that for example L_2 decreases when U is increased. This result is in perfect agreement with the result of the dimensional analysis of Example 2.7.

5.6.3 Laminar flow in Cartesian coordinates: due to gravity

In this example, it concerns a laminar flow of a Newtonian liquid between two very large vertical plates. The distance between the two plates (of width B) is referred to now as $2D$ (for reasons that will later become apparent). The flow here occurs only under the influence of gravity (there is no difference in pressure driving the flow). This steady-state situation is shown in Figure 5.20. Different to the earlier situations, the flow is now in the (negative) y-direction!

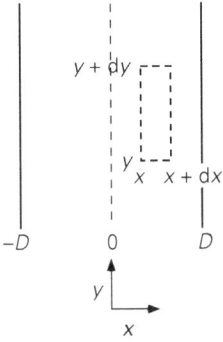

Figure 5.20

Here, too, the shear stress profile is determined first. And again, the y momentum balance for the control volume $Bdxdy$ reduces to the **force balance** $\Sigma F_y = 0$:

$$0 = B\,dy \cdot \tau_{xy}\big|_x + B\,dy \cdot \left[-\tau_{xy}\big|_{x+dx}\right] - B\,dx\,dy \cdot \rho g \tag{5.95}$$

in which – in comparison with equation (5.76) – the role of the two pressure terms is taken over by gravity. Gravity acts on the mass of the slice: $F = M{\cdot}g$ (in kg·m/s² or in kg·N/kg) and again supplies momentum (in Ns) per unit of time and mass. From equation (5.95) the following differential equation for the shear stress is obtained:

$$\frac{d\tau_{xy}}{dx} = -\rho g \tag{5.96}$$

with the solution being:

$$\tau_{xy}(x) = -\rho g x + C_1 \tag{5.97}$$

Here, the specific weight ρg, sometimes denoted by γ, pops up as the driving force for flow (see also the discussion after Example 2.6). The velocity profile then follows by using the fact that the liquid is Newtonian:

$$\tau_{xy} = -\mu \frac{d}{dx} v_y \rightarrow \frac{d}{dx} v_y = \frac{\rho g}{\mu} x - \frac{C_1}{\mu} \tag{5.98}$$

and by integrating equation (5.98):

$$v_y = \frac{\rho g}{\mu} \frac{1}{2} x^2 - \frac{C_1}{\mu} x + C_2 \tag{5.99}$$

The two integration constants can be determined with the help of the two obvious boundary conditions for the velocity, that is: $x = \pm D \rightarrow v_y = 0$. This gives $C_1 = 0$ and $2C_2 = -\rho\,g\,D^2/\mu$, and therefore:

$$v_y(x) = -\frac{\rho g}{2\mu}\left(D^2 - x^2\right) \tag{5.100}$$

The velocity profile therefore has the form of a parabola. The liquid flows down of course – note the minus sign! The maximum velocity is found at $x = 0$. Figure 5.21 shows the shear stress and velocity profiles.

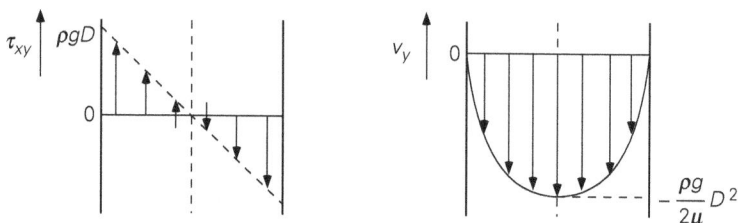

Figure 5.21

The integration constant C_1 can be more directly determined in this case by using at an earlier stage the symmetry that is present in the problem. This symmetry is hidden above, in the two velocity-related boundary conditions: both plates are motionless walls with the same effect on the flow (just like in Section 5.6.2). This symmetry also means that transport of y momentum through plane $x = 0$ is not possible at all, because then the two 'halves' of the flow would be or become different. This means that both 'halves' exert no force (shear stress) on each other: $x = 0 \rightarrow \tau_{xy} = 0$. It is precisely for this reason that in such a symmetrical situation the $x = 0$ axis is selected halfway between the two plates.

Also, it is precisely because of the link between shear stress and velocity gradient that the velocity at $x = 0$ is also at its maximum. The symmetry condition therefore makes it possible to directly conclude here too that with equation (5.97), $C_1 = 0$, so that without having to go the roundabout route via the velocity profile it follows directly that:

$$\tau_{xy}(x) = -\rho g x \tag{5.101}$$

Example 5.11. Film condensation II.
With the help of the above theory, it is possible to further elaborate equation (3.193) from Example 3.16 concerning film condensation. To do this, it is necessary to find out how ϕ_v changes in the flow direction x (now selected as downwards). This flow rate depends on both the thickness of the layer and the velocity profile in the layer. The dependence can be found by determining the velocity profile for a certain thickness δ and the resulting liquid flow rate associated with that δ.

In the situation in Figure 3.32, then, consider a thin slice of thickness dy at some distance from the wall (this is actually half of the aforementioned case of flow between two thin vertical slices). From a force balance for this slice it follows that, analogous to equation (5.97), but thanks to the other choice of coordinate:

$$\tau_{yx} = -\rho\, g\, y + C_1 \tag{5.102}$$

With boundary condition $y = \delta \rightarrow \tau_{yx} = 0$ (because the maximum velocity is at the surface of the film and virtually no momentum is transferred to, or shear stress exerted on, the air: compare a similar assumption concerning the interface of liquid and air in Example 3.10) it follows that

$$\tau_{yx} = \rho\, g(\delta - y) \tag{5.103}$$

For a Newtonian fluid, substituting Newton's viscosity law, integration and the use of the boundary condition $y = 0 \rightarrow v = 0$ leads to

$$v = \frac{\rho g}{\mu}\left(\frac{1}{2}y^2 - y\delta\right) \tag{5.104}$$

Integrating this velocity profile for thickness δ results in the following expression for the flow rate across width b:

$$\phi_V = \frac{\rho g \delta^3 b}{3\mu} \tag{5.105}$$

Differentiating versus δ and substituting the result into equation (3.193) gives

$$\Delta h_V\, \rho^2 g \delta^3\, d\delta = \mu\, \lambda(T - T_w)\, dx \tag{5.106}$$

Integration between the limits $x = 0 \rightarrow \delta = 0$ and $x = L \rightarrow \delta = \delta_L$ leads to

$$\delta_L = \left(\frac{4\mu\,\lambda(T - T_w)L}{\rho^2 g\, \Delta h_V}\right)^{1/4} \tag{5.107}$$

An expression for the heat transfer coefficient $\langle h \rangle$ averaged over the height L follows from a heat balance for the whole film between $x = 0$ and $x = L$:

$$\langle h \rangle (T - T_w)\, bL = \phi_V|_{x=L} \cdot \rho \Delta h_V \tag{5.108}$$

Combining equations (5.105), (5.107), and (5.108) produces equation (3.194).

5.6.4 Laminar flow in cylindrical coordinates: due to a pressure gradient

Finally, we will discuss an example of a flow involving a cylindrical geometry: steady-state laminar flow of a Newtonian fluid through a very long horizontal straight cylinder under the influence of a pressure gradient. A **force balance** can be drawn up for a cylindrical case within the liquid of thickness dr (between r and $r + dr$) and length dx (see Figure 5.22).

A force $p|_x \cdot 2\pi r\, dr$ is being exerted in the positive x direction on the left-hand end of the cylindrical case (at $x = x$). Similarly, a force in the negative x direction, that is, $-p|_{x+dx} \cdot 2\pi r\, dr$, works on the right-hand end from outside (at $x = x + dx$). Shear stress $\tau_{rx}|_r$ is being exerted on the inside of the cylindrical case on surface $2\pi r\, dx$, while the

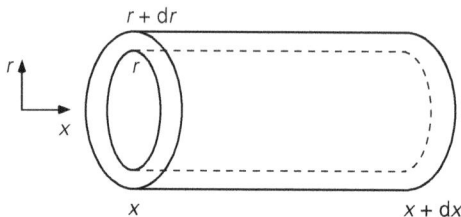

Figure 5.22

stress $-\tau_{rx}|_{r+dr}$ works on the outside on surface $2\pi(r + dr)dx$. This produces the following force balance:

$$0 = p|_x \, 2\pi r \, dr + \left(-p|_{x+dx} \, 2\pi r \, dr\right) +$$
$$+ \tau_{rx}|_r \cdot 2\pi r \, dx + \left[-\tau_{rx}|_{r+dr} \cdot 2\pi(r + dr) \, dx\right] \tag{5.109}$$

This equation can also be written as

$$0 = p|_x \, 2\pi r \, dr - p|_{x+dx} \, 2\pi r \, dr +$$
$$+ 2\pi(r\tau_{rx})|_r dx - 2\pi(r\tau_{rx})|_{r+dr} dx \tag{5.110}$$

where in the third and fourth terms of the right-hand side, the product of radius and shear stress is given in brackets because the perimeter (or surface area) and shear stress both depend on r (see also, e.g. Section 3.1.3). Equation (5.110) can be easily rewritten (divide all terms by $2\pi r dr dx$) to the differential equation for the shear stress:

$$\frac{1}{r}\frac{d}{dr}(r\tau_{rx}) = -\frac{dp}{dx} = \Gamma \tag{5.111}$$

Now the pressure gradient is a given constant, so – with for $[-dp/dx]$ again the notation Γ (see Section 5.6.2) – the equation can be solved fairly easily:

$$\frac{d}{dr}(r\tau_{rx}) = \Gamma r \rightarrow r\tau_{rx} = \frac{1}{2}\Gamma r^2 + C_1$$
$$\rightarrow \tau_{rx} = \frac{1}{2}\Gamma r + \frac{C_1}{r} \tag{5.112}$$

Again, on the basis of symmetry considerations, it is the case that the shear stress on the axis of the cylinder must equal zero: $r = 0 \rightarrow \tau_{rx} = 0$. This means that the integration constant C_1 is zero. The shear stress profile is then:

$$\tau_{rx}(r) = \frac{1}{2}\Gamma r \tag{5.113}$$

Remember that $\Gamma > 0$: there is always a pressure drop involved, because the friction on the wall and internally between the layers of liquid has to be overcome. Equation (5.113) shows that, in accordance with the definition of shear stress, $\tau_{rx} \geq 0$.

The velocity profile now follows by including the fact that the liquid is Newtonian; as a result:

$$\frac{d}{dr} v_x = -\frac{\Gamma}{2\mu} r \tag{5.114}$$

The solution to this is

$$v_x = -\frac{\Gamma}{4\mu} r^2 + C^2 \tag{5.115}$$

Integration constant C_2 can be determined using the boundary condition: $r = R \rightarrow v_x = 0$. This gives

$$v_x(r) = \frac{\Gamma}{4\mu} \left(R^2 - r^2 \right) = \frac{R^2 \Gamma}{4\mu} \left(1 - \frac{r^2}{R^2} \right) \tag{5.116}$$

Again, this is a parabola with the maximum velocity on the axis of the cylinder ($r = 0$). The shear stress profile – equation (5.113) – and the velocity profile – equation (5.116) – are shown in Figure 5.23. Note that the centre-line $r = 0$ is an axis of symmetry!

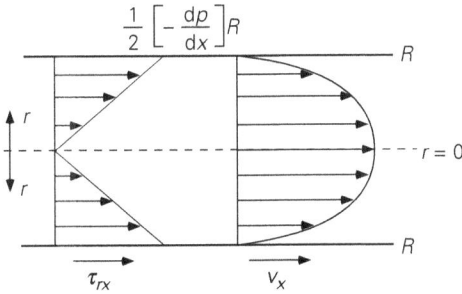

Figure 5.23

The mean liquid velocity is

$$\langle v \rangle = \frac{\int_0^R v_x(r) 2\pi r \, dr}{\int_0^R 2\pi r \, dr} = \frac{R^2 \Gamma}{8\mu} = \frac{R^2}{8\mu} \left(-\frac{dp}{dx} \right) \tag{5.117}$$

This is precisely half the maximum velocity in the cylindrical tube: $v_{max} = 2 \langle v \rangle$.

Equation (5.117) can also be written in another form, namely as an equation between the pressure drop Δp over a tube length L and the mean velocity:

$$\Delta p = \frac{8\mu L}{R^2} \langle v \rangle = \frac{32\mu L}{D^2} \langle v \rangle = \frac{64\mu}{\rho \langle v \rangle D} \cdot \frac{L}{D} \cdot \frac{1}{2} \rho \langle v \rangle^2 \tag{5.118}$$

The latter term of this equation is deliberately written in a form that was also used for modelling energy dissipation due to wall friction (e_{fr}). This is easy to understand:

$$\frac{\Delta p}{\rho} = e_{fr} = \frac{64\mu}{\rho \langle v \rangle D} \cdot \frac{L}{D} \cdot \frac{1}{2} \langle v \rangle^2 \tag{5.119}$$

This therefore proves that in the case of laminar flow through a horizontal cylindrical tube, the Fanning friction factor $4f$ is equal to 64/Re indeed: see expression (5.28).

Finally, the flow rate that flows under influence of a pressure drop Δp through a pipeline with a circular cross section and of length L follows from equation (5.116) through integration:

$$\phi_V = \int_0^R v_x(r) 2\pi r \, dr = \frac{\pi R^4}{8\mu} \frac{\Delta p}{L} \tag{5.120}$$

This equation is called the Hagen–Poiseuille law; the associated laminar flow is referred to as just Poiseuille flow.

The calculation of the velocity profile in tubes with a non-circular cross section is a good deal more complicated. For a tube with a square-shaped cross section with edge a, an equation applies that is very similar to equation (5.120), but with a factor $a^4/28.6$ instead of $\pi R^4/8$. Notice that the flow rate always remains proportional to the typical dimension to the power of 4.

! **Summary**

For steady-state laminar flow, the velocity profile (=the radial momentum–concentration distribution) is easy to calculate if the geometry is sufficiently simple. This is analogous to the situation regarding molecular transport of heat or mass. To do this, it is necessary to draw up a momentum balance which, for simple geometries, is reduced to a force balance. The shear stress profile can be determined using this force balance.

This profile is **independent** of the type of liquid. It is only when determining the velocity profile that the link between shear stress and velocity gradient is important: does the fluid obey Newton's viscosity law or not? In many cases, the integration constants that appear during the integration process of force balance into velocity profile can be determined most safely from the boundary conditions related to the velocity.

For the steady-state laminar flow of a Newtonian liquid through a cylindrical tube, it has been derived that the shear stress is zero at the axis and increases in a radial direction, that the velocity profile is parabolic, and that the Hagen–Poiseuille law applies to the flow rate:

$$\phi_V = \frac{\pi R^4}{8\mu} \frac{\Delta p}{L}$$

but shear stress and velocity profiles have also been derived for other geometries.

5.7 Laminar flow of non-Newtonian liquids

In the previous section, we looked only at fluids that obeyed Newton's viscosity law – see equation (2.4) in Section 2.1.2 – the standards of example of which are air and water. However, there are very many liquids that comply only slightly with the law, or indeed not at all. Some examples of liquids that display **non-Newtonian behaviour** are:

- Polymers
- Rubbers
- Paint
- Wet cement

- Toothpaste
- Blood
- Peanut butter
- Margarine

Non-Newtonian behaviour is more the rule than the exception. This is because Newton's viscosity law applies only to spherical molecules which, apart from collisions, can move among each other relatively unimpeded. The molecules of most substances are not spherical, but instead consist of either longer or branched chains of atoms or groups of atoms; moreover, the molecules influence each other in terms of their movements, often through charge distributions within the molecules (polar groups). To pretend that the molecules behave like marbles is a long way removed from the truth.

Under the influence of shear forces that are imposed upon them, the molecules may rearrange themselves and/or intermolecular bonds between polar groups or branches may be broken to a greater or lesser degree. Only then perhaps does Newton's viscosity law apply.

Non-Newtonian liquids can be divided into difference categories, each of which has their own rule for the link between the shear stress and the velocity gradient. The term for this link is the **rheology** of the liquid. We will discuss a number of commonly occurring categories in this section.

5.7.1 Power law liquids

A large number of liquids obey the Ostwald–De Waele model for the link between shear stress and velocity gradient:

$$\tau_{xy} = - K \left|\frac{dv_y}{dx}\right|^{n-1} \cdot \frac{dv_y}{dx} \tag{5.121}$$

Equation (5.121) is also known as the **power law**; any liquid that obeys this law is re-ferred to as a power law liquid. The constant K is called the **consistency** and n the **flow index**. Index n is non-dimensional, while K does of course have a dimension, which is dependent on the value of n. The vertical bars in this power law denote that the abso-lute value (or modulus) of the derivative between the bars should be taken here.

Note that Newtonian liquids form a special case of a power law liquid, namely when $n = 1$. There is another distinction in this category of non-Newtonian liquids: power law liquids with an index $n < 1$ are called **pseudoplastic**. An example of this is a 4 wt% pulp in water for which $n = 0.575$ and $K = 20.0$ Ns$^{0.575}$/m^2 are found. If $n > 1$, we refer to a **dilatant** liquid, an example of which is wet cement.

Often, an **apparent viscosity** coefficient or **effective viscosity** coefficient μ_e is used which is defined as

$$\mu_e = K \cdot \left|\frac{dv_y}{dx}\right|^{n-1} = K \cdot \dot{\gamma}^{n-1} \tag{5.122}$$

This **apparent viscosity** coefficient μ_e depends not only on the physical property K, but also on the (local) velocity gradient or **shear rate** $\dot{\gamma}$, and expresses that $\dot{\gamma}$ leads to rearrangement of supramolecular structures and/or to (partial) breaking of bonds be-tween polar groups or branches of the (polymer) molecules and in this way affects the flow properties of the liquid. Notice that it is only the magnitude (denoted by the modulus bars), not the sign of the velocity gradient that influences the intermolecular interaction. With this definition of an apparent viscosity, Expression (5.121) is con-verted into an equation which again resembles Newton's viscosity law but now with a flow dependent viscosity coefficient.

The solution to laminar, steady-state flow problems runs entirely analogously to that of Newtonian liquids, but μ_e can have a different value anywhere in the flow do-main, depending on the magnitude of the local velocity gradient. This effect should be factored in. This will be demonstrated by means of an example.

Example 5.12. A power law liquid between two vertical plates.
Consider the laminar steady-state flow of a power law liquid between two very large vertical plates (see Figure 5.24). The flow is solely influenced by gravity; no pressure difference is imposed. The question is to derive an expression for the velocity profile.

Solution
From the force balance $\sum F_y = 0$ comes

$$0 = B\,dx \cdot \tau_{xy}|_x + B\,dx \cdot \left(-\tau_{xy}|_{x+dx}\right) - B\,dx\,dy \cdot \rho g \tag{5.123}$$

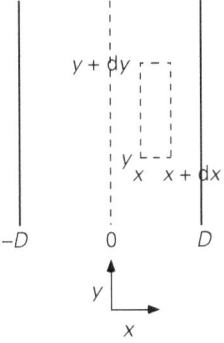

Figure 5.24

Of course, this equation again produces the shear stress profile of equation (5.101):

$$\tau_{xy}(x) = -\rho g x \tag{5.124}$$

where, as before, use has been made of the symmetry plane at $x = 0$ midway between the plates. Remember that the rheology of the liquid has no influence on the shear stress at all. In order now to determine the velocity profile, the fact that the liquid in question is a power law liquid **should** be taken into account:

$$\tau_{xy} = -K \left|\frac{dv_y}{dx}\right|^{n-1} \frac{dv_y}{dx} \tag{5.125}$$

First, the two vertical bars in equation (5.125) need to be disposed of, for which we need the velocity gradient sign. Consider the domain $0 \le x \le D$ and remember that – because of the symmetry in this problem – the velocity profile in the domain $-D \le x \le 0$ can be obtained later through reflection of the $(x = 0)$ line. In the domain $0 \le x \le D$, velocity v_y becomes less negative as x increases; in other words, the derivative $dv_y/dx > 0$. On that case, the modulus lines can be dropped.

[In the other case – that is: with $dv_y/dx < 0$ – $|dv_y/dx|$ should be replaced by $(-dv_y/dx)$ to the effect that $|dv_y/dx|^{\frac{1}{n}}$ becomes $(-dv_y/dx)^{\frac{1}{n}}$.]

Combining equations (5.124) and (5.125) then produces for the domain $0 \le x \le D$

$$-K\left(\frac{dv_y}{dx}\right)^n = -\rho g x \;\rightarrow\; \frac{dv_y}{dx} = \left(\frac{\rho g}{K}\right)^{\frac{1}{n}} x^{\frac{1}{n}} \tag{5.126}$$

Integration gives

$$v_y(x) = \left(\frac{\rho g}{K}\right)^{\frac{1}{n}} \frac{n}{n+1} x^{\frac{n+1}{n}} + C_2 \tag{5.127}$$

The integration constant, C_2, can be determined with the help of the boundary condition $x = D \rightarrow v_y = 0$. This means the velocity profile in the domain $0 \le x \le D$ ultimately becomes

$$v_y(x) = -\frac{n}{n+1} \left(\frac{\rho g}{K}\right)^{\frac{1}{n}} \left(D^{\frac{n+1}{n}} - x^{\frac{n+1}{n}}\right) \tag{5.128}$$

Notice that for $n = 1$, the solution for Newtonian liquids, like the one determined in Section 5.6.3, appears again. Figure 5.25 shows the velocity profiles for two cases: $n = \frac{1}{3}$ and $n = 3$.

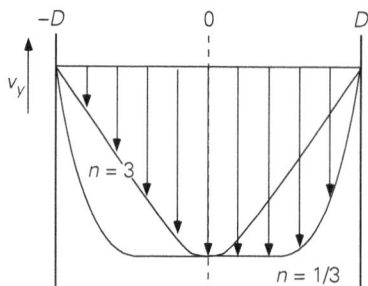

Figure 5.25

There are two other interesting limit cases:

i) $n \to \infty$: $v_y = -\left(\frac{\rho g}{K}\right)^{\frac{1}{n}}(D-x) \to -(D-x)$: see Figure 5.26.
ii) $n \to 0$: the profile v_y approaches that of ideal plug flow; the latter assertion has been made visible in Figure 5.27 by drawing cases $n = 1/10$, $1/100$, $1/1000$.

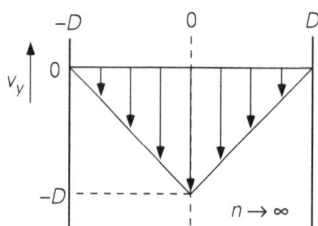

Figure 5.26

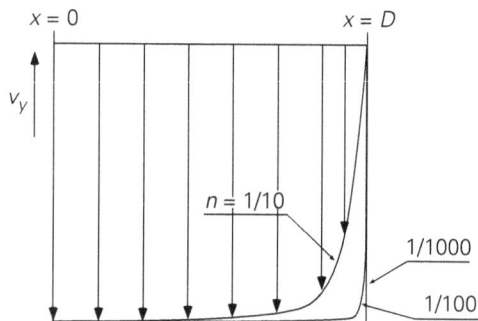

Figure 5.27

5.7.2 Bingham liquids

There are also materials that only start to flow once the shear stress exerted on them exceeds a certain value. Examples of these are clay and toothpaste. The behaviour of toothpaste in particular is well known: if the tube (with the cap removed) is held upside down, nothing happens. It is only by squeezing (which actually means increasing the pressure in the tube) that the liquid flows out. Liquids that demonstrate this kind of behaviour are known as **Bingham liquids**. The link between the shear stress and the velocity gradient in the case of a Bingham liquid is

$$\left|\tau_{xy}\right| - \tau_0 = \mu\left|\frac{dv_y}{dx}\right| \quad \text{for} \quad \left|\tau_{xy}\right| \geq \tau_0$$
$$\frac{dv_y}{dx} = 0 \quad \text{for} \quad \left|\tau_{xy}\right| < \tau_0$$

(5.129)

The quantity τ_0 is called the **yield stress** and is a physical property of the compound involved: it is only when the tension being exerted exceeds a critical value – this yield stress – that a liquid may develop and exhibit velocity gradients because the intermolecular network then gives way. We will illustrate the use of a yield stress of this kind with an example.

Example 5.13. A Bingham liquid between two vertical plates.
Consider again the situation in which a liquid is flowing downwards between two vertical and motionless plates under the influence of gravity. The Bingham liquid in this example has a yield stress that is numerically equal to $\rho\, gD/2$ (Note: increasing or decreasing the distance D between the plates does not change the yield stress; here, it concerns just the value of the stress yield of the liquid considered.)
   Now, the question is about the velocity profile.
   It follows still from a force balance for a typical volume element that:

$$\tau_{xy} = -\rho g x \tag{5.130}$$

– cf. Equations (5.101) and (5.124). In order to solve the velocity profile, the rheology of the liquid has to be specified. Because of the symmetry, only the domain $0 \le x \le D$ has to be analysed. Here, the following applies:

$$|\tau_{xy}| = \rho g x \tag{5.131}$$

It is then necessary – due to equation (5.129) – to make a distinction between the domains where $|\tau_{xy}| \ge \tau_0$ and where $|\tau_{xy}| < \tau_0$. Consider first the domain where $|\tau_{xy}| \ge \tau_0$. Here,

$$|\tau_{xy}| \ge \tau_0 \;\; \rightarrow \;\; \rho g x \ge \frac{1}{2}\rho g D \;\; \rightarrow \;\; x \ge \frac{1}{2}D \tag{5.132}$$

Because of $dv_y/dx \ge 0$ for $x \ge 0$, it follows from equations (5.129) to (5.132) that

$$\rho g x - \tau_0 = \mu \frac{dv_y}{dx} \tag{5.133}$$

From this, it follows for the velocity gradient

$$\frac{dv_y}{dx} = \frac{\rho g}{\mu}x - \frac{\tau_0}{\mu} = \frac{\rho g}{\mu}\left(x - \frac{1}{2}D\right) \tag{5.134}$$

where the given expression for τ_0 has also been entered. The solution to this differential equation, thanks to the boundary condition $x = D \rightarrow v_y = 0$, is

$$v_y(x) = \frac{\rho g}{2\mu}\left(x^2 - Dx\right) \tag{5.135}$$

Then, the profile for the range $0 \le x < D/2$ has to be worked out. Here, given expressions (5.129), the following applies: $|\tau_{xy}| < \tau_0$, and therefore

$$\frac{dv_y}{dx} = 0 \;\; \rightarrow \;\; v_y = \text{constant} = v_y\left(\frac{1}{2}D\right) = -\frac{\rho g}{8\mu}D^2 \tag{5.136}$$

In this latter equation, a boundary condition has been used whereby the value of the velocity must be taken as $x = D/2$, as follows from the first part of the solution.

In Figure 5.28, the velocity profile of the Bingham liquid is shown after it has been mirrored with respect to the symmetry axis. In the range $-D/2 < x < D/2$, it is not the case that $v_y = 0$, but $v_y =$ constant $\neq 0$! In this range, genuine plug flow occurs.

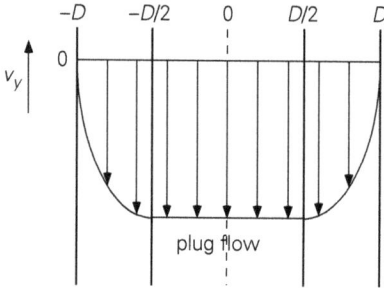

Figure 5.28

What would the solution to the velocity profile have been if the yield stress value had been greater than $\rho g D$?

In that case, $|\tau_{xy}| < \tau_0$ would have applied throughout the liquid and therefore everywhere the velocity $v_y =$ constant. Applying the boundary condition $x = D \rightarrow v_y = 0$ then means that the following applies to the whole liquid layer: $v_y = 0$!

5.7.3 Casson liquids

The rheology of Casson liquids is described by

$$
\begin{aligned}
|\tau_{xy}|^{1/2} - \tau_c^{1/2} = \mu_c^{1/2} \left|\frac{dv_y}{dx}\right|^{1/2} \quad &\text{if} \quad |\tau_{xy}| \geq \tau_c, \\
\frac{dv_y}{dx} = 0 \quad &\text{if} \quad |\tau_{xy}| < \tau_c
\end{aligned}
\tag{5.137}
$$

in which τ_c again is a yield stress. Examples of liquids that fall into this category are yoghurt and blood.

5.7.4 Viscoelastic liquids

Depending on the circumstances (imposed stresses), some liquids also have **elastic properties**; the behaviour of liquids of this kind are often satisfactorily described by

$$
\tau_{xy} + \lambda \frac{d\tau_{xy}}{dt} = -\mu \frac{dv_y}{dx}
\tag{5.138}
$$

where λ is the time constant that describes the elastic behaviour.

The remarkable thing about this is that the liquid behaves in part like a 'normal' viscous liquid: in a steady-state condition, equation (5.138) is reduced to the Newtonian equation. However, if the shear stress varies in time, the substance will display elastic properties. This behaviour too can be traced back to the formation of networks between strongly branched and rolled up molecules and the reaction of these networks to imposed stresses.

A very good illustration of this **viscoelastic** behaviour can be demonstrated through shark fin soup. If a spoon is used to stir the soup in the same direction for a longer period of time, a steady-state situation will arise. If the spoon is then removed from the liquid, then it will continue to rotate for a short time and come to a standstill as a result of viscosity. Now, however, the situation is no longer steady and the other properties of the liquid become clearly visible. The soup does not slowly come to a standstill, but actually abruptly reverses the direction of flow! It is as if the liquid (actually the polymer molecules in the soup) were stretched like an elastic band and, now that the spoon has been taken out, that the band wishes to resume its state of balance. This is a simple but effective test for finding out whether shark fin soup is in fact 'genuine' ($\lambda \approx 1$ s).

At the end of this section on non-Newtonian liquids, Figure 5.29 shows, by way of summary, the link between the shear stress and the velocity gradient for two Newtonian fluids, a dilatant liquid, a pseudoplastic liquid, a Bingham liquid and a Casson liquid.

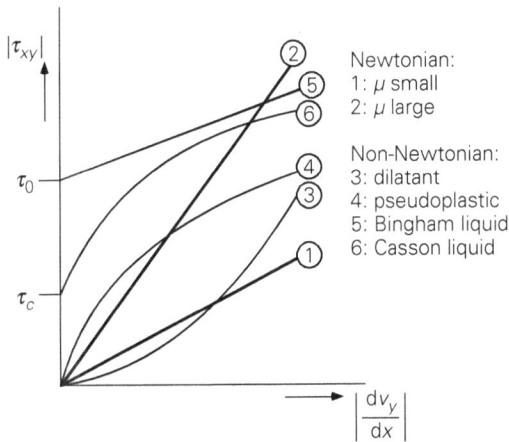

Figure 5.29

Summary

The link between the velocity gradient and the shear stress is not shown for all liquids by Newton's law. On the contrary, most liquids that are used in industry and indeed in everyday life do not fall into this category. There are many different categories of liquid, each with their own link between shear stress and velocity gradient (see Figure 5.29). The field that is involved with this is an entirely separate part of fluid mechanics and is called **rheology**. The link between the shear stress and the velocity gradient is also referred to as the rheology of the liquid.

Remember that the shear stress profile does not depend on the rheology of the liquid, but the velocity profile does. Velocity profiles have been derived for power law liquids and Bingham liquids. Casson liquids and viscoelastic liquids have also been introduced. Sometimes, the flow starts to resemble plug flow.

5.8 The general equations of motion

5.8.1 Fluid mechanics and transport phenomena

In Chapters 1 and 2 of this book we discussed the techniques of drawing up balances and dimensional analysis. Additionally, all kinds of concepts (such as mechanical energy, residence time distribution, and shear stress), non-dimensional numbers, and several phenomenological laws have been introduced. We then looked at heat and mass transport in Chapters 3 and 4, focusing first on molecular transport before dealing with the use of transfer coefficients for convective transport. With the description, development, application, and upscaling of process equipment, chemical engineers have come a long way on the basis of the aforementioned approach, which has been substantiated by extensive empirical information. The analogy between heat and mass transfer was apposite and extremely helpful.

Fluid mechanics, which is actually momentum transport or momentum transfer, can also be considered analogously to heat and mass transfer. This was done in Sections 2.1 and 5.6 for laminar flow (molecular transport) and in Section 5.3.3 for friction drag (momentum transfer at solid walls) under turbulent flow conditions. Nonetheless, the sequence of the topics covered in this chapter on fluid mechanics has been largely different to that in Chapters 3 and 4. Chapter 5 starts with the phenomenological approach to friction and with pressure drop calculations, with molecular transport coming only thereafter. There are four reasons for this inverted sequence.

First, for many chemical engineers, pressure drop calculations across pipeline systems and packed beds are highly relevant, with the mechanical energy balance playing a major role. This balance falls slightly outside the classic analogy of momentum, heat, and mass transfer; note that the friction factor (Section 5.3.1) and the loss coefficient of a fitting (Section 5.4) are defined in a very different way to the heat and mass transfer coefficients (in Sections 3.5 and 4.5, respectively), although the analogy for the friction factor is restored again in Section 5.3.3. With engineering practice in mind, the chapter therefore began with engineering fluid mechanics for pressure drops that concurs with the coverage of mechanical energy balances and Bernoulli's law (in Section 1.3.3 and Examples 1.20 and 1.21) and that of drag force (in Section 2.3).

Second, apart from the analogy with heat and mass transport, there is – seen from the industrial engineering practice – no good reason for Chapter 5 to start with molecular momentum transport. The simplest version of molecular momentum transport is one-dimensional laminar flow.[21] This is not particularly relevant for pressure drop calculations in most of engineering fluid mechanics, while on the contrary molecular heat transport for example is important with regard to the description of convective heat transport (because of the term 'resistance to heat transport', the Nusselt number, and the overall heat transfer coefficient). This makes it possible to postpone the coverage of molecular momentum transport.

Third, reference was already made in Chapters 3 and 4 to the fact that a more precise alternative **is** available for the phenomenological approach to convective heat and mass transfer. In Sections 3.4 and 4.4, after all, the generally valid micro-balances that lead to the transport equations for time-dependent three-dimensional heat and mass transport were presented on the basis of the **cubic volume element method**. It was pointed out here that solving this transport equation requires an equally detailed knowledge of the (time-dependent three-dimensional) velocity field.

The so-called Navier–Stokes equations of motion, which describe the flow (the momentum housekeeping) in a three-dimensional domain, will be looked at now; this links up to the coverage of the one-dimensional flows in Sections 5.6 and 5.7. The coverage of the Navier–Stokes equations once again highlights the analogy between heat, mass, and momentum transport in all its glory: the transport equations for momentum (3x), heat, and mass are also completely identical, mathematically.

Finally, whereas the Navier–Stokes equations used to be something of an oddity and the be-all and end-all of theory for many classically educated chemical engineers, they now serve as the starting point for modern chemical engineering, which uses numerical techniques in order to gain an inside picture of local transport and transfer phenomena in process equipment. This makes it much easier to understand and manage all kinds of effects of geometry and scale on the yield and selectivity of chemical reactions, for example, on the intensity and effectiveness of heat and mass transfer, and on separation processes (through variations in flow patterns, residence time distributions, contact times, etc.). The chapter ends with this highly promising future prospect.

5.8.2 The continuity equation

Before deriving the aforementioned Navier–Stokes equations for three-dimensional momentum transport in a flow domain, we will first use the **cubic volume element**

---

[21] One-dimensional laminar flows occur in micro-fluidics (micro-reactors, lab on a chip), in polymer technology (extrusion and coating processes), in the life sciences, and in the food industry; in the three latter fields, non-Newtonian liquid properties often play a dominant role.

method in order to derive an equation on the basis of an overall mass balance for a small cube, with which every flow field must comply. Once again, we will use only a Cartesian coordinate system, for the sake of simplicity (see Figure 5.30, which is identical to Figure 3.22). The control volume *dxdydz* is located at a random position somewhere in the field of flow, which is three-dimensional and time-dependent. This means that the three velocity components and the density ρ are all a function of both time and the x, y, and z coordinates.

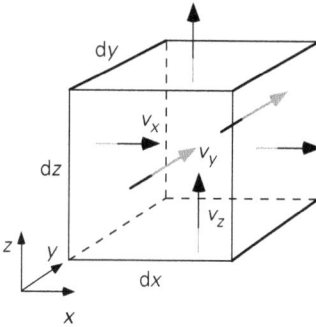

Figure 5.30

The micro-balance for the total mass for the cube in Figure 5.30 contains, in addition to the accumulation term,

$$\frac{\partial}{\partial t} \rho \, dxdydz \tag{5.139}$$

only the convective transport terms through each of the six planes. As a result of convective transport, therefore,

$$[\rho v_x]|_{x,y,z} \, dydz \tag{5.140}$$

of mass enters through the left-hand plane, while

$$[\rho v_x]|_{x+dx,y,z} \, dydz \tag{5.141}$$

of mass leaves again through the right-hand plane. This produces a net contribution of convective mass transport in the direction of x of

$$[\rho v_x]|_{x,y,z} \, dydz - [\rho v_x]|_{x+dx,y,z} \, dydz = -\frac{\partial}{\partial x}(\rho v_x) \, dxdydz \tag{5.142}$$

In an identical manner, the convective mass flows through the rear and front planes produce a net contribution

$$[\rho v_y]|_{x,y,z} \, dxdz - [\rho v_y]|_{x,y+dy,z} dxdz = -\frac{\partial}{\partial y}(\rho v_y) dxdydz \tag{5.143}$$

and through the top and bottom planes

$$[\rho v_z]|_{x,y,z} \, dxdy - [\rho v_z]|_{x,y,z+dz} \, dxdy = -\frac{\partial}{\partial z}(\rho v_z) \, dxdydz \tag{5.144}$$

Merging equations (5.139), (5.142), (5.143), and (5.144), and dividing this by the fixed magnitude $dxdydz$ of the cube produces what is known as the **continuity equation**

$$\frac{\partial \rho}{\partial t} = -\frac{\partial \rho v_x}{\partial x} - \frac{\partial \rho v_y}{\partial y} - \frac{\partial \rho v_z}{\partial z} \tag{5.145}$$

This continuity equation is a direct result of the **law of conservation of mass**. This is still the case if the density is constant, and equation (5.145) therefore changes to

$$0 = \frac{\partial v_x}{\partial x} + \frac{\partial v_y}{\partial y} + \frac{\partial v_z}{\partial z} \tag{5.146}$$

What enters through the left-hand plane, for example, and does not exit through the right-hand plane must, where density is constant, exit through one or more of the other planes. This is the case at all times! Notice that the density (the mass) no longer features in equation (5.146).

Summary
The continuity equation has been derived with the help of the cubic volume element method, with which every velocity field must comply at all times: the direct consequence of the law of conservation of mass.

5.8.3 The Navier–Stokes equations

For the derivation of the general micro-balances for momentum transport, or the Navier–Stokes equations, it is advisable to again use the term **momentum concentration** that was introduced in Section 1.4: the velocity component v_x (in m/s) in the direction of x is actually a concentration of x-momentum (in Ns/kg). This is also why ρv_x represents the momentum concentration on the basis of volume (in Ns/m³).

The actual velocity distributions, or the field of velocity, in a flow domain come about because momentum is redistributed across the domain by convective and molecular momentum flows, while forces are exerted on the mass and on the boundaries of the domain. Once again, the **cubic volume element method** is used in order to identify this momentum housekeeping.

Because of the vector character of momentum and of forces, a separate momentum balance for the cube in Figure 5.30 should be drawn up for every direction.

Below, only the derivative for the x-momentum balance is presented. First, there is an inventory of the individual contributions to this micro-balance.

Accumulation

The unsteady-state term is

$$\frac{\partial}{\partial t}\rho v_x \; dxdydz \tag{5.147}$$

Convective transport

A volume flow rate of $v_x(x,y,z,t) \cdot dydz$ enters the left-hand plane of the cube, while the x-momentum concentration in this flow is $\rho v_x(x,y,z,t)$. In general, the local velocity vector does not happen to be perpendicular to the left-hand plane. This vector is decomposed into its components in the selected axis system; the x-component of the velocity is the only one to bring mass into the cube through this left-hand plane, and this mass carries (among other things) x-momentum with it. Convectively, therefore,

$$[v_x \cdot \rho v_x]|_{x,y,z} \; dydz \tag{5.148}$$

of x-momentum enters the cube through the left-hand plane

$$[v_x \cdot \rho v_x]|_{x+dx,y,z} \; dydz \tag{5.149}$$

and exits the cube through the right-hand plane, where both v_x and ρv_x have different values on the right to those they have on the left. These flows make a net contribution in the direction of x of

$$[v_x \cdot \rho v_x]|_{x,y,z} \; dydz - [v_x \cdot \rho v_x]|_{x+dx,y,z} \; dydz = -\frac{\partial}{\partial x}(v_x\rho v_x) \; dxdydz \tag{5.150}$$

to the accumulation of x-momentum in the cube. Similarly, the y-component of the velocity at the front plane of the cube brings in mass that carries (among other things) x-momentum with it:

$$[v_y \cdot \rho v_x]|_{x,y,z} \; dxdz \tag{5.151}$$

Together with a similar term at the rear plane of the cube, the net contribution of the convective transport in the direction of y to the accumulation of the x-momentum is obtained as follows:

$$[v_y \cdot \rho v_x]|_{x,y,z} dxdz - [v_y \cdot \rho v_x]|_{x,y+dy,z} \; dxdz = -\frac{\partial}{\partial y}(v_y\rho v_x) \; dxdydz \tag{5.152}$$

Analogously, the net contribution by the top and bottom planes is

$$[v_z \cdot \rho v_x]_{x,y,z} \, dxdy - [v_z \cdot \rho v_x]_{x,y,z+dz} \, dxdy = -\frac{\partial}{\partial z}(v_z \rho v_x) \, dxdydz \qquad (5.153)$$

Molecular transport

Providing that Newton's law applies, the molecular x-momentum flow through the front side can be written as follows:

$$-\mu \left[\frac{\partial v_x}{\partial y}\right]_{x,y,z} dxdz \qquad (5.154)$$

and through the rear plane

$$-\mu \left[\frac{\partial v_x}{\partial y}\right]_{x,y+dy,z} dxdz \qquad (5.155)$$

So therefore a net contribution by the molecular transport of x-momentum in the direction of y to the accumulation results that is given by

$$-\mu \left[\frac{\partial v_x}{\partial y}\right]_{x,y,z} dxdz - \left\{-\mu \left[\frac{\partial v_x}{\partial y}\right]_{x,y+dy,z}\right\} dxdz = \frac{\partial}{\partial y}\left[\mu\frac{\partial v_x}{\partial y}\right] dxdydz \qquad (5.156)$$

The left- and right-hand planes make a net contribution to the accumulation given by

$$\frac{\partial}{\partial z}\left[\mu\frac{\partial v_x}{\partial z}\right] dxdydz \qquad (5.157)$$

It should be pointed out here that this concerns **molecular** transport of x-momentum to each of the two planes associated with the local values of the velocity gradient $\partial v_x/\partial x$. This gradient expresses whether, on average, molecules in the direction of x **accelerate** or **decelerate**, while $\mu \, \partial v_x/\partial x$ is a measure for the local net transfer of x-momentum by collisions of **individual** molecules, **given this average acceleration or deceleration**. This net transfer is different on the left and right, and it is in this way that the contribution of expression (5.157) results.

Molecular transport of x-momentum in the direction of z at the top and bottom planes results in the following net contribution:

$$\frac{\partial}{\partial z}\left[\mu\frac{\partial v_x}{\partial z}\right] dxdydz \qquad (5.158)$$

In the above derivations of the molecular transport terms, it has been assumed that, on each plane of a cube, either only a shear stress or only a normal stress prevails to each of which Newton's law applies. This representation – that was chosen because of the classical analogy between heat, mass, and momentum transport (see Sections 3.4 and 4.4) – is in fact erroneous, takes no account of the simultaneous occurrence and mutual relationship between shear and normal stresses in every transforming fluid element, and is at odds with a number of laws of physics. The eventual result, equation (5.161), is correct, however. For the correct derivation, in physics terms, which uses a linear connection between stresses in a fluid and transformation velocities of fluid elements, readers should refer to e.g. Munson.[22]

Production
Production of x-momentum is made by the x-components of all the forces that are exerted on the cube. As far as the pressure is concerned, in the direction of x it is exerted only on the left- and right-hand planes of the cube. The net result is

$$p|_x \, dydz + \{-p|_{x+dx} \, dydz\} = -\frac{\partial p}{\partial x} \, dxdydz \qquad (5.159)$$

A physically correct interpretation of this net contribution to the total x-momentum balance is that $[-\partial p/\partial x]$ is a source term expressing that per unit of volume and per unit of time a certain amount of x-momentum (in Ns) is supplied to the volume element. Finally, the x-component g_x of gravity affects the *mass* in the cube by supplying x-momentum (in Ns) per unit of mass per unit of time:

$$g_x \rho \, dxdydz \qquad (5.160)$$

This means that gravity is a typical example of a **body force** (in N/kg) where pressure (in N/m^2) is exerted on a plane.

The equations of motion
Merging expressions (5.147), (5.150), (5.152), (5.153), and (5.156) through (5.160), dividing by $dxdydz$, and considering the case of constant μ provides the x-momentum balance:

$$\frac{\partial \rho v_x}{\partial t} = -\frac{\partial}{\partial x}(v_x \rho v_x) - \frac{\partial}{\partial y}(v_y \rho v_x) - \frac{\partial}{\partial z}(v_z \rho v_x) +$$
$$+ \mu \left\{ \frac{\partial^2 v_x}{\partial x^2} + \frac{\partial^2 v_x}{\partial y^2} + \frac{\partial^2 v_x}{\partial z^2} \right\} - \frac{\partial p}{\partial x} + \rho g_x \qquad (5.161)$$

This equation is completely comparable and analogous to the equations (3.135) and (4.72).

The x-momentum balance can also be derived in terms of shear stresses, which means it is valid for non-Newtonian fluids as well. This derivation contains no molecular transport terms of the kind of expression (5.154), but the shear stresses on the six planes should be included under the section **Production**. Consider the fluid flow in the direction of x along the front plane of the cube in Figure 5.30; because of the local shear stress as a result of the local velocity gradient $\partial v_x/\partial y$, this fluid exerts a force

---

22 Munson, B.R., D.F. Young & T.H. Okiishi, *Fundamentals of Fluid Mechanics*, Wiley, 1994, Section 6.1.

$$\tau_{yx}\big|_{x,y,z} \, dxdz \qquad (5.162)$$

on the fluid in the cube. There is a similar force in the direction of x on the rear plane as a result of the fluid that flows on the outside the cube there:

$$-\tau_{yx}\big|_{x,y+dy,z} \, dxdz \qquad (5.163)$$

Together, these forces produce a net force in the direction of x

$$-\frac{\partial \tau_{yx}}{\partial y} \, dxdydz \qquad (5.164)$$

on the cube. Similar forces are exerted in the direction of x on the other four planes, giving the following net contributions:

$$-\frac{\partial \tau_{xx}}{\partial x} \, dxdydz \qquad (5.165)$$

due to spatial variations in the normal stress τ_{xx}, and

$$-\frac{\partial \tau_{zx}}{\partial z} \, dxdydz \qquad (5.166)$$

These latter two contributions to the x-momentum balance relate to the local velocity gradients $\partial v_x/\partial x$ and $\partial v_x/\partial z$.

Merging expressions (5.147), (5.150), (5.152), (5.153), (5.159), and (5.160) with (5.164) through (5.166) and dividing by $dxdydz$ produces

$$\frac{\partial \rho v_x}{\partial t} = -\frac{\partial}{\partial x}(v_x \rho v_x) - \frac{\partial}{\partial y}(v_y \rho v_x) - \frac{\partial}{\partial z}(v_z \rho v_x) +$$
$$-\frac{\partial \tau_{xx}}{\partial x} - \frac{\partial \tau_{yx}}{\partial y} - \frac{\partial \tau_{zx}}{\partial z} - \frac{\partial p}{\partial x} + \rho g_x \qquad (5.167)$$

The x-momentum balance is often presented in another form, for which the left-hand side of equations (5.161) and (5.167) are written as

$$\rho \frac{\partial v_x}{\partial t} + v_x \frac{\partial \rho}{\partial t} \qquad (5.168)$$

This means also that, for example, the second convective term on the right-hand side of equations (5.161) and (5.167) is elaborated to

$$\rho v_y \frac{\partial v_x}{\partial y} + v_x \frac{\partial \rho v_y}{\partial y} \qquad (5.169)$$

By doing the same thing with the two remaining convective terms, six terms are created on the right-hand side of equations (5.161) and (5.167). Thanks to the **continuity**

equation (5.145), three of these cancel against the second term of equation (5.168). This means that equation (5.161) can be written as follows:

$$\rho \frac{\partial v_x}{\partial t} + \rho v_x \frac{\partial v_x}{\partial x} + \rho v_y \frac{\partial v_x}{\partial y} + \rho v_z \frac{\partial v_x}{\partial z} =$$

$$= \mu \left\{ \frac{\partial^2 v_x}{\partial x^2} + \frac{\partial^2 v_x}{\partial y^2} + \frac{\partial^2 v_x}{\partial z^2} \right\} - \frac{\partial p}{\partial x} + \rho g_x$$

(5.170)

where μ has again been assumed to be constant.

Similar equations can be derived for the two remaining velocity components:

$$\rho \frac{\partial v_y}{\partial t} + \rho v_x \frac{\partial v_y}{\partial x} + \rho v_y \frac{\partial v_y}{\partial y} + \rho v_z \frac{\partial v_y}{\partial z} =$$

$$= \mu \left\{ \frac{\partial^2 v_y}{\partial x^2} + \frac{\partial^2 v_y}{\partial y^2} + \frac{\partial^2 v_y}{\partial z^2} \right\} - \frac{\partial p}{\partial y} + \rho g_y$$

(5.171)

$$\rho \frac{\partial v_z}{\partial t} + \rho v_x \frac{\partial v_z}{\partial x} + \rho v_y \frac{\partial v_z}{\partial y} + \rho v_z \frac{\partial v_z}{\partial z} =$$

$$= \mu \left\{ \frac{\partial^2 v_z}{\partial x^2} + \frac{\partial^2 v_z}{\partial y^2} + \frac{\partial^2 v_z}{\partial z^2} \right\} - \frac{\partial p}{\partial z} + \rho g_z$$

(5.172)

The equations (5.170) – (5.172). are denoted as the equations of motion, govern and describe – along with the continuity equation, see equation (5.145) – the flow of any Newtonian fluid (of constant μ), and are the cornerstone of the field of fluid mechanics. The equations of motion are named after Navier (1785–1836) and Stokes (1819–1903), and are therefore also denoted as the Navier–Stokes equations.

In vector notation, the set reads very compactly as follows:

$$\rho \frac{\partial \boldsymbol{v}}{\partial t} + \rho \boldsymbol{v} \cdot \nabla \boldsymbol{v} = \mu \nabla^2 \boldsymbol{v} - \nabla p + \rho \boldsymbol{g}$$

(5.173)

In terms of shear stresses – going forward from equation (5.167) – the following is true:

$$\rho \frac{\partial \boldsymbol{v}}{\partial t} + \rho \boldsymbol{v} \cdot \nabla \boldsymbol{v} = -\nabla \cdot \boldsymbol{\tau} - \nabla p + \rho \boldsymbol{g}$$

(5.174)

For flows where viscous effects can be disregarded, the Euler equation applies:

$$\rho \frac{\partial \boldsymbol{v}}{\partial t} + \rho \boldsymbol{v} \cdot \nabla \boldsymbol{v} = -\nabla p + \rho \boldsymbol{g}$$

(5.175)

For viscous (laminar) flows in which just the inertial terms may be ignored, and for steady-state conditions, equation (5.172) simplifies to

$$0 = \mu \nabla^2 \mathbf{v} - \nabla p + \rho \mathbf{g} \tag{5.176}$$

It falls beyond the scope of this book to show the form of these sets of equations for cylinder or sphere coordinates. Readers are referred to e.g. Bird et al.[23]

The set of Equations (5.170), (5.171), and (5.172), or the vector Equations (5.173) and (5.174), apply in a very general sense to time-dependent three-dimensional flows. An equation for a simpler type of flow, of the kind dealt with in Section 5.6 for example, can easily be extracted from a general system. For the time-independent, one-dimensional, vertical, incompressible flow of a layer of uniform thickness, for example, equation (5.96) can be obtained from equation (5.174) by using the following assumptions:

$$v_x = 0, \ v_z = 0, \ \frac{\partial v_y}{\partial y} = 0, \ \frac{\partial}{\partial t} = 0, \ g_x = g_z = 0, \ g_y = g$$

plus all pressure gradients being zero and all shear stresses being zero except τ_{xy}. Similarly, equation (5.82) can be obtained from equation (5.176) in the absence of gravity.

In a way similar to that of the redefinition of equation (5.161) to the end-result of equation (5.170), that is, with the help of the continuity equation – see equation (5.145) – also the earlier equation (3.135) governing heat transport can be rewritten as

$$\rho \frac{\partial c_p T}{\partial t} + \rho v_x \frac{\partial c_p T}{\partial x} + \rho v_y \frac{\partial c_p T}{\partial y} + \rho v_z \frac{\partial c_p T}{\partial z} =$$
$$= \lambda \left\{ \frac{\partial^2 T}{\partial x^2} + \frac{\partial^2 T}{\partial y^2} + \frac{\partial^2 T}{\partial z^2} \right\} + q \tag{5.177}$$

In exactly the same way, equation (4.72) can be rearranged:

$$\frac{\partial c}{\partial t} + v_x \frac{\partial c}{\partial x} + v_y \frac{\partial c}{\partial y} + v_z \frac{\partial c}{\partial z} = = \text{ID} \left\{ \frac{\partial^2 c}{\partial x^2} + \frac{\partial^2 c}{\partial y^2} + \frac{\partial^2 c}{\partial z^2} \right\} + r \tag{5.178}$$

It should be pointed out that in equation (5.177) the unit of combination $c_p T$ is J/kg and can therefore be regarded as an **energy concentration**, as explained previously. Finally, equation (5.177) for constant c_p changes to

23 Bird, R.B., W.E. Stewart & E.N. Lightfoot, Transport Phenomena, Wiley, 2nd *Ed.*, 2002.

$$\frac{\partial T}{\partial t} + v_x \frac{\partial T}{\partial x} + v_y \frac{\partial T}{\partial y} + v_z \frac{\partial T}{\partial z} = a\left\{\frac{\partial^2 T}{\partial x^2} + \frac{\partial^2 T}{\partial y^2} + \frac{\partial^2 T}{\partial z^2}\right\} + \frac{q}{\rho c_p} \qquad (5.179)$$

Find out yourself how equation (3.72), for example, can be obtained from equation (5.179).

Equations (5.170) through (5.172), (5.177), and (5.178) perfectly illustrate in their form the analogy between momentum, heat, and species transport. Together with equation (5.145) – the continuity equation – a system of six partial differential equations is available from which, in principle, with sufficient boundary and initial conditions, the spatial distributions of the three velocity components, pressure, temperature, and species concentration can be calculated as a function of time. This system of six differential equations is strongly linked internally and, thanks to the convective terms, is also strongly nonlinear.

! **Summary**
With the help of the **cubic volume element method**, it has been demonstrated how the Navier–Stokes equations can be derived for the x-momentum: for each plane of a cube, both convective and molecular transport terms have to be formulated, and all relevant forces have to be taken into consideration. The molecular transport terms can also be represented as shear stresses. The continuity equation can be used for rewriting the momentum balances, and the same thing applies to the heat and species transport equations.

The classic analogy between momentum, heat, and species transport is apparent from the similarity between the various transport equations. Finally, it is possible to obtain the specific equations for specific simple cases from the general transport equations by omitting non-relevant terms.

5.8.4 Computational fluid dynamics

Systems like equations (5.170) through (5.172), (5.177), (5.178), and (5.145) form the basis for what is nowadays known as computational fluid dynamics (CFD). Modern chemical engineering exploits CFD in order to achieve better descriptions and designs of process equipment, for example, on the basis of information about **local** variables (velocities, pressure, temperature, concentrations). The state of computer technology means that the phenomenological approach to transport phenomena can be replaced to an increasing degree by CFD simulations. In one sense, though, there is nothing new – it is still all about the concepts, the laws, and the technique of drawing up balances, as covered in Chapters 1 and 2.

The systems of transport equations that have been mentioned are all derived using a small cube (Figure 5.30). The dimensions of the cube have not been specified, but they should preferably be small in comparison with the dimensions of the flow domain (the equipment) in order to arrive at local values for the variable. The absolute lower limit for the dimensions of the cube is the free path length of the molecules: if the dimensions approach this, then continuum terms like mean velocity and convective transport lose their significance. As long as this is not the case, the velocities, the temperature and the concentrations that are allocated in the **cubic volume element** method to a cube can be regarded as point properties at the centre of the cube.

For CFD calculations based on the **finite volume method**, the flow domain is divided up into a very large number of small volume elements (small cubes in a Cartesian coordinate system). In practice, the number of 'cells' is limited at the lower end by the desired degree of detail (resolution), especially where steep gradients can be expected, and at the upper end by capacity, memory, and speed of the computer.

Mass, momentum, heat and/or species mass balances are drawn up for each of the volume elements. That is equivalent to solving the system (5.145), (5.170) through (5.172), (5.177) and (5.178), for example, at the discrete (grid) points – or nodes – of a grid – or lattice – that spans the whole flow domain; the variables only have values at the grid points which are in the heart of every cell. All terms with derivatives in the (micro)-balances for the cells are discretised with the help of the variables in the adjacent lattice points; a large number of so-called differential schemes are available for this. This helps the partial differential equations to change into algebraic equations and to link the values of any variable to those at adjacent points. This procedure has been illustrated in Section 3.1.5 for the simple case of two-dimensional heat conduction in a Cartesian coordinate system.

Every cell produces as many algebraic equations as there are balances set up for that cell. For the whole flow domain with its many cells, this actually means a very large number of algebraic equations that have to be solved simultaneously, or iteratively. There is a wide choice of methods for solving this. The iterations are continued until all the micro-balances in the domain have been complied with throughout, to a predetermined level of accuracy.

It should be repeated here that most flows (both in process equipment and in nature) are **turbulent**. The lattice on which the flow should be calculated should be fine enough in order for the smallest eddies to be visible. The state of computer technology has not yet advanced to the stage that it is possible to allow this for higher Reynolds numbers. (The smallest eddies become smaller as the Reynolds number increases.) Moreover, most technologists are not interested in the instantaneous velocity field, including all its eddies. Generally, knowledge of mean quantities and their variations about the flow domain is more than sufficient.

This means that calculations are made using a coarser grid and that the **momentum transport** caused by the **eddies** is modelled as a separate contribution to the momentum housekeeping on a scale between convective transport by the (mean) flow and molecular transport. During their existence, all the eddies – individually and collectively – transport momentum, among other things (via convection). (See also Section 3.5.2.) This turbulent momentum transport can, in a large number of cases, be modelled sufficiently successfully in terms of two core concepts from turbulence mechanics:

- The concentration of **turbulent kinetic energy**, denoted by k and with unit m^2/s^2, or J/kg: a measure for magnitude and strength of all current velocity fluctuations as a result of the whole spectrum of turbulent eddies.
- The rate ε at which k is dissipated into heat caused by viscosity inside the smallest eddies (the so-called **Kolmogorov eddies**); this means that ε has the unit m^2/s^3 or W/kg. The image that goes with ε is that of the number density of the smallest eddies in which the **energy dissipation** takes place.

Both k and ε can clearly be regarded as concentrations; also, both variables appear to be able to vary considerably across a flow domain. In relation to local and transient momentum transport by eddies, it is therefore usually a good idea to calculate k and ε as well. Transport equations can be derived for both variables, which actually stand for micro-balances for k and ε and, as far as their form is concerned (number and type of terms), display a strong similarity with the Navier–Stokes equations. However, it should be pointed out that k and ε are **not conserved quantities** (see the discussion in Section 1.3.3). All of this means that the number of transport equations that have to be solved in this approach is greater still than for laminar flows.

If classical chemical engineering, which uses phenomenological concepts like mass and heat transfer coefficients, is compared to modern, CFD-based chemical engineering, then it is noticeable that

- both rely strongly on the drawing up of balances;
- both use the same concepts and laws;
- the phenomenological approach leads very quickly to a fairly accurate result and in the past has led to major successes;
- CFD produces much more detailed information and therefore promises much for the future;
- that further training in fluid mechanics (including turbulence) and numerical analysis is desirable for CFD.

It should be pointed out that, with the current state of CFD technology, experimental verification of simulation results is still very much required, especially in the case of turbulent flows. Partly for this reason, the phenomenological approach to transport phenomena **and** CFD simulations will, for the time being, continue to exist side-by-side as valuable and complementary techniques.

! Summary

The principle of CFD, based on the general transport equations (micro-balances) for momentum, heat and mass, has been set out. The procedure for finding solutions as laid down in the finite-volume method has been described in brief. For turbulent flows, two new variables – the turbulent kinetic energy k and the rate ε at which k is dissipated – has been introduced and their use explained. They help in the modelling of momentum transport through the eddies of the turbulent velocity field. Finally, it was stated that simulations created with the help of CFD are a welcome and highly promising extension to the toolbox of chemical engineers.

List of symbols

Only those symbols that are used throughout the textbook are reported here.

		Units
a	Absorption coefficient	–
a	Specific area	m^2/m^3
a	Thermal diffusivity	m^2/s
A	Area	m^2
b	Width	m
c	Concentration	kg/m^3 of $kmol/m^3$
c^*	Saturation concentration, or solubility	kg/m^3 of $kmol/m^3$
c_p	Specific heat at constant pressure	$J/kg\ K$
c_v	Specific heat at constant volume	$J/kg\ K$
C_d	Discharge coefficient	–
C_D	Drag coefficient	–
$C(\theta)$	C-function	–
d	Diameter	m
D	Diameter	m
D_h	Hydraulic diameter	m
ID	Diffusion coefficient	m^2/s
e	Energy per unit of mass	J/kg
e_{diss}	Specific total energy dissipation	J/kg
e_{fr}	Specific energy dissipation in pipes (friction)	J/kg
e_L	Specific energy dissipation in fittings	J/kg
e	Emission coefficient	–
E	Energy	J
E_q	Heat extraction coefficient	–
$E(\theta)$	E-function	–
f	Friction factor	–
f_D	Stefan's correction factor	–
F_D	Drag force	N
F	Force	N
$F(\theta)$	F-function	–
g	Gravitational acceleration	m/s^2
h	Height	m
h	Heat transfer coefficient	J/sm^2K
h	Enthalpy per unit of mass	J/kg of J/mol
Δh_v	Heat of evaporation	J/kg
H	Enthalpy	J
H	Henry coefficient	Pa
$\mathcal{H}$	Absolute humidity	kg/kg
j_D	Mass transfer number	–
j_H	Heat transfer number	–
k	Mass transfer coefficient	m/s
k	Concentration of turbulent kinetic energy	J/kg
k	Constant	–
k_B	Boltzmann constant	J/K
k_r	Reaction rate constant	(varying)
K	Overall mass transfer coefficient	m/s

		Units
K	Consistency (power law liquid)	(varying)
K_L	Loss coefficient	–
l	Length	m
L	Length (scale)	m
m	Mass of a molecule	kg
m	Partition coefficient	–
M	(Total) mass	kg
M	Molar mass	kg/mol
n	Number	–
n	Flow index (power law)	–
N	Number of tanks in series	–
N	Number of impeller revolutions per unit of time	s^{-1}
N_{avo}	Number of Avogadro	mol^{-1}
p	Pressure	N/m^2
p^*	Equilibrium vapour pressure	N/m^2
P	Power	J/s
P	Production	s^{-1}
P	Power	Ns
q	Rate of specific heat production	J/m^3s
r	Radial coordinate	m
r	Chemical reaction rate	$kmol/m^3s$
r	Reflection coefficient	–
R	Radius	m
R	Gas constant	J/mol K
s	Distance	m
S	(Wetted) perimeter	m
t	Time	s
t	Transmission coefficient	–
T	Temperature	K
u	Internal, or thermal, energy per unit of mass	J/kg
U	Internal, or thermal, energy	J
U	Overall heat transfer coefficient	J/m^2 s K
v	Velocity	m/s
v_r	Relative velocity	m/s
v_o	Superficial velocity	m/s
V	Volume	m^3
w	Impeller blade height	m
w	Width	m
W	Work	J/s
W	Length dimension	m
x	Mass fraction	kg/kg
x, y	Spatial coordinates	m
z	Vertical coordinate	m
β	Thermal expansion coefficient	K^{-1}
δ	Film, gap, or coating thickness	m
Δ	Difference, or interval	–
ε	Rate of dissipation of turbulent kinetic energy	J/kg s
ε	Porosity	m^3/m^3
ε	Wall roughness	m
γ	Specific weight (= ρg)	kg/m^2s^2

		Units
θ	Non-dimensional time	–
θ	Angle	rad
ϕ	Flow rate	s^{-1}
ϕ'	Flow rate per unit of width	$s^{-1}\,m^{-1}$
ϕ''	Flux (= flow rate per unit of area)	$s^{-1}\,m^{-2}$
ϕ_w	Work	J/s
λ	Thermal conductivity coefficient	J/msK
λ	Time constant for viscoelastic fluids	s
λ	Wavelength	m
μ	Dynamic viscosity coefficient	Ns/m^2
υ	Kinematic viscosity coefficient	m^2/s
ρ	Density	kg/m^3
σ	Diameter of a molecule	m (or J/m^2)
σ	Interfacial, or surface, tension	N/m
σ	Stefan–Boltzmann constant	J/m^2sK^4
τ	Residence time	s
τ_o	Yield stress	N/m^2
τ_{xy}	Shear stress	N/m^2

Subscripts

a	Ambient
A	Component A
buoy	Buoyancy
B	Component B
c	Value in centre
diss	Dissipation
e	Energy
fr	Friction
g	Gravity
h	Hydraulic
I	Interface
l	Liquid
L	Position L
m	Measured
m	Mass
mol	Mol
p	Momentum
p	Particle
q	Heat
r	Relative
s	Slip
tot	Total
u	Internal energy
v	Vapour
V	Volume
w	Wall
x,y,z	x,y,z-Direction

X	Component X
0	Start; entrance
0	Wet bulb
1, 2	At positions 1, 2
$\perp$	Normal to direction of flow or motion
⟨ ⟩	Refers to spatial averaging
–	Refers to averaging over time
Bi	Biot number
Br	Brinkman number
Fo	Fourier number
Gr	Grashof number
Gz	Graetz number
Le	Lewis number
Nu	Nusselt number
Pe	Péclet number
Po	Power number
Pr	Prandtl number
Re	Reynolds number
Sc	Schmidt number
Sh	Sherwood number
Vi	Viscosity number

Index

www.ingramcontent.com/pod-product-compliance
Lightning Source LLC
Chambersburg PA
CBHW080935220326
41598CB00034B/5788